CONDENSED-PHASE MOLECULAR SPECTROSCOPY AND PHOTOPHYSICS

CONDENSED-PHASE MOLECULAR SPECTROSCOPY AND PHOTOPHYSICS

ANNE MYERS KELLEY

WILEY

A JOHN WILEY & SONS, INC., PUBLICATION

Published by John Wiley & Sons, Inc., Hoboken, New Jersey
Published simultaneously in Canada

For general information on our other products and services or for technical support, please contact our Customer Care Department within the United States at (800) 762-2974, outside the United States at (317) 572-3993 or fax (317) 572-4002.

Wiley also publishes its books in a variety of electronic formats. Some content that appears in print may not be available in electronic formats. For more information about Wiley products, visit our web site at www.wiley.com.

Library of Congress Cataloging-in-Publication Data is available
ISBN 9780470946701

Printed in the United States of America

10 9 8 7 6 5 4 3 2 1

CONTENTS

PREFACE xi

1 REVIEW OF TIME-INDEPENDENT QUANTUM MECHANICS 1

1.1 States, Operators, and Representations / 1
1.2 Eigenvalue Problems and the Schrödinger Equation / 4
1.3 Expectation Values, Uncertainty Relations / 6
1.4 The Particle in a Box / 7
1.5 Harmonic Oscillator / 9
1.6 The Hydrogen Atom and Angular Momentum / 12
1.7 Approximation Methods / 15
1.8 Electron Spin / 18
1.9 The Born–Oppenheimer Approximation / 22
1.10 Molecular Orbitals / 22
1.11 Energies and Time Scales; Separation of Motions / 25
Further Reading / 26
Problems / 27

2 ELECTROMAGNETIC RADIATION 31

2.1 Classical Description of Light / 31
2.2 Quantum Mechanical Description of Light / 35
2.3 Fourier Transform Relationships Between Time and Frequency / 38

2.4 Blackbody Radiation / 40
2.5 Light Sources for Spectroscopy / 42
References and Further Reading / 44
Problems / 44

3 RADIATION–MATTER INTERACTIONS **47**

3.1 The Time-Dependent Schrödinger Equation / 47
3.2 Time-Dependent Perturbation Theory / 50
3.3 Interaction of Matter with the Classical Radiation Field / 54
3.4 Interaction of Matter with the Quantized Radiation Field / 59
References and Further Reading / 63
Problems / 64

4 ABSORPTION AND EMISSION OF LIGHT **67**

4.1 Einstein Coefficients for Absorption and Emission / 67
4.2 Other Measures of Absorption Strength / 69
4.3 Radiative Lifetimes / 72
4.4 Oscillator Strengths / 73
4.5 Local Fields / 73
Further Reading / 74
Problems / 75

5 SYSTEM–BATH INTERACTIONS **79**

5.1 Phenomenological Treatment of Relaxation and Lineshapes / 79
5.2 The Density Matrix / 86
5.3 Density Matrix Methods in Spectroscopy / 90
5.4 Exact Density Matrix Solution for a Two-Level System / 95
References and Further Reading / 98
Problems / 98

6 SYMMETRY CONSIDERATIONS **103**

6.1 Qualitative Aspects of Molecular Symmetry / 103
6.2 Introductory Group Theory / 104
6.3 Finding the Symmetries of Vibrational Modes of a Certain Type / 109
6.4 Finding the Symmetries of All Vibrational Modes / 111
Further Reading / 113
Problems / 113

7 MOLECULAR VIBRATIONS AND INFRARED SPECTROSCOPY 115

7.1 Vibrational Transitions / 115
7.2 Diatomic Vibrations / 117
7.3 Anharmonicity / 118
7.4 Polyatomic Molecular Vibrations: Normal Modes / 121
7.5 Symmetry Considerations / 127
7.6 Isotopic Shifts / 130
7.7 Solvent Effects on Vibrational Spectra / 130
References and Further Reading / 135
Problems / 135

8 ELECTRONIC SPECTROSCOPY 139

8.1 Electronic Transitions / 139
8.2 Spin and Orbital Selection Rules / 141
8.3 Spin–Orbit Coupling / 143
8.4 Vibronic Structure / 143
8.5 Vibronic Coupling / 148
8.6 The Jahn–Teller Effect / 151
8.7 Considerations in Large Molecules / 152
8.8 Solvent Effects on Electronic Spectra / 154
Further Reading / 159
Problems / 160

9 PHOTOPHYSICAL PROCESSES 163

9.1 Jablonski Diagrams / 163
9.2 Quantum Yields and Lifetimes / 166
9.3 Fermi's Golden Rule for Radiationless Transitions / 167
9.4 Internal Conversion and Intersystem Crossing / 167
9.5 Intramolecular Vibrational Redistribution / 173
9.6 Energy Transfer / 179
9.7 Polarization and Molecular Reorientation in Solution / 182
References and Further Reading / 186
Problems / 186

10 LIGHT SCATTERING 191

10.1 Rayleigh Scattering from Particles / 191
10.2 Classical Treatment of Molecular Raman and Rayleigh Scattering / 193

10.3 Quantum Mechanical Treatment of Molecular Raman and Rayleigh Scattering / 195
10.4 Nonresonant Raman Scattering / 204
10.5 Symmetry Considerations and Depolarization Ratios in Raman Scattering / 206
10.6 Resonance Raman Spectroscopy / 207
References and Further Reading / 211
Problems / 211

11 NONLINEAR AND PUMP–PROBE SPECTROSCOPIES 215

11.1 Linear and Nonlinear Susceptibilities / 215
11.2 Multiphoton Absorption / 216
11.3 Pump–Probe Spectroscopy: Transient Absorption and Stimulated Emission / 219
11.4 Vibrational Oscillations and Impulsive Stimulated Scattering / 225
11.5 Second Harmonic and Sum Frequency Generation / 227
11.6 Four-Wave Mixing / 232
11.7 Photon Echoes / 232
References and Further Reading / 234
Problems / 234

12 ELECTRON TRANSFER PROCESSES 239

12.1 Charge–Transfer Transitions / 239
12.2 Marcus Theory / 243
12.3 Spectroscopy of Anions and Cations / 247
References and Further Reading / 248
Problems / 248

13 COLLECTIONS OF MOLECULES 251

13.1 Van der Waals Molecules / 251
13.2 Dimers and Aggregates / 252
13.3 Localized and Delocalized Excited States / 253
13.4 Conjugated Polymers / 256
References / 259
Problems / 259

14 METALS AND PLASMONS 263

14.1 Dielectric Function of a Metal / 263
14.2 Plasmons / 266

14.3 Spectroscopy of Metal Nanoparticles / 268
14.4 Surface-Enhanced Raman and Fluorescence / 270
References and Further Reading / 274
Problems / 275

15 CRYSTALS **277**

15.1 Crystal Lattices / 277
15.2 Phonons in Crystals / 281
15.3 Infrared and Raman Spectra / 284
15.4 Phonons in Nanocrystals / 286
References and Further Reading / 287
Problems / 287

16 ELECTRONIC SPECTROSCOPY OF SEMICONDUCTORS **291**

16.1 Band Structure / 291
16.2 Direct and Indirect Transitions / 296
16.3 Excitons / 296
16.4 Defects / 298
16.5 Semiconductor Nanocrystals / 298
Further Reading / 302
Problems / 302

APPENDICES

A PHYSICAL CONSTANTS, UNIT SYSTEMS, AND CONVERSION FACTORS 305
B MISCELLANEOUS MATHEMATICS REVIEW 309
C MATRICES AND DETERMINANTS 313
D CHARACTER TABLES FOR SOME COMMON POINT GROUPS 317
E FOURIER TRANSFORMS 321

INDEX 323

PREFACE

Faculty members teaching advanced or graduate level courses in their specialty tend to have well-defined ideas about how such courses should be taught—what material should be covered and in what depth, how the topics should be organized, and what "point of view" should be adopted. There are probably as many different ideas about how to teach any graduate course as there are individuals doing the teaching. It is therefore no surprise that many faculty cannot find any single textbook that truly meets their needs for any given course. Most simply accept this situation, and either choose the book they like best (or dislike least) and supplement it with other textbooks and/or notes, or adopt no primary text at all. A few become sufficiently unhappy with the lack of a suitable text that they decide to write one themselves. This book is the outcome of such a decision.

Most chemistry departments offer a course in molecular spectroscopy for graduate students and advanced undergraduates. Some very good textbooks have been written for traditional molecular spectroscopy courses, but my favorite and that of many others, Walter Struve's *Fundamentals of Molecular Spectroscopy*, is now out of print. Furthermore, most traditional textbooks focus on high-resolution spectroscopy of small molecules in the gas phase, and topics such as rotational spectroscopy and rotation–vibration interactions that are not the most relevant to much current research in chemistry programs. A great deal of modern research involves the interaction of radiation with condensed-phase systems, such as molecules in liquids, solids, and more complex media, molecular aggregates, metals and semiconductors, and their composites. There is a need for a graduate-level textbook that covers the basics of traditional molecular spectroscopy but takes a predominantly condensed-phase perspective and also addresses optical processes in extended systems, such as metals, semiconductors, and conducting polymers.

This book aims to provide a treatment of radiation–matter interactions that is useful for molecules in condensed phases, as well as supramolecular structures and nanostructures. The book is written at a level appropriate for advanced undergraduates or beginning graduate students in physical or materials chemistry. Much of the organization and topic selection is similar to that of a traditional graduate-level molecular spectroscopy text, but atomic spectroscopy, rotational spectroscopy, and other topics relevant mainly to gas-phase systems are omitted entirely, and there is much more emphasis on the molecule–environment interactions that strongly influence spectra in condensed phases. Additions often not found in molecular spectroscopy texts include the spectroscopy and photophysics of molecular aggregates and molecular solids and of metals and semiconductors, with particular emphasis on nanoscale size regimes. Spin-resonance methods (e.g., NMR and ESR) are not covered, nor are x-ray or electron spectroscopies. Experimental techniques are addressed only to the extent needed to understand spectroscopic data.

Chapters 1 through 10 address spectroscopic fundamentals that should probably be included in any spectroscopy course or in prerequisite courses. Chapters 11 through 16 are less fundamental, and some will probably need to be omitted in a one-semester spectroscopy course, and certainly in a one-quarter course. The instructor may choose to omit all of these remaining chapters or may pick and choose among them to tailor the course. It is assumed that students using this book have already taken a course in basic quantum mechanics and many instructors may choose to omit explicit coverage of Chapter 1, but it is included for review and reference as it is essential to understanding what comes later. Instructors may also wish to omit Chapter 6 if it is assumed that students have been introduced to group theory in undergraduate inorganic chemistry or quantum chemistry courses.

Textbooks of this type inevitably reflect the experiences and biases of their authors, and this one is no exception. Group theory is discussed in less depth and is used less extensively than in most spectroscopy textbooks. Raman and resonance Raman scattering are developed in considerable detail, while most other multiphoton spectroscopies are treated only at a rather superficial level. Spectroscopies that involve circularly polarized light, such as circular dichroism, are ignored completely. These all represent choices made in an effort to keep this book a reasonable length while trying to optimize its usefulness for a wide range of students. Only time will tell how well I have succeeded.

The general structure and emphasis of this book developed largely from discussions with David Kelley, whose encouragement pushed me over the edge from thinking about writing a book to actually doing it. He would not have written the same book I did, but his input proved valuable at many stages. I would also like to acknowledge the graduate students in my Spring 2012 Molecular Spectroscopy course (Gary Abel, Joshua Baker, Ke Gong, Cheetar Lee, and Xiao Li) for patiently pointing out numerous typographical errors, notational inconsistencies, and confusing explanations present in the first draft of this book. Their input was very helpful in bringing this project to its conclusion.

ANNE MYERS KELLEY

Merced, CA
May 2012

CHAPTER 1

REVIEW OF TIME-INDEPENDENT QUANTUM MECHANICS

While some spectroscopic observations can be understood using purely classical concepts, most molecular spectroscopy experiments probe explicitly quantum mechanical properties. It is assumed that students using this text have already taken a course in basic quantum mechanics, but it is also recognized that there are likely to be some holes in the preparation of most students and that all can benefit from a brief review. As this is not a quantum mechanics textbook, many results in this chapter are given without proof and with minimal explanation. Students seeking a deeper treatment are encouraged to consult the references given at the end of this chapter.

This chapter, like most introductory quantum chemistry courses, focuses on solutions of the time-independent Schrödinger equation. Because of the importance of time-dependent quantum mechanics in spectroscopy, that topic is discussed further in Chapter 2.

1.1. STATES, OPERATORS, AND REPRESENTATIONS

A quantum mechanical system consisting of N particles (usually electrons and/or nuclei) is represented most generally by a state function or state vector Ψ. The state vector contains, in principle, all information about the quantum mechanical system.

Condensed-Phase Molecular Spectroscopy and Photophysics, First Edition. Anne Myers Kelley.

In order to be useful, state vectors have to be expressed in some basis. In the most commonly used position basis, the state vector is called the wavefunction, written as $\Psi(\mathbf{r}_1, \mathbf{r}_2, \ldots \mathbf{r}_N)$, where $\mathbf{r}_i$ is the position in space of particle i. The position $\mathbf{r}$ may be expressed in Cartesian coordinates (x, y, z), spherical polar coordinates (r, θ, φ), or some other coordinate system. Wavefunctions may alternatively be expressed in the momentum basis, $\Psi(\mathbf{p}_1, \mathbf{p}_2, \ldots \mathbf{p}_N)$, where $\mathbf{p}_i$ is the momentum of particle i. Some state vectors cannot be expressed as a function of position, such as those representing the spin of an electron. But there's always a state function that describes the system, even if it's not a "function" of ordinary spatial coordinates.

The wavefunction itself, also known as the probability amplitude, is not directly measurable and has no simple physical interpretation. However, the quantity $|\Psi(\mathbf{r}_1, \mathbf{r}_2, \ldots \mathbf{r}_N)|^2\, d\mathbf{r}_1 d\mathbf{r}_2 \ldots d\mathbf{r}_N$ gives the probability that particle 1 is in some infinitesimal volume element around $\mathbf{r}_1$, and so on. Integration over a finite volume then gives the probability that the system is found within that volume. A "legal" wavefunction has to be single valued, continuous, differentiable, and normalizable.

The scalar product or inner product of two wavefunctions Ψ and Φ is given by $\int \Psi^*\Phi$, where the asterisk means complex conjugation and the integration is performed over all of the coordinates of all the particles. The inner product is not a function but a number, generally a complex number if the wavefunctions are complex. In Dirac notation, this inner product is denoted $\langle\Psi|\Phi\rangle$. The absolute square of the inner product, $|\langle\Psi|\Phi\rangle|^2$, gives the probability (a real number) that a system in state Ψ is also in state Φ. If $\langle\Psi|\Phi\rangle = 0$, then Ψ and Φ are said to be orthogonal. Reversing the order in Dirac notation corresponds to taking the complex conjugate of the inner product: $\langle\Phi|\Psi\rangle = \int \Phi^*\Psi$, while $\langle\Psi|\Phi\rangle = \int \Psi^*\Phi = (\int \Phi^*\Psi)^*$.

The inner product of a wavefunction with itself, $\langle\Psi|\Psi\rangle = \int \Psi^*\Psi$, is always real and positive. Usually, wavefunctions are chosen to be normalized to $\int \Psi^*\Psi = 1$. This means that the probability of finding the system somewhere in space is unity.

The quantities we are used to dealing with in classical mechanics are represented in quantum mechanics by operators. Operators act on wavefunctions or state vectors to give other wavefunctions or state vectors. Operator $\mathbb{A}$ acting on wavefunction Ψ to give wavefunction Φ is written as $\mathbb{A}\Psi = \Phi$. The action of an operator can be as simple as multiplication, although many (not all) operators involve differentiation.

Quantum mechanical operators are linear, which means that if λ_1 and λ_2 are numbers (not states or operators), then $\mathbb{A}(\lambda_1\Phi_1 + \lambda_2\Phi_2) = \lambda_1\mathbb{A}\Phi_1 + \lambda_2\mathbb{A}\Phi_2$, and $(\mathbb{AB})\Phi = \mathbb{A}(\mathbb{B}\Phi) = \mathbb{AB}\Phi$. However, it is *not* true in general that $\mathbb{AB}\Phi = \mathbb{BA}\Phi$; the order in which the operators are applied often matters. The quantity $\mathbb{AB} - \mathbb{BA}$ is called the commutator of $\mathbb{A}$ and $\mathbb{B}$ and is symbolized $[\mathbb{A}, \mathbb{B}]$, and it is zero for some pairs of operators but not for all. Most of what is "interesting" (i.e., nonclassical) about quantum mechanical systems arises from the noncommutation of certain operators.

A representation is a set of basis vectors, which may be discrete (finite or infinite) or continuous. An example of a finite discrete basis is the eigenstates of the z-component of spin for a spin-1/2 particle (two states, usually called α and β). An example of a discrete infinite basis is the set of eigenstates of a one-dimensional harmonic oscillator, $\{\psi_v\}$, where v must be an integer but can go from 0 to ∞. An example of a continuous basis is the position basis $\{\mathbf{r}\}$, where $\mathbf{r}$ can take on any real value. To be a representation, a set of basis vectors must obey certain extra conditions. One is orthonormality: $\langle u_i|u_j\rangle = \delta_{ij}$ (the Kronecker delta) for a discrete basis, or $\langle w_\alpha|w_{\alpha'}\rangle = \delta(\alpha\text{-}\alpha')$ (the Dirac delta function) for a continuous basis. The Kronecker delta is defined by $\delta_{ij} = 1$ if $i = j$, $\delta_{ij} = 0$ if $i \neq j$. The Dirac delta function $\delta(\alpha\text{-}\alpha')$ is a hypothetical function of the variable α that is infinitely sharply peaked around $\alpha = \alpha'$ and has an integrated area of unity. Three useful properties of the Dirac delta function are:

$$\int_{-\infty}^{\infty} dke^{ikx} = 2\pi\delta(x) \tag{1.1}$$

$$\int_{-\infty}^{\infty} dxf(x)\delta(x-a) = f(a) \tag{1.2}$$

$$\delta(ax) = \delta(x)/|a|, \tag{1.3}$$

where a is a constant.

A set of vectors in a particular state space is a basis if every state in that space has a unique expansion, such that $\Psi = \Sigma_i c_i u_i$ (discrete basis) or $\Psi = \int d\alpha c(\alpha) w_\alpha$ (continuous basis), where the c's are (complex) numbers. "In a particular state space" means, for example, that if we want to describe only the spin state of a system, the basis does not have to include the spatial degrees of freedom. Or, the states of position in one dimension $\{x\}$ can be a basis for a particle in a one-dimensional box, but not a two-dimensional box, which requires a two-dimensional position basis $\{(x,y)\}$. An important property of a representation is closure:

$$\sum_i \langle\Phi|u_i\rangle\langle u_i|\Psi\rangle = \langle\Phi|\Psi\rangle \text{ or } \int d\alpha\langle\Phi|w_\alpha\rangle\langle w_\alpha|\Psi\rangle = \langle\Phi|\Psi\rangle. \tag{1.4}$$

Representations of states and operators in discrete bases are often conveniently written in matrix form (see Appendix C). A state vector is represented in a basis by a column vector of numbers:

$$\begin{pmatrix} \langle u_1|\Psi\rangle \\ \langle u_2|\Psi\rangle \\ \vdots \end{pmatrix},$$

and its complex conjugate by a row vector: $(\langle\Phi|u_1\rangle\langle\Phi|u_2\rangle \cdots)$. The inner product is then obtained by the usual rules for matrix multiplication as

$$\langle\Phi|\Psi\rangle = (\langle\Phi|u_1\rangle\langle\Phi|u_2\rangle\cdots)\begin{pmatrix}\langle u_1|\Psi\rangle\\ \langle u_2|\Psi\rangle\\ \vdots\end{pmatrix} = \text{a number.}$$

An operator is represented by a square matrix having elements $A_{ij} = \langle u_i|\mathbb{A}|u_j\rangle$:

$$\begin{pmatrix}\langle u_1|\mathbb{A}|u_1\rangle & \langle u_1|\mathbb{A}|u_2\rangle & \cdots\\ \langle u_2|\mathbb{A}|u_1\rangle & \langle u_2|\mathbb{A}|u_2\rangle & \\ \vdots & & \ddots\end{pmatrix}$$

For Hermitian operators, $A^*_{ji} = A_{ij}$. It follows that the diagonal elements must be real for Hermitian operators, since only then can $A^*_{ii} = A_{ii}$.

The operator expression $\mathbb{A}\Psi = \Phi$ is represented in the $\{u_i\}$ basis as the matrix equation

$$\begin{pmatrix}\langle u_1|\mathbb{A}|u_1\rangle & \langle u_1|\mathbb{A}|u_2\rangle & \cdots\\ \langle u_2|\mathbb{A}|u_1\rangle & \langle u_2|\mathbb{A}|u_2\rangle & \\ \vdots & & \ddots\end{pmatrix}\begin{pmatrix}\langle u_1|\Psi\rangle\\ \langle u_2|\Psi\rangle\\ \vdots\end{pmatrix} = \begin{pmatrix}\langle u_1|\Phi\rangle\\ \langle u_2|\Phi\rangle\\ \vdots\end{pmatrix}$$

1.2. EIGENVALUE PROBLEMS AND THE SCHRÖDINGER EQUATION

The state Ψ is an eigenvector or eigenstate of operator $\mathbb{A}$ with eigenvalue λ if $\mathbb{A}\Psi = \lambda\Psi$, where λ is a number. That is, operating on Ψ with $\mathbb{A}$ just multiplies Ψ by a constant. The eigenvalue λ is nondegenerate if there is only one eigenstate having that eigenvalue. If more than one distinct state (wavefunctions that differ from each other by more than just an overall multiplicative constant) has the same eigenvalue, then that eigenvalue is degenerate.

To every observable (measurable quantity) in classical mechanics, there corresponds a linear, Hermitian operator in quantum mechanics. Since observables correspond to measurable things, this means all observables have only real eigenvalues. It can be shown from this that eigenfunctions of the same observable having different eigenvalues are necessarily orthogonal (orthonormal if we require they be normalized).

In Dirac notation, using basis $\{u_i\}$, the eigenvalue equation is $\langle u_i|\mathbb{A}|\Psi\rangle = \lambda\langle u_i|\Psi\rangle$. Inserting closure gives $\Sigma_j\langle u_i|\mathbb{A}|u_j\rangle\langle u_j|\Psi\rangle = \lambda\langle u_i|\Psi\rangle$, or in a shorter form $\Sigma_j A_{ij}c_j = \lambda c_i$, or in an even more compact form, $\Sigma_j\{A_{ij} - \lambda\delta_{ij}\}c_j = 0$. This is a system of N equations (one for each i) in N unknowns, which has a nontrivial solution if and only if the determinant of the coefficients is zero: $|\mathbb{A} - \lambda\mathbb{1}| = 0$, where $\mathbb{1}$ is an $N \times N$ unit matrix (1's along the diagonal). So to find the eigenvalues, we need to set up the determinantal equation

$$\begin{vmatrix}A_{11}-\lambda & A_{12} & \cdots\\ A_{21} & A_{22}-\lambda & \\ \vdots & & \ddots\end{vmatrix} = 0,$$

and solve for the N roots λ, then find the eigenvector corresponding to each eigenvalue $\lambda^{(i)}$ by solving the matrix equation

$$\begin{pmatrix} A_{11} & A_{12} & \cdots \\ A_{21} & A_{22} & \\ \vdots & & \ddots \end{pmatrix} \begin{pmatrix} c_1^{(i)} \\ c_2^{(i)} \\ \vdots \end{pmatrix} = \lambda^{(i)} \begin{pmatrix} c_1^{(i)} \\ c_2^{(i)} \\ \vdots \end{pmatrix}$$

(see Appendix C).

Any measurement of the observable associated with operator $\mathbb{A}$ can give only those values that are eigenvalues of $\mathbb{A}$. If the system is in an eigenstate of the operator, then every measurement will yield the same value, the eigenvalue. If the system is not in an eigenstate, different measurements will yield different values, but each will be one of the eigenvalues.

A particularly important observable is the one associated with the total energy of the system. This operator is called the Hamiltonian, symbolized $\mathbb{H}$, the sum of the kinetic and potential energy operators. The eigenstates of the Hamiltonian are therefore eigenstates of the energy, and the associated eigenvalues represent the only values that can result from any measurement of the energy of that system.

To find the energy eigenvalues and eigenstates, one must first write down the appropriate Hamiltonian for the problem at hand, which really amounts to identifying the potential function in which the particles move, since the kinetic energy is straightforward. One then solves the eigenvalue problem $\mathbb{H}\psi_n = E_n\psi_n$, which is the time-independent Schrödinger equation. For most Hamiltonians, there are many different pairs of wavefunctions ψ_n and energies E_n that can satisfy the equation.

Two observables of particular importance in quantum mechanics are the position $\mathbb{Q}$ and the linear momentum $\mathbb{P}$ along the same coordinate (e.g., x and p_x). The commutator is $[\mathbb{Q}, \mathbb{P}] = i\hbar$ and the action of $\mathbb{P}$ in the q representation is $-i\hbar(\partial/\partial q)$. That is, $\mathbb{P}\Psi(q) = -i\hbar(\partial/\partial q)\Psi(q)$, where $\Psi(q)$ is the wavefunction as a function of the coordinate q. The operator $\mathbb{P}^2 = \mathbb{P}\mathbb{P}$ in the q representation is $-\hbar^2(\partial^2/\partial q^2)$.

The Schrödinger equation in the position basis, $\mathbb{H}\Psi(q) = E\Psi(q)$, can therefore be written for a particle moving in only one dimension as

$$[(p^2/2m) + V(q)]\Psi(q) = E\Psi(q), \tag{1.5}$$

or

$$-(\hbar^2/2m)\{\partial^2\Psi(q)/\partial q^2\} + V(q)\Psi(q) = E\Psi(q). \tag{1.6}$$

The position operator in three dimensions is a vector, $= \hat{\boldsymbol{x}}x + \hat{\boldsymbol{y}}y + \hat{\boldsymbol{z}}z$, where $\hat{\boldsymbol{x}}$, $\hat{\boldsymbol{y}}$, and $\hat{\boldsymbol{z}}$ are unit vectors along x, y, and z directions. The momentum operator is also a vector, which in the position basis is

$$p = -i\hbar(\hat{\boldsymbol{x}}\partial/\partial x + \hat{\boldsymbol{y}}\partial/\partial y + \hat{\boldsymbol{z}}\partial/\partial z) = -i\hbar\boldsymbol{\nabla}, \tag{1.7}$$

where $\boldsymbol{\nabla}$ is called the del or grad operator. The square of the momentum is

$$\begin{aligned} \boldsymbol{p}^2 = \boldsymbol{p}\cdot\boldsymbol{p} &= -\hbar^2(\hat{\boldsymbol{x}}\partial/\partial x + \hat{\boldsymbol{y}}\partial/\partial y + \hat{\boldsymbol{z}}\partial/\partial z)\cdot(\hat{\boldsymbol{x}}\partial/\partial x + \hat{\boldsymbol{y}}\partial/\partial y + \hat{\boldsymbol{z}}\partial/\partial z) \\ &= -\hbar^2(\partial^2/\partial x^2 + \partial^2/\partial y^2 + \partial^2/\partial z^2) = -\hbar^2\nabla^2, \end{aligned} \tag{1.8}$$

where ∇^2 is called the Laplacian. Notice that while the momentum is a vector, the momentum squared is not.

1.3. EXPECTATION VALUES, UNCERTAINTY RELATIONS

When a system is in state Ψ, the mean value or expectation value of observable $\mathbb{A}$ is defined as the average of a large number of measurements. It is given by $\langle\mathbb{A}\rangle = \langle\Psi|\mathbb{A}|\Psi\rangle = \int\Psi^*\mathbb{A}\Psi$. If Ψ is an eigenfunction of $\mathbb{A}$, then we'll always measure the same number, the eigenvalue, for the observable a. If Ψ is not an eigenfunction of $\mathbb{A}$, then each measurement may yield a different value, but we can calculate its average with complete certainty from the previous expression. Note that if $\mathbb{A}$ involves just multiplication, such as q to some power, we can just write this as $\langle\mathbb{q}^n\rangle = \int|\Psi(q)|^2\, q^n\, dq$. But if $\mathbb{A}$ does something like differentiation, then we have to make sure it operates on Ψ(q) only, not also $\Psi^*(q)$. For example, for momentum, $\mathbb{p} = -i\hbar(\partial/\partial q)$, we have $\langle\mathbb{p}\rangle = -i\hbar\int\Psi^*(q)\{(\partial/\partial q)\Psi(q)\}dq$.

The *root-mean-square deviation* of the value of operator $\mathbb{A}$ in state Ψ is

$$\Delta A = \sqrt{\langle(\mathbb{A} - \langle\mathbb{A}\rangle)^2\rangle} \tag{1.9a}$$

that is, $\Delta A = \sqrt{\langle(\mathbb{A} - A_{\text{avg}})^2\rangle}$. Note that if Ψ is an eigenstate of $\mathbb{A}$, then $\Delta A = 0$, since every measurement of $\mathbb{A}$ gives the same result. An alternative and sometimes preferable form for ΔA can be written by noting that since $\langle\mathbb{A}\rangle$ is a number, $\langle(\mathbb{A} - \langle\mathbb{A}\rangle)^2\rangle = \langle\mathbb{A}^2 - 2\mathbb{A}\langle\mathbb{A}\rangle + \langle\mathbb{A}\rangle^2\rangle = \langle\mathbb{A}^2\rangle - 2\langle\mathbb{A}\rangle^2 + \langle\mathbb{A}\rangle^2 = \langle\mathbb{A}^2\rangle - \langle\mathbb{A}\rangle^2$. So,

$$\Delta A = \sqrt{\langle\mathbb{A}^2\rangle - \langle\mathbb{A}\rangle^2} \tag{1.9b}$$

is an alternative equivalent form.

The product of the root-mean-square deviations of two operators $\mathbb{A}$ and $\mathbb{B}$, in any state, obeys the relationship

$$\Delta A\Delta B \geq \frac{1}{2}|\langle[\mathbb{A}, \mathbb{B}]\rangle|. \tag{1.10}$$

A particularly important example is for the various components of position and momentum; since $[\mathbb{Q}, \mathbb{P}] = i\hbar$, then $\Delta Q\Delta P \geq \hbar/2$. This is often known as the Heisenberg uncertainty relation between position and momentum. The interpretation is that you cannot simultaneously know both the position and the momentum along the same direction (e.g., $\mathbb{x}$ and $\mathbb{p}_x$) to arbitrary accuracy; their uncertainty product, ΔQΔP, has a finite nonzero value. Only for operators that commute can the uncertainty product vanish (although it is not *necessarily* zero in any state). Note that since different components of position and momentum (e.g., $\mathbb{x}$ and $\mathbb{p}_y$) do commute, their uncertainty product can be zero.

Because commuting observables can have a zero uncertainty product, they are said to be compatible observables. This means that if $[\mathbb{A},\mathbb{B}] = 0$, and $\mathbb{A}\Psi = a\Psi$, (Ψ is an eigenstate of the operator $\mathbb{A}$), then the state given by ($\mathbb{B}\Psi$) is also an eigenstate of $\mathbb{A}$ with the same eigenvalue. The result one gets from a measurement of $\mathbb{A}$ is not affected by having previously measured $\mathbb{B}$. One can have simultaneous eigenstates of $\mathbb{A}$ and $\mathbb{B}$. In general, this is not the case if the operators do not commute.

1.4. THE PARTICLE IN A BOX

The "particle in a box," in one dimension, refers to a particle of mass m in a potential defined by $V(x) = 0$ for $0 \leq x \leq a$, and $V(x) = \infty$ everywhere else. In one dimension, this may be used to model an electron in a delocalized molecular orbital, for example, the pi-electron system of a linear polyene molecule, conjugated polymer, or porphyrin. In three dimensions, it may be used to model the electronic states of a semiconductor nanocrystal.

Since the potential energy becomes infinite at the walls, the boundary condition is that the wavefunction must go to zero at both walls. Thus, the relevant Schrödinger equation becomes (in 1-D)

$$-(\hbar^2/2m)(d^2/dx^2)\psi(x) = E\psi(x) \quad \text{for } 0 \leq x \leq a. \tag{1.11}$$

The solutions to this equation are

$$\psi_n(x) = B\sin(n\pi x/a) \quad \text{and} \quad E_n = h^2n^2/8ma^2 \quad \text{for } n = 1, 2, \ldots. \tag{1.12}$$

The function will be a solution for any value of the constant B. We find B by requiring that $\psi_n(x)$ be normalized: that is, the total probability of finding the particle somewhere in space must be unity. This means that $\int_0^a dx|\psi_n(x)|^2 = 1$, and we find $B = \sqrt{2/a}$. These solutions are plotted in Figure 1.1.

The particle in a box in two or three dimensions is a simple extension of the 1-D case. The Schrödinger equation for the 3-D case is

$$\begin{aligned} &-(\hbar^2/2m)(\partial^2/\partial x^2 + \partial^2/\partial y^2 + \partial^2/\partial z^2)\psi(x, y, z) = -(\hbar^2/2m)\nabla^2\psi(x, y, z) \\ &= E\psi(x, y, z), \end{aligned} \tag{1.13}$$

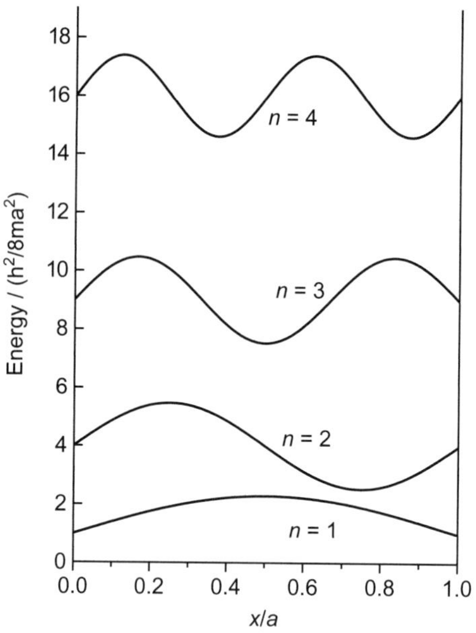

Figure 1.1. The one-dimensional particle in a box potential and its first four eigenfunctions.

for $0 \le x \le a$, $0 \le y \le b$, $0 \le z \le c$. Since the Hamiltonian for this system is a simple sum of operators in each of the three spatial dimensions, the total energy of the system is also a simple *sum* of contributions from x, y, and z motions, and the wavefunctions that are eigenfunctions of the Hamiltonian are just *products* of wavefunctions for x, y, and z individually:

$$\Psi(x, y, z) = \{(2/a)^{1/2} \sin(n_x \pi x/a)\}\{(2/b)^{1/2} \sin(n_y \pi y/b)\}\{(2/c)^{1/2} \sin(n_z \pi z/c)\} \tag{1.14a}$$

$$E(n_x, n_y, n_z) = (h^2/8m)\{(n_x^2/a^2) + (n_y^2/b^2) + (n_z^2/c^2)\} \tag{1.14b}$$

$$n_x, n_y, n_z = 1, 2, 3, \ldots \tag{1.14c}$$

A separable Hamiltonian has eigenfunctions that are products of the solutions for the individual dimensions and energies that are sums. This result is general and important.

If all three dimensions a, b, and c are different and not multiples of each other, generally, each set of quantum numbers (n_x, n_y, n_z) will have a different energy. But if any of the box lengths are the same or integer multiples, then certain energies will correspond to more than one state, that is, more than one combination of quantum numbers. This *degeneracy* is a general feature of quantum mechanical systems that have some symmetry.

1.5. HARMONIC OSCILLATOR

Classically, the relative motion of two masses m_1 and m_2, connected by a Hooke's law spring of force constant k, is described by Newton's law as

$$\mu d^2x/dt^2 + kx = 0, \tag{1.15}$$

where the reduced mass is $\mu = m_1m_2/(m_1 + m_2)$ (μ has units of mass). This is just $F = ma$, where the force $F = -kx$ and the acceleration $a = d^2x/dt^2$. Here x is the deviation of the separation from its equilibrium value, $x = (x_1 - x_2) - x_0$, and the motion of the center of mass, $M = m_1 + m_2$, has been factored out. Remember the force is just $-dV/dx$, so $V(x) = (1/2)kx^2 + C$; the potential energy is a quadratic function of position. The classical solution to this problem is $x(t) = c_1\sin\omega t + c_2\cos\omega t$, where the constants depend on the initial conditions, and the frequency of oscillation in radians per second is $\omega = (k/\mu)^{1/2}$.

The form $V(x) = (1/2)kx^2$ is the lowest-order nonconstant term in the Taylor series expansion of any potential function about its minimum:

$$\begin{aligned} V(\ell) &= V(\ell_0) + (dV/d\ell)_0(\ell - \ell_0) + (1/2)(d^2V/d\ell^2)_0(\ell - \ell_0)^2 \\ &\quad + (1/6)(d^3V/d\ell^3)_0(\ell - \ell_0)^3 + \ldots. \end{aligned} \tag{1.16}$$

But if ℓ_0 is the potential minimum, then by definition $(dV/d\ell)_0 = 0$, so the first nonzero term besides a constant is the quadratic one. Redefining $x = \ell - \ell_0$, we get

$$\begin{aligned} V(x) &= C + (1/2)(d^2V/dx^2)_0 x^2 + \text{higher ``anharmonic'' terms} \\ &= C + (1/2)kx^2 + \text{anharmonic terms}. \end{aligned} \tag{1.17}$$

Since the anharmonic terms depend on higher powers of x, they will be progressively less important for very small displacements x. So oscillations about a minimum can generally be described well by a harmonic oscillator as long as the amplitude of motion is small enough.

The quantum mechanical harmonic oscillator has a Schrödinger equation given by

$$-(\hbar^2/2\mu)(d^2/dx^2)\Psi(\mathrm{x}) + (1/2)kx^2\Psi(x) = E\Psi(x). \tag{1.18}$$

The solutions to this differential equation involve a set of functions called the Hermite polynomials, H_v. The eigenfunctions and eigenvalues are:

$$\Psi_v(x) = N_v[H_v(\alpha^{1/2}x)]\exp(-\alpha x^2/2),\ E_v = \hbar\omega(v + 1/2),\ v = 0, 1, 2, \ldots, \tag{1.19}$$

with $\alpha = (k\mu/\hbar^2)^{1/2}$ and $\omega = (k/\mu)^{1/2}$ as in the classical case. Note the units: k is a force constant in $\mathrm{N{\cdot}m^{-1}} = (\mathrm{kg{\cdot}m{\cdot}s^{-2}}){\cdot}\mathrm{m^{-1}} = \mathrm{kg{\cdot}s^{-2}}$; μ is a mass in kg; so ω has units of $\mathrm{s^{-1}}$ as it must. $\hbar$ has units of $\mathrm{J{\cdot}s} = (\mathrm{kg{\cdot}m^2{\cdot}s^{-2}})\ \mathrm{s} = \mathrm{kg{\cdot}m^2{\cdot}s^{-1}}$ so α has units of inverse length squared, making $(\alpha^{1/2}x)$ a unitless quantity. This quantity is often renamed as q, a *dimensionless coordinate*. N_v is a normalization constant that depends on the quantum number:

$$N_v = (\alpha/\pi)^{1/4}(2^v v!)^{-1/2}, \tag{1.20}$$

[recall $v! = (v)(v-1)(v-2)\ldots(2)(1)$, and $0! = 1$ by definition]. The Hermite polynomial H_v is a vth order polynomial; the number of nodes equals the quantum number v. The first four of them are

$$H_0(q) = 1, \qquad H_1(q) = 2q, \qquad H_2(q) = 4q^2 - 2, \qquad H_3(q) = 8q^3 - 12q.$$

The quantity $\alpha^{1/2}$ scales the displacement coordinate to the force constant and masses involved.

The energy levels of the quantum mechanical harmonic oscillator are equally spaced by an energy corresponding to the classical vibrational frequency, $\hbar\omega$. There is a zero-point energy, $\hbar\omega/2$, arising from confinement to a finite region of coordinate space.

The lowest-energy wavefunction is

$$\Psi_0(x) = (\alpha/\pi)^{1/4} \exp(-\alpha x^2/2) \tag{1.21}$$

Its modulus squared, $|\Psi_0(x)|^2 = (\alpha/\pi)^{1/2}\exp(-\alpha x^2)$, gives the probability of finding the oscillator at a given position x. This probability has the form of a Gaussian function—it has a single peak at $x = 0$. This means that if this is a model for a vibrating diatomic molecule, the most probable bond length is the equilibrium length. Higher vibrational eigenstates do not all have this property. Note that the probability never goes to zero even for very large positive or negative displacements. In particular, this means that there is finite probability to find the oscillator at a position where the potential energy exceeds the total energy. This is an example of quantum mechanical tunneling.

For $v = 1$, the probability distribution is

$$|\Psi_1(x)|^2 = \left|(\alpha/\pi)^{1/4} 2^{-1/2} 2\alpha^{1/2} x \exp(-\alpha x^2/2)\right|^2 = 2\alpha^{3/2}\pi^{-1/2} x^2 \exp(-\alpha x^2), \tag{1.22}$$

which has a node at the exact center and peaks on either side. The first few harmonic oscillator wavefunctions are plotted in Figure 1.2.

A classical oscillator is most likely to be found (i.e., spends most of its time) at the turning point, the outside edge where the total energy equals the potential energy. This is not true for the quantum oscillator, except at high quantum numbers. In many ways, quantum systems act like classical ones only in the limit of large quantum numbers.

All of the Hermite polynomials are either even or odd functions. The ones with even v are even and those with odd v are odd. The Gaussian function is always even. Thus, $\Psi_v(x)$ is even for even v and odd for odd v. Therefore, $|\Psi_v(x)|^2$ is always an even function, and

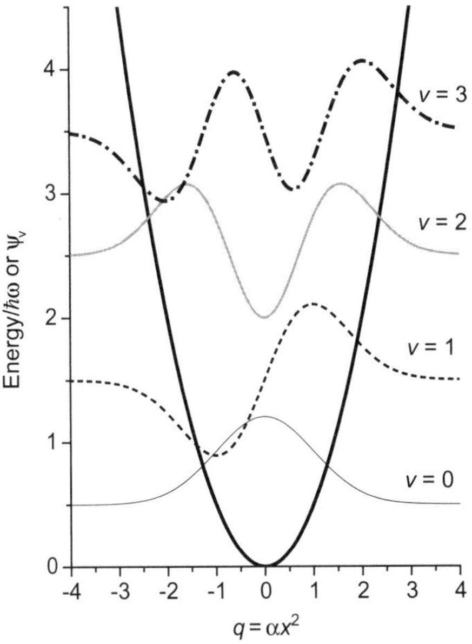

Figure 1.2. The harmonic oscillator potential energy function and its first four eigenfunctions.

$$\langle x\rangle = \int_{-\infty}^{\infty} dx \Psi_v^*(x) x \Psi_v(x) = \int_{-\infty}^{\infty} dx x |\Psi_v(x)|^2 = 0,$$

since the product of an even and an odd function is odd. The *average* position is always at the center (potential minimum) of the oscillator. Similarly,

$$\langle p\rangle = \int_{-\infty}^{\infty} dx \Psi_v^*(x)\left(-i\hbar\frac{d}{dx}\right)\Psi_v(x) = 0,$$

since the derivative of an even function is always odd and vice versa, so the integrand is always odd.

It is often more convenient to work in the reduced coordinates

$$q = \sqrt{\frac{\mu\omega}{\hbar}}x \quad \text{and} \quad P = \frac{1}{\sqrt{\mu\omega\hbar}}p.$$

Notice that the operators $\mathbb{q}$ and $\mathbb{P}$ obey the same commutation relations as $\mathbb{x}$ and $\mathbb{p}$ without the factor of $\hbar$: $[\mathbb{q}, \mathbb{P}] = (1/\hbar)[\mathbb{x}, \mathbb{p}] = i$. The harmonic oscillator Hamiltonian can be written in these reduced coordinates as

$$\mathbb{H} = \frac{1}{2}\hbar\omega\,(\mathbb{P}^2 + \mathbb{q}^2) \tag{1.23}$$

We then introduce the raising and lowering operators, $\mathbb{a}^\dagger$ and $\mathbb{a}$, where

$$\mathbb{a} = \frac{1}{\sqrt{2}}(\mathbb{q} + i\mathbb{P}) \tag{1.24a}$$

$$\mathbb{a}^\dagger = \frac{1}{\sqrt{2}}(\mathbb{q} - i\mathbb{P}), \tag{1.24b}$$

so the reduced position and momentum operators can be written in terms of the raising and lowering operators as

$$\mathbb{q} = \frac{1}{\sqrt{2}}(\mathbb{a} + \mathbb{a}^\dagger) \tag{1.25a}$$

$$\mathbb{P} = -\frac{i}{\sqrt{2}}(\mathbb{a} - \mathbb{a}^\dagger). \tag{1.25b}$$

The raising and lowering operators act on the harmonic oscillator eigenstates $|v>$ as follows:

$$\mathbb{a}|v\rangle = \sqrt{v}\,|v-1\rangle \qquad (v > 0) \tag{1.26a}$$

$$\mathbb{a}|0\rangle = 0 \tag{1.26b}$$

$$\mathbb{a}^\dagger|v\rangle = \sqrt{v+1}\,|v+1\rangle. \tag{1.26c}$$

Also note

$$\mathbb{a}^\dagger\mathbb{a} = \frac{1}{2}(\mathbb{P}^2 + \mathbb{q}^2 + i\mathbb{q}\mathbb{P} - i\mathbb{P}\mathbb{q}) = \frac{1}{2}(\mathbb{P}^2 + \mathbb{q}^2 + i[\mathbb{q}, \mathbb{P}]) = \frac{1}{2}(\mathbb{P}^2 + \mathbb{q}^2 - 1),$$

so we can rewrite the Hamiltonian as

$$\mathbb{H} = \hbar\omega\left(\mathbb{a}^\dagger\mathbb{a} + \frac{1}{2}\right) = \hbar\omega\left(\mathbb{N} + \frac{1}{2}\right). \tag{1.27}$$

$\mathbb{N} = \mathbb{a}^\dagger\mathbb{a}$ is called the number operator because the eigenstates of the Hamiltonian are also eigenstates of $\mathbb{N}$ with eigenvalues given by the quantum numbers.

1.6. THE HYDROGEN ATOM AND ANGULAR MOMENTUM

The hydrogen atom consists of one electron and one proton, interacting through a Coulombic potential. If we assume that the nucleus is fixed in space, the Hamiltonian consists of the kinetic energy of the electron (mass m_e) plus the Coulombic attraction between the electron and the nucleus:

$$\mathbb{H} = -(\hbar^2/2m_e)\nabla^2 - e^2/(4\pi\varepsilon_0 r). \tag{1.28}$$

The problem is most readily solved in spherical polar coordinates. Transforming the Laplacian operator into spherical polar coordinates gives the Schrödinger equation,

$$\begin{aligned}\mathbb{H} = -(\hbar^2/2m_e)\{(1/r^2)(\partial/\partial r)[(r^2(\partial/\partial r)] + (1/r^2\sin\theta)(\partial/\partial\theta)[\sin\theta\partial/\partial\theta)] \\ + (1/r^2\sin^2\theta)(\partial^2/\partial\varphi^2)\}\Psi(r,\theta,\varphi) - (e^2/4\pi\varepsilon_0 r)\Psi(r,\theta,\varphi) = E\Psi(r,\theta,\varphi).\end{aligned} \tag{1.29}$$

The solution to this equation is described in nearly all basic quantum mechanics textbooks. The wavefunctions are products of an r-dependent part and a (θ,φ)-dependent part, and they depend on three quantum numbers, n, ℓ, and m:

$$\Psi_{n\ell m}(r,\theta,\phi) = R_{n\ell}(r)Y_\ell^m(\theta,\phi), \tag{1.30}$$

with allowed values for the quantum numbers of

$$n = 1, 2, 3, \ldots \tag{1.31a}$$

$$\ell = 0, 1, 2, \ldots (n-1) \tag{1.31b}$$

$$m = -\ell, (-\ell+1), \ldots (\ell-1), \ell. \tag{1.31c}$$

The quantum number m is often designated m_ℓ to distinguish it from the spin quantum number m_s (see Section 1.8).

The associated energy eigenvalues depend only on n and are given by

$$E_n = -e^2/(8\pi\varepsilon_0 a_0 n^2). \tag{1.32}$$

The quantity $a_0 = 4\pi\varepsilon_0\hbar^2/(m_e e^2)$ is the Bohr radius. It has the numerical value 0.529 Å. It defines an intrinsic length scale for the H-atom problem, much as the constant α does for the harmonic oscillator.

The angle-dependent functions $Y_\ell^m(\theta,\phi)$ are known as the spherical harmonics. They are normalized and orthogonal. The first few spherical harmonics are

$$Y_0^0 = \frac{1}{\sqrt{4\pi}} \tag{1.33a}$$

$$Y_1^0 = \sqrt{\frac{3}{4\pi}}\cos\theta \tag{1.33b}$$

$$Y_1^{\pm 1} = \mp\sqrt{\frac{3}{8\pi}}\sin\theta e^{\pm i\phi}. \tag{1.33c}$$

The quantum number ℓ refers to the total angular momentum, while m refers to its projection onto an arbitrary space-fixed axis. The spherical harmonics are eigenfunctions of both the total angular momentum (or its square, $\mathbb{L}^2$) and the z-component of the angular momentum, $\mathbb{L}_z$, but not of $\mathbb{L}_x$ or $\mathbb{L}_y$. Also note that the energy depends only on $\mathbb{L}^2$. So the z-component of angular momentum is quantized, but this quantization affects only the *degeneracy* of each energy level. The spherical harmonics satisfy the eigenvalue equations

$$\mathbb{L}^2 Y_\ell^m(\theta,\phi) = \hbar^2 \ell(\ell+1) Y_\ell^m(\theta,\phi) \tag{1.34a}$$

$$\mathbb{L}_z Y_\ell^m(\theta,\phi) = \hbar m Y_\ell^m(\theta,\phi). \tag{1.34b}$$

The radial part of the wavefunction is given by

$$R_{n\ell}(r) = -\sqrt{\left(\frac{2}{na_0}\right)^3 \frac{(n-\ell-1)!}{2n[(n+\ell)!]^3}} \left(\frac{2r}{na_0}\right)^\ell \exp\left(-\frac{r}{na_0}\right) L_{n+\ell}^{2\ell+1}\left(\frac{2r}{na_0}\right), \tag{1.35}$$

where the $L_{n+\ell}^{2\ell+1}(2r/na_0)$ are called the associated Laguerre polynomials. The functions $R_{n\ell}(r)$ are normalized with respect to integration over the radial coordinate, such that $\int_0^\infty r^2 dr |R_{n\ell}(r)|^2 = 1$. Note that the radial part of the volume element in spherical polar coordinates is $r^2 dr$.

The H-atom wavefunctions depend on three quantum numbers. The principal quantum number n, the only one on which the energy depends, mainly determines the overall size of the wavefunction (larger n gives a larger average distance from the nucleus). The angular momentum quantum number ℓ determines the overall shape of the wavefunction; $\ell = 0, 1, 2, 3, \ldots$ correspond to s, p, d, f . . . orbitals. The magnetic quantum number m determines the orientation of the orbital and causes each degenerate energy level to split into $2\ell + 1$ different energies in the presence of a magnetic field.

Recall that in general, the more nodes in a wavefunction, the higher its energy. The number of radial nodes [nodes in $R_{nl}(r)$] is given by $(n - \ell - 1)$, and the number of nodal planes in angular space is given by ℓ, so the total number of nodes is $(n - 1)$, and the energy goes up with increasing n. For $\ell = 0$, there is no (θ, ϕ) dependence. The orbital is spherically symmetric and is called an s orbital. For $\ell = 1$, there is one nodal plane, and the orbital is called a p orbital. Since in spherical polar coordinates $z = r\cos\theta$, Y_1^0 points along the z-direction and we refer to the $\ell = 1, m = 0$ orbitals as p_z orbitals. Y_1^1 and Y_1^{-1} are harder to interpret because they are complex. However, since they're degenerate, any linear combination of them will also be an eigenstate of the Hamiltonian. Therefore, it's traditional to work with the linear combinations

$$p_x = 2^{-1/2}(Y_1^1 + Y_1^{-1}) = (3/4\pi)^{1/2} \sin\theta\cos\varphi \tag{1.36a}$$

$$p_y = -i2^{-1/2}(Y_1^1 - Y_1^{-1}) = (3/4\pi)^{1/2} \sin\theta\sin\varphi, \tag{1.36b}$$

which are real and point along the x- and y-directions, respectively. For $\ell = 2$, we can similarly make linear combinations of $m = \pm 1$ and $m = \pm 2$ to create five real d orbitals.

The H-atom wavefunctions are orthonormal. Remember that in spherical polar coordinates, the ranges of the coordinates are

$$r = 0 \text{ to } \infty, \theta = 0 \text{ to } \pi, \varphi = 0 \text{ to } 2\pi,$$

and the volume element is $r^2 \sin\theta dr d\theta d\varphi$. So the orthonormality condition can be written as

$$\int_0^\infty r^2 dr \int_0^\pi \sin\theta d\theta \int_0^{2\pi} d\phi \Psi^*_{n'\ell'm'}(r,\theta,\phi)\Psi_{n\ell m}(r,\theta,\phi) = \delta_{nn'}\delta_{\ell\ell'}\delta_{mm'}.$$

Note that if we are using the previous real linear combinations of the $Y_\ell^{\pm m}$, p_x and p_y instead of Y_1^1 and Y_1^{-1}, the orthonormality relations still hold because we defined p_x and p_y to be orthogonal linear combinations of Y_1^1 and Y_1^{-1}. The same holds for the d orbitals.

For future reference, here we give the complete normalized wavefunctions for $n = 1$ and $n = 2$:

$$\Psi_{100} = \frac{1}{\sqrt{\pi}}\left(\frac{1}{a_0}\right)^{3/2} e^{-r/a_0} \tag{1.37a}$$

$$\Psi_{200} = \frac{1}{\sqrt{\pi}}\left(\frac{1}{2a_0}\right)^{3/2}\left(1 - \frac{r}{2a_0}\right)e^{-r/2a_0} \tag{1.37b}$$

$$\Psi_{21-1} = \frac{1}{8\sqrt{\pi}}\left(\frac{1}{a_0}\right)^{5/2} re^{-r/2a_0}\sin\theta e^{-i\varphi} \tag{1.37c}$$

$$\Psi_{210} = \frac{1}{\sqrt{\pi}}\left(\frac{1}{2a_0}\right)^{5/2} re^{-r/2a_0}\cos\theta \tag{1.37d}$$

$$\Psi_{211} = \frac{1}{8\sqrt{\pi}}\left(\frac{1}{a_0}\right)^{5/2} re^{-r/2a_0}\sin\theta e^{i\varphi}. \tag{1.37e}$$

For one-electron ions that have a nuclear charge greater than 1, the wavefunctions are the same except that a_0 is replaced by a_0/Z, where Z is the nuclear charge.

1.7. APPROXIMATION METHODS

There are two basic ways to handle quantum mechanical problems that do not have an exact analytical solution (i.e., all the really interesting ones). The variational method is the most general and flexible, and lies at the heart of most techniques for calculating the electronic structure of multielectron atoms and molecules. The perturbation method is most useful when the real problem of interest is very similar to one of the simple problems that can be solved exactly.

The variational method expresses the unknown wavefunction as a function of one or more parameters, and then tries to find the best value(s) of those parameter(s) that make the resulting function and its energy as close as possible to the real one. But how do we find "best" if, by definition, we do not know the exact answer?

We take advantage of a theorem called the variational principle. Let $\Phi(\alpha,\beta,\gamma,\ldots)$ be this arbitrary function of one or more parameters, and let Ψ_0 be the true ground-state wavefunction of the Hamiltonian of interest, $\mathbb{H}$. We assume that we know $\mathbb{H}$ exactly but we cannot solve the resulting Schrödinger equation exactly to find Ψ_0 and its energy eigenvalue, E_0. Here we will not assume that Φ is normalized. If the system is in state Φ, then the expectation value of its energy is given by

$$E_\Phi = \int \Phi^* \mathbb{H} \Phi d\tau \Big/ \int \Phi^* \Phi d\tau \qquad (1.38)$$

where the integrations are over all space. The variational principle says that for any function Φ, it is always true that $E_\Phi \geq E_0$; that is, the energy we get by using any approximate function can approach the true energy as a lower bound, but can never be lower. Thus, the lower the energy we get by varying the parameters in the function $\Phi(\alpha,\beta,\gamma,\ldots)$, the closer this energy is to the true ground-state energy, and by assumption, the closer Φ is to the real wavefunction.

The way we actually implement this idea is to *minimize* E_Φ with respect to each of the parameters $\alpha,\beta,\gamma,\ldots$. that is, set $\partial E_\Phi/\partial\alpha = 0$ and solve for α, and so on. Of course, how good the resulting energy and wavefunction are depends on how good a choice we made for the general functional form of $\Phi(\alpha,\beta,\gamma,\ldots)$. If we choose a different functional form, or one that depends on more parameters, and get a lower value for E_Φ, then that is a better approximation to the real wavefunction. Notice that if we "accidentally" choose a trial function that has the same functional form as the real ground-state wavefunction, then this procedure will result in the exact wavefunction and energy.

While using a trial function that depends in a nonlinear way on one or more variational parameters can be quite useful for analytical calculations, for numerical (computer) calculations, it turns out to be much more efficient to choose a trial function that depends linearly on the variational parameters. That is, one assumes

$$\Phi = c_1 f_1 + c_2 f_2 + c_3 f_3 \ldots = \sum\nolimits_n c_n f_n, \qquad (1.39a)$$

where the f_n are known functions that one tries to select intelligently, and the c_n are parameters whose best values are to be determined using the variational principle. For just two functions,

$$\Phi = c_1 f_1 + c_2 f_2. \tag{1.39b}$$

Minimizing the energy with respect to both of the parameters gives two equations in two unknowns:

$$c_1(H_{11} - ES_{11}) + c_2(H_{12} - ES_{12}) = 0 \tag{1.40a}$$

$$c_1(H_{21} - ES_{21}) + c_2(H_{22} - ES_{22}) = 0, \tag{1.40b}$$

where the matrix elements of the Hamiltonian are defined as $H_{ij} = \langle f_i | \mathbb{H} | f_j \rangle$, and the overlap integrals are defined as $S_{ij} = \langle f_i | f_j \rangle$ (this is just the normalization integral for $i = j$). By definition, it is clear that $S_{ij} = S_{ji}^*$, and using the Hermitian property of quantum mechanical operators, one can also show that $H_{ij} = H_{ji}^*$. In order for this system of two equations in two unknowns to have a nontrivial solution, the determinant of their coefficients must vanish:

$$\begin{vmatrix} H_{11} - ES_{11} & H_{12} - ES_{12} \\ H_{21} - ES_{21} & H_{22} - ES_{22} \end{vmatrix} = 0. \tag{1.41}$$

Expanding this 2×2 determinant gives a quadratic equation in E, which has two roots. The lower value of the two corresponds to the lower bound on the energy. We then plug this value of E back into the original pair of equations,

$$c_1(H_{11} - ES_{11}) + c_2(H_{12} - ES_{12}) = 0 \tag{1.42a}$$

$$c_1(H_{21} - ES_{21}) + c_2(H_{22} - ES_{22}) = 0, \tag{1.42b}$$

and solve for the parameters c_1 and c_2. Actually, we can only solve for one of them in terms of the other (both equations will give the same relationship between c_1 and c_2), and the value of the remaining parameter is determined by normalization (see also Appendix C).

Generally, the more functions one combines (the larger is the basis set), the more accurate the result will be. This is because we are trying to approximate the (unknown) actual wavefunction by a sum of other functions, and the more free parameters this fitting function has, the more closely it can approximate the actual one. It is similar to trying to fit an arbitrary curve by a polynomial; the higher order the polynomial we use, in general, the better we can fit the arbitrary function. Using a linear combination of N functions yields an $N \times N$ determinant. For large N, this gets pretty ugly to solve by hand, but computers can manipulate determinants and matrices very efficiently.

The alternative approach to solving a quantum mechanical problem approximately, perturbation theory, is most useful for situations where the problem of interest is only slightly different from a problem that we do know how to solve exactly. What constitutes "slightly different" is a matter of experience and judgment.

Assume we can write the complete Hamiltonian whose solutions we are interested in as the sum of a part we know how to solve exactly, $\mathbb{H}^{(0)}$, and another part, $\mathbb{H}^{(1)}$, so that $\mathbb{H} = \mathbb{H}^{(0)} + \mathbb{H}^{(1)}$. Let the normalized eigenfunctions and eigenvalues of $\mathbb{H}^{(0)}$ be $\Psi_v^{(0)}$ and $E_v^{(0)}$, where v represents a generalized quantum number. We now assume that the eigenfunctions and eigenvalues of H can be written in the form of an expansion,

$$\Psi_v = \Psi_v^{(0)} + \Psi_v^{(1)} + \Psi_v^{(2)} + \ldots \tag{1.43a}$$

$$E_v = E_v^{(0)} + E_v^{(1)} + E_v^{(2)} + \ldots, \tag{1.43b}$$

where each term is assumed to be successively smaller than the previous one. Expressions can be derived for each of these corrections, but here we will address only the first-order corrections. The first-order correction to the energy, $E_v^{(1)}$, is simply the expectation value of the perturbation part of the Hamiltonian, $\mathbb{H}^{(1)}$, when the system is in the zeroth-order wavefunction, $\Psi_v^{(0)}$:

$$E_v^{(1)} = \langle \Psi_v^{(0)} | \mathbb{H}^{(1)} | \Psi_v^{(0)} \rangle. \tag{1.44}$$

The first-order correction to the wavefunction, $\Psi_v^{(1)}$, is

$$|\Psi_v^{(1)}\rangle = \sum_{n \neq v} \frac{\langle \Psi_n^{(0)} | \mathbb{H}^{(1)} | \Psi_v^{(0)} \rangle}{E_v^{(0)} - E_n^{(0)}} |\Psi_n^{(0)}\rangle. \tag{1.45}$$

That is, the first-order correction to the wavefunction for the vth state involves a sum of contributions from all the other zeroth-order wavefunctions, each weighted by a matrix element of the perturbation Hamiltonian divided by an energy denominator. The energy denominator has the effect of weighting most heavily the states that are closest in energy to the state of interest.

1.8. ELECTRON SPIN

You should remember from freshman chemistry the aufbau or "building-up" process for determining the electron configurations of atoms beyond He: order the hydrogenlike atomic orbitals by energy, 1s, 2s, 2p, 3s, and so on, and then start putting electrons into these orbitals, two electrons per orbital, until all the electrons are used up. The reason only two electrons can go into each orbital is that each electron can be in either of two spin states, which are usually labeled α and β (or + and –, or "up" and "down"), and the Pauli Exclusion principle says that it is not possible for two electrons to be in exactly the same quantum state. Thus, if two electrons are in the same orbital (same *spatial* part of the wavefunction), they must be in different spin states.

All subatomic particles (electrons and nuclei) have associated with them a quality called spin, which acts in some ways like a spinning top having a certain angular momentum, but really does not have a classical analog. Spin is quantized and can have either integer (including zero) or half-integer values. Particles having integer spins are called bosons, and they do not have any exclusion principle; any number of bosons can occupy the same quantum state. Particles having half-integer spin are called fermions, and no two fermions can occupy the same quantum state. Electrons have spin 1/2 and are fermions.

We define operators for the square of the total spin angular momentum, $\mathbb{S}^2$, and the z-component of the spin angular momentum, $\mathbb{S}_z$, which are entirely analogous to those for ordinary ("orbital") angular momentum, $\mathbb{L}^2$ and $\mathbb{L}_z$, except that they have different eigenvalues:

$$\mathbb{S}^2\alpha = \left(\frac{1}{2}\right)\left(\frac{1}{2}+1\right)\hbar^2\alpha \tag{1.46a}$$

$$\mathbb{S}^2\beta = \left(\frac{1}{2}\right)\left(\frac{1}{2}+1\right)\hbar^2\beta \tag{1.46b}$$

$$\mathbb{S}_z\alpha = (\hbar/2)\alpha \tag{1.46c}$$

$$\mathbb{S}_z\beta = (-\hbar/2)\beta. \tag{1.46d}$$

That is, the two spin states, α and β, are both eigenstates of the total spin angular momentum, $\mathbb{S}^2$, with eigenvalue $\hbar^2 s(s+1)$, where $s = 1/2$; s is analogous to the quantum number ℓ for orbital angular momentum, but it has half-integer rather than integer values. α and β are also both eigenstates of the z-component of the total angular momentum with eigenvalue $\hbar m_s$, where $m_s = +(1/2)$ for α and $-(1/2)$ for β ("spin-up" and "spin-down"). The eigenvalue $\hbar m_s$ refers to the projection of the spin angular momentum onto the z-axis. The operators $\mathbb{S}^2$ and $\mathbb{S}_z$ are Hermitian, so their eigenstates corresponding to different eigenvalues must be orthogonal and are chosen to be normalized:

$$\langle\alpha|\alpha\rangle = \langle\beta|\beta\rangle = 1 \tag{1.47a}$$

$$\langle\alpha|\beta\rangle = \langle\beta|\alpha\rangle = 0, \tag{1.47b}$$

where the brackets imply integration over a spin variable σ that has no classical analog.

We assume that the spatial and spin degrees of freedom are independent, so the whole wavefunction for an electron is just a product of the spatial and spin parts:

$$\Psi(r,\theta,\varphi,\sigma) = \psi(r,\theta,\varphi)\alpha(\sigma) \text{ or } \psi(r,\theta,\varphi)\beta(\sigma). \tag{1.48}$$

The complete one-electron function $\Psi(\mathrm{r},\theta,\varphi,\sigma)$ is called a spin orbital or spinorbital. For the H atom, the spinorbitals depend on four quantum numbers, m_s as well as n, ℓ, and m. The two lowest-energy spinorbitals combine the 1s spatial orbital with the α or β spin function:

$$\Psi_{100\alpha}(r,\theta,\varphi,\sigma) = \psi_{1s}(r,\theta,\varphi)\alpha(\sigma) = \pi^{-1/2}e^{-r}\alpha \tag{1.49a}$$

$$\Psi_{100\beta}(r,\theta,\varphi,\sigma) = \psi_{1s}(r,\theta,\varphi)\beta(\sigma) = \pi^{-1/2}e^{-r}\beta. \tag{1.49b}$$

These functions are written in atomic units where the electron mass, electron charge, $\hbar$, $4\pi\varepsilon_0$, and a_0 are all equal to unity. Spinorbitals for other values of n, ℓ, and m are constructed in the same way. Since the spatial and spin parts are individually orthonormal and they represent independent sets of coordinates, the spinorbitals are also orthonormal, for example

$$\begin{aligned}\langle \Psi_{100\alpha}|\Psi_{100\alpha}\rangle &= \int_0^\infty r^2 dr \int_0^\pi \sin\theta d\vartheta \int_0^{2\pi} d\phi \psi_{1s}^*(r,\theta,\varphi)\psi_{1s}(r,\theta,\varphi) \\ &\quad \cdot \int d\sigma \alpha^*(\sigma)\alpha(\sigma) = (1)(1) = 1\end{aligned} \tag{1.50a}$$

$$\begin{aligned}\langle \Psi_{100\alpha}|\Psi_{100\beta}\rangle &= \int_0^\infty r^2 dr \int_0^\pi \sin\theta d\vartheta \int_0^{2\pi} d\phi \psi_{1s}^*(r,\theta,\varphi)\psi_{1s}(r,\theta,\varphi) \\ &\quad \cdot \int d\sigma \alpha^*(\sigma)\beta(\sigma) = (1)(0) = 0.\end{aligned} \tag{1.50b}$$

It might seem reasonable to guess that an atom or molecule with two electrons could be described by a wavefunction like $\Psi(1,2) = \psi(1)\alpha(1)\psi(2)\beta(2)$, that is, both electrons in the same spatial orbital ψ, but having different spin states. There's only one problem with this wavefunction: it implies that electron 1 and electron 2 are *distinguishable* by having different spins. But in fact, they are not; they are the same kind of particle and occupy the same region of space, and they hold no "labels" to tell them apart. None of the physically observable properties associated with the wavefunction can be changed by simply changing the labels of the electrons. This requirement can be satisfied by constructing wavefunctions that are either symmetric or antisymmetric combinations of the previous function with the labels reversed, that is, $\Psi_\pm(1,2) = \psi(1)\alpha(1)\psi(2)\beta(2) \pm \psi(2)\alpha(2)\psi(1)\beta(1)$. It turns out that for electrons, which are fermions, only the *antisymmetric* (minus) combination is legal. This actually turns out to be a more general statement of the Pauli exclusion principle: Wave functions must be antisymmetric under the exchange of any two fermions. A legal wavefunction for a two-electron atom or molecule (unnormalized) is thus given by

$$\begin{aligned}\Psi(1,2) &= \psi(1)\alpha(1)\psi(2)\beta(2) - \psi(2)\alpha(2)\psi(1)\beta(1) \\ &= \psi(1)\psi(2)\{\alpha(1)\beta(2) - \beta(1)\alpha(2)\}.\end{aligned} \tag{1.51}$$

In the example with only two electrons, it is easy to just write down a wavefunction that has the proper behavior of changing its sign when the two electrons are interchanged. It is not so obvious how to do this for a system with more than two electrons. Fortunately, there is a well-defined way to do this which makes use of one of the properties of determinants: exchanging any two rows or columns changes the sign of the determinant. For N electrons occupying N spinorbitals, the wavefunction is

$$\Psi(1,2,\ldots N)=\frac{1}{\sqrt{N!}}\begin{vmatrix} u_1(1) & \cdots & u_N(1) \\ \vdots & \ddots & \vdots \\ u_1(N) & \cdots & u_N(N) \end{vmatrix}, \tag{1.52}$$

where $u_i(j)$ represents the jth electron in the ith spinorbital. Each column represents a particular spinorbital, and each row represents a particular electron. This formula automatically makes the wavefunction normalized (by the $1/\sqrt{N!}$) and antisymmetric with respect to interchange of any two electrons.

Another nice property of determinants is that if any two rows or columns are identical, the determinant is zero. This means that if we try to put two electrons into the same spinorbital, the determinant will vanish. Thus, this determinantal form for the wavefunction (also called a Slater determinant) automatically obeys the Pauli exclusion principle that every electron must occupy a different spinorbital.

Equation (1.51) describes a wavefunction that is symmetric with respect to the spatial part of the wavefunction and antisymmetric with respect to the spin part. This is the only acceptable overall wavefunction that places both electrons in the lowest-energy spatial wavefunction. If we instead consider states in which the two electrons are in two different spatial functions, ψ and ϕ, it turns out there are four different total wavefunctions we can write that satisfy the requirement to be antisymmetric with respect to interchange of electrons:

$$\Psi(1,2)=\frac{1}{2}\{\psi(1)\phi(2)+\phi(1)\psi(2)\}\{\alpha(1)\beta(2)-\beta(1)\alpha(2)\} \tag{1.53}$$

$$\Psi(1,2)=\frac{1}{2}\{\psi(1)\phi(2)-\phi(1)\psi(2)\}\{\alpha(1)\beta(2)+\beta(1)\alpha(2)\} \tag{1.54a}$$

$$\Psi(1,2)=\frac{1}{\sqrt{2}}\{\psi(1)\phi(2)-\phi(1)\psi(2)\}\alpha(1)\alpha(2) \tag{1.54b}$$

$$\Psi(1,2)=\frac{1}{\sqrt{2}}\{\psi(1)\phi(2)-\phi(1)\psi(2)\}\beta(1)\beta(2). \tag{1.54c}$$

Equation (1.53) has a symmetric spatial part and an antisymmetric spin part, and there is only one such function. This is referred to as a singlet state, as is the function in Equation (1.51). Equations (1.54) have antisymmetric spatial parts and symmetric spin parts, and there are three such functions. These are referred to a triplet states. The three triplet states are degenerate in the absence of external magnetic fields, and they are nearly always lower in energy than the corresponding singlet. Note that only two of these four wavefunctions can be represented by a single Slater determinant; the other two are linear combinations of more than one Slater determinant.

1.9. THE BORN–OPPENHEIMER APPROXIMATION

The Born-Oppenheimer approximation is a very useful and usually quite good approximation made in finding the electronic energy levels and wavefunctions of molecules. It may be most easily introduced by reference to the simplest possible diatomic molecule, H_2^+: two nuclei and only one electron. Its full Hamiltonian is

$$\mathbb{H}_{H_2^+} = -(\hbar^2/2M)(\nabla_A^2 + \nabla_B^2) - (\hbar^2/2m_e)\nabla_1^2 - (e^2/4\pi\varepsilon_0)(1/r_{1A} + 1/r_{1B}) + e^2/4\pi\varepsilon_0 R. \tag{1.55}$$

Here, A and B label the two nuclei, R is the distance between them, and 1 labels the electron. M is the nuclear mass and m_e is the electron mass. The four terms represent, respectively, the nuclear kinetic energy, electronic kinetic energy, electron–nuclear attraction, and nuclear–nuclear repulsion.

We simplify this equation by recognizing that since nuclei are much heavier than electrons (three orders of magnitude or more), they move much more slowly than the electrons. As far as the electrons are concerned, the nuclei appear nearly stationary, and the electrons are always able to "solve" their own wavefunction for each value of the nuclear position. Therefore, to find out what the electrons are doing, it is a good approximation to neglect the nuclear kinetic energy operator in the H_2^+ Hamiltonian and leave the internuclear separation R as a *parameter* in the Schrödinger equation for the electronic motion. The energy will depend on this parameter R. If we then want to deal with the motions of the nuclei explicitly, we will have to find the dependence of this energy on the internuclear separation. For example, to calculate the vibrational spectrum of a molecule, we need its force constant, which means we need to find both the value of R that minimizes the energy and the second derivative of the energy with respect to R.

Within the Born–Oppenheimer approximation and in atomic units, the electronic Hamiltonian for the H_2^+ molecule is

$$\mathbb{H}_{H_2^+}(1) = -\frac{1}{2}\nabla_1^2 - 1/r_{1A} - 1/r_{1B} + 1/R, \tag{1.56}$$

where the (1) just emphasizes that the Hamiltonian is a *function* of only the coordinates of electron 1, with R being a parameter. Because there is only one electron, it is possible to solve this problem exactly by transforming to elliptic coordinates. The exact solution will not be discussed further because it is not possible to treat any molecule with more than one electron exactly.

1.10. MOLECULAR ORBITALS

The simplest way to think about H_2^+ is to consider it as two H nuclei with the electron on one or the other nucleus. We know that the lowest-energy solution for a single H atom is for the electron to occupy the 1s orbital. So one might guess that a reasonable variational trial function for H_2^+ is

$$\psi = c_A 1s_A + c_B 1s_B,$$

where $1s_A$ and $1s_B$ are 1s orbitals centered at nuclei A and B, respectively, and c_A and c_B are variational parameters. To solve this problem, we first need to solve the determinantal equation,

$$\begin{vmatrix} H_{AA} - ES_{AA} & H_{AB} - ES_{AB} \\ H_{BA} - ES_{BA} & H_{BB} - ES_{BB} \end{vmatrix} = 0, \tag{1.57}$$

where

$$H_{AA} = H_{BB} = \langle 1s_A | \mathbb{H} | 1s_A \rangle = \langle 1s_B | \mathbb{H} | 1s_B \rangle \tag{1.58a}$$

$$H_{AB} = H_{BA} = \langle 1s_A | \mathbb{H} | 1s_B \rangle = \langle 1s_B | \mathbb{H} | 1s_A \rangle \tag{1.58b}$$

$$S_{AA} = S_{BB} = \langle 1s_A | 1s_A \rangle = \langle 1s_B | 1s_B \rangle = 1 \text{ for normalized 1s orbitals} \tag{1.59a}$$

$$S_{AB} = S_{BA} = \langle 1s_A | 1s_B \rangle = \langle 1s_B | 1s_A \rangle = \mathrm{S} \text{ (the "overlap integral").} \tag{1.59b}$$

Plugging in the H_2^+ Hamiltonian and doing the integrals gives, in atomic units,

$$H_{AA} = H_{BB} = E_{1s} + J \tag{1.60a}$$

$$H_{AB} = H_{BA} = E_{1s}S + K, \tag{1.60b}$$

where

$$E_{1s} = \int d\mathbf{r} 1s_A(\mathbf{r})\{-\nabla^2/2 - 1/r_A\} 1s_A(\mathbf{r}) \quad \text{the ground-state H atom energy}$$

$$J = \int d\mathbf{r} 1s_A(\mathbf{r})\{-1/r_B + 1/R\} 1s_A(\mathbf{r}) \quad \text{the "Coulomb integral"} \tag{1.61a}$$

$$K = \int d\mathbf{r} 1s_B(\mathbf{r})\{-1/r_B + 1/R\} 1s_A(\mathbf{r}) \quad \text{the "exchange integral."} \tag{1.61b}$$

Therefore, the determinant to be solved becomes

$$\begin{vmatrix} E_{1s} + J - E & E_{1s}S + K - ES \\ E_{1s}S + K - ES & E_{1s} + J - E \end{vmatrix} = 0, \tag{1.62}$$

which we can solve to find two values of E,

$$E_{\pm} = E_{1s} + (J \pm K)/(1 \pm S), \tag{1.63}$$

with E_+ being lower than E_-. Going back and solving for the coefficients, we find $|c_A| = |c_B|$ as required by the symmetry of the problem, and the wavefunctions are $\psi_{\pm} = c_{\pm}(1s_A \pm 1s_B)$, where $c_{\pm}$ are overall normalization constants that are different for the two solutions. These are referred to as molecular orbitals (MOs), which have the explicit form of linear combinations of atomic orbitals (LCAOs). The wavefunction corresponding to the lower energy, ψ_+, has no nodes and is called a bonding orbital; ψ_- has a node halfway between the two nuclei and is called an antibonding orbital. The normalized solutions are

$$\psi_\pm = [2(1 \pm S)]^{-1/2}(1s_A \pm 1s_B). \tag{1.64}$$

ψ_+ is the first approximation for the spatial wavefunction of H_2^+.

The simplest model for H_2 puts two electrons, spin paired, into the same kind of bonding molecular orbital that we found as the lowest-energy solution for the one-electron H_2^+ molecular ion. The wavefunction must be antisymmetric with respect to interchange of electrons, and is most easily written as a Slater determinant. Expanding this determinant gives a wavefunction that factors into a product of spin and space parts,

$$\Psi(1,2) = \psi_+(1)\psi_+(2)\{\alpha(1)\beta(2) - \beta(1)\alpha(2)\}/\sqrt{2}, \tag{1.65}$$

where ψ_+ is the properly normalized sum of two 1s orbitals. The ground-state energy of H_2 at this level of approximation is given by $E = \int dr_1 dr_2 \Psi(1,2)\mathbb{H}_{H_2}\Psi(1,2)$ with $\mathbb{H}_{H2}$ being the Hamiltonian for H_2, given within the Born–Oppenheimer approximation in atomic units by

$$\mathbb{H}_{H_2} = -\nabla_1^2/2 - \nabla_2^2/2 - 1/r_{1A} - 1/r_{1B} - 1/r_{2A} - 1/r_{2B} + 1/r_{12} + 1/R. \tag{1.66}$$

The result for the energy, after doing the integrals, is that the binding energy (the difference in energy between the H_2 molecule and two isolated H atoms) is $260\,kJ\,mol^{-1}$, and the equilibrium nuclear separation is $R_e = 1.61a_0 = 0.85\,Å$, compared with experimental values of $457\,kJ\,mol^{-1}$ and $R_e = 0.74\,Å$. This is a pretty bad estimate for the energy, for two reasons. First, the best functional form for the one-electron molecular orbital is not a simple sum of two 1s atomic orbitals. This is also a problem for H_2^+. But H_2 has a second problem, and that is that the real wavefunction is not just a Slater determinant, where the spatial wavefunction is a product of two one-electron orbitals. In the real wavefunction, there is correlation between the motions of the two electrons due to their repulsive interaction. A number of different methods are used to handle this electron correlation in quantum chemistry calculations.

A useful and reasonable approximation to the ground-state wavefunctions of other homonuclear diatomic molecules is obtained by constructing molecular orbitals as plus and minus combinations of the hydrogenlike atomic orbitals. The MOs found for H_2^+, ψ_+, and ψ_-, are the two lowest-energy such orbitals, being constructed from the 1s hydrogenlike AOs. A similar pair of MOs can be made by taking the plus and minus combinations of the 2s hydrogenlike orbitals. Six more MOs can be formed from each of the three pairs of 2p orbitals. For symmetry reasons, p_x orbitals combine only with p_x, p_y with p_y, and p_z with p_z. This is because one type of p orbital on one atom has zero overlap integral with a different type of p orbital on the other atom. Thus, we build up a set of MOs, ordered according to their energies. First approximations to the ground state wavefunctions of higher diatomic molecules are then constructed by putting two electrons with opposite spin into each MO, starting from the lowest-energy MOs and working up.

In principle, one could form MOs from linear combinations of different AOs—for example, a 1s orbital on one atom and a 2s on another atom. However, because the isolated 1s and 2s orbitals have quite different energies when the nuclear charge is the same, they do not "couple" strongly to make MOs—the MOs formed from them turn out to be nearly the same as the original AOs, and the plus (bonding) combination is higher in energy than the plus combination formed from two 1s orbitals.

To get the best possible wavefunction, one uses a trial function that is a linear combination of a very large number of basis functions—hydrogenlike orbitals, Slater-type orbitals, Gaussian-type orbitals, and so on—and uses the variational method to find the best values of the coefficients. For example, for a homonuclear diatomic, one might use a trial function of the type

$$\psi = c_1 1s_A + c_2 1s_B + c_3 2s_A + c_4 2s_B + c_5 2p_{zA} + c_6 2p_{zB} + c_7 3s_A + c_8 3s_B.$$

Employing the variational method then gives eight different energies, corresponding to the orbital energies, and eight different sets of coefficients $\{c_n\}$ describing the orbitals. What is found is that the lowest-energy orbital is mostly $1s_A + 1s_B$, with some other stuff mixed in; the next-lowest is mostly $1s_A - 1s_B$, *etc.*

1.11. ENERGIES AND TIME SCALES; SEPARATION OF MOTIONS

A molecule consists of a collection of nuclei and electrons. The first step in describing its spectroscopy is generally to factor out the center of mass (CM) motion, defining a new coordinate system that has its origin at the molecule's center of mass. The total Hamiltonian can then be written as a sum of the kinetic energy of motion of the center of mass and the kinetic and potential energies of motion relative to that CM. The CM motion is not very interesting and for most purposes not very important and we will not consider it further. From now on, we will treat the molecule's center of mass as fixed (in other words, let the coordinate system move with the CM).

One then typically makes the Born–Oppenheimer approximation as mentioned previously. This consists of assuming that the total position–space wavefunction for the electrons and the nuclei can be factored into a function of the electronic positions **r** and a function of the nuclear positions **R**:

$$\Psi(\mathbf{r},\mathbf{R}) \approx \psi_{\text{nuc}}(\mathbf{R})\psi_{\text{elec}}(\mathbf{r};\mathbf{R}). \tag{1.67}$$

The notation (**r**; **R**) means that ψ_{elec} is explicitly a function of **r** only but depends parametrically on the nuclear coordinates **R**. The Born–Oppenheimer approximation is based on an approximate separation of time scales between nuclear and electronic motions. The electronic wavefunction does depend on the nuclear positions, but as the nuclei move, the electrons are able to keep up with that motion such that the electronic wavefunction is always at "equilibrium" with the nuclear motion. The idea that the electronic wavefunction adjusts itself instantaneously to keep up with the nuclei is also known as the adiabatic approximation. The nuclear wavefunction, on the other hand, depends on what the electrons are doing but not *explicitly* on their positions, since the electrons move so fast that all the nuclei see is their blurred-out average charge distribution. The combination of the nuclear–nuclear repulsions and the attraction of the nuclei to this averaged electron distribution produces a potential energy surface, $V(\mathbf{R})$, on which the nuclei move. It is different for each electronic state.

This separation of time scales implies a hierarchy of states and also of energy separations. Different electronic states are widely spaced in energy (thousands to tens of thousands of cm^{-1}), and each state has an energy that depends on the internal coordinates **Q**; this energy dependence, together with the internuclear repulsions, defines a potential surface $V(\mathbf{Q})$ for vibrational motion. Each electronic state has its own set of vibrational states, which are usually much more closely spaced in energy (10–3000 cm^{-1}). Therefore, one can sensibly talk about making transitions between different vibrational levels of the same electronic state. However, when considering a transition that changes the electronic state, we also have to consider what happens to the vibrations, since the effective Hamiltonians for nuclear motion are often quite different in different electronic states.

FURTHER READING

A. Das and A. C. Melissinos, *Quantum Mechanics: A Modern Introduction* (Gordon and Breach, New York1986.

M. D. Fayer, *Elements of Quantum Mechanics* (Oxford University Press, New York, 2001).

I. N. Levine, *Quantum Chemistry*, 6th edition (Prentice Hall, Upper Saddle River, NJ, 2009).

D. A. McQuarrie, *Quantum Chemistry*, 2nd edition (University Science Books, Mill Valley, CA, 2008).

J. J. Sakurai, *Modern Quantum Mechanics*, revised edition (Addison-Wesley, New York, 1993).

PROBLEMS

1. Consider a quantum-mechanical system in the state

$$\Psi = \frac{1}{\sqrt{5}}(\phi_1 + 2\phi_2),$$

where ϕ_1 and ϕ_2 are normalized eigenstates of the Hamiltonian with energy eigenvalues E and $2E$, respectively. Calculate the expectation value of the energy.

2. Find the expectation value of the energy of the state defined by $\Psi = \mathbb{q}\phi_v$, where ϕ_v is an eigenstate of the one-dimensional harmonic oscillator (arbitrary v), and q is the reduced position operator, $\mathbb{q} = (\mu\omega/\hbar)^{1/2}x$. Hint: Write $\mathbb{q}$ in terms of the harmonic oscillator raising and lowering operators.

3. Consider a simple model for an electron in a symmetric linear triatomic molecule, A—B—C, and assume that the electron can be described by a basis $\{\varphi_A, \varphi_B, \varphi_C\}$, denoting three orthonormal states localized on atoms A, B, and C. When we neglect "tunneling" of the electron from one center to another, its energy is described by the zero-order Hamiltonian $\mathbb{H}_0$ whose eigenstates and eigenvalues are

$$\mathbb{H}_0\varphi_A = \mathbb{E}_0\varphi_A, \qquad \mathbb{H}_0\varphi_B = \mathbb{E}_0\varphi_B, \qquad \mathbb{H}_0\varphi_C = \mathbb{E}_0\varphi_C.$$

The coupling is described by an additional Hamiltonian $\mathbb{H}$ defined by

$$\mathbb{W}\varphi_A = -\varepsilon\varphi_B, \qquad \mathbb{W}\varphi_B = -\varepsilon(\varphi_A + \varphi_C), \qquad \mathbb{W}\varphi_C = -\varepsilon\varphi_B,$$

where ε is a real positive constant having units of energy. Find the eigenvalues and eigenvectors of the total Hamiltonian, $\mathbb{H}_{total} = \mathbb{H}_0 + \mathbb{H}$.

4. Find the expectation values $\langle \mathbb{p}_i \rangle$ and $\langle \mathbb{p}_i^2 \rangle$ for each of the three components of the linear momentum ($i = x, y,$ or z) for an arbitrary energy eigenstate of a three-dimensional particle in a box having sides of length a, b, and c.

5. What are the degeneracies of the first four energy levels of a three-dimensional particle in a box with sides of length $a = b = 1.5c$?

6. Use raising and lowering operators to find general expressions for the expectation values $\langle \mathbb{x}^2 \rangle$ and $\langle \mathbb{x}^4 \rangle$ for a one-dimensional harmonic oscillator. Express your results in terms of the inverse length parameter $\alpha = (k\mu/\hbar^2)^{1/2}$ and the quantum number v.

7. The force constant of $^{79}Br^{79}Br$ is $240\,N{\cdot}m^{-1}$, and the equilibrium bond length is $2.28\,Å$. Calculate the fundamental vibrational frequency ω in radians per second, the period of one full vibration in picoseconds, and the zero-point energy in joules per mole, electron volts, and wavenumbers. Also, find the root-mean-square displacement, $\langle x^2 \rangle^{1/2}$, of this oscillator in its $v = 0$ state and express it as a percentage of the equilibrium bond length.

8. Show that the sum of the probability densities of the three 2p hydrogen atom orbitals is spherically symmetric by using the explicit forms for the complex wavefunctions. This is a special case of the general theorem

$$\sum_{m=-\ell}^{\ell} \left| Y_\ell^m(\theta, \varphi) \right|^2 = \text{constant},$$

known as Unsöld's theorem. Use this result to show that the sum of the probability densities for all the $n = 2$ hydrogen atom wavefunctions is spherically symmetric. Do you expect this to hold for other values of n? Explain.

9. Calculate the expectation value of the electron-nuclear separation, $\langle r \rangle$, for the hydrogen atom in states $(n,\ell,m) = (2,1,1)$, $(2,1,0)$, and $(2,0,0)$.

10. The Hamiltonian for a hydrogen atom in an external magnetic field along the z-axis is given by

$$\mathbb{H} = \mathbb{H}_0 + (\beta_B B_z/\hbar)\mathbb{L}_z,$$

where $\mathbb{H}_0$ is the Hamiltonian for the hydrogen atom in the absence of the external field. β_B is the Bohr magneton, a constant having units of joules per tesla ($J{\cdot}T^{-1}$), and B_z is the strength of the external magnetic field in units of Tesla. Find the eigenfunctions and eigenvalues of $\mathbb{H}$.

11. Consider the anharmonic oscillator defined by the potential

$$V(x) = \frac{1}{2}kx^2 + \frac{1}{6}\gamma x^3 + \frac{1}{24}\delta x^4,$$

where γ and δ are small constants.

(a) Use perturbation theory to estimate the ground-state energy to first order in the perturbation.

(b) Use the variational method to estimate the ground-state energy, taking a trial function of the form $\phi(x) = c_0\psi_0(x) + c_2\psi_2(x)$, where ψ_0 and ψ_2 are the $v = 0$ and $v = 2$ eigenfunctions of the corresponding harmonic oscillator, and c_0 and c_2 are variational parameters.

12. In terms of the two spatial molecular orbitals

$$\psi_+ = [2(1+S)]^{-1/2}(1s_A + 1s_B) \text{ and } \psi_- = [2(1-S)]^{-1/2}(1s_A - 1s_B),$$

and the spin functions α and β,

(a) Write the wavefunction $\Psi(1,2)$ that corresponds to the ground state of the H_2 molecule as a normalized Slater determinant.

(b) Write the three components of the wavefunction that correspond to the lowest energy triplet state of the H_2 molecule as normalized Slater determinants. In some cases, a linear combination of Slater determinants may be required.

(c) Write the two equivalent Slater determinants that correspond to the lowest energy singlet excited configurations of H_2. These differ only in the association of spin states with spatial functions.

13. Using your result for Problem 12(a), calculate the probability that when an H_2 molecule is in its ground electronic state, both electrons are found on the same atom (i.e., in either $1s_A$ or $1s_B$). Does this probability depend on the internuclear separation R? Discuss what this means physically.

14. Consider a particle of mass m in the one-dimensional potential defined as follows:

$$V(x) = \infty \text{ for } x < -L/2 \text{ or } x > L/2$$

$$V(x) = 0 \text{ for } -L/2 < x < 0$$

$$V(x) = V_0 \text{ for } 0 < x < L/2$$

Under what conditions do there exist one or more solutions having energy less than V_0? You do not need to calculate the energies, but simplify the expression you would need to solve as far as possible.

15. A variation on the particle in a box problem is the particle on a ring, where a particle of mass m is constrained to move in the xy plane on a ring of radius R. The position of the particle is defined by a single angular variable φ. The Hamiltonian is given by

$$\hat{\mathbb{H}} = -\frac{\hbar^2}{2I}\frac{d^2}{d\varphi^2},$$

where $I = mR^2$ is the moment of inertia.

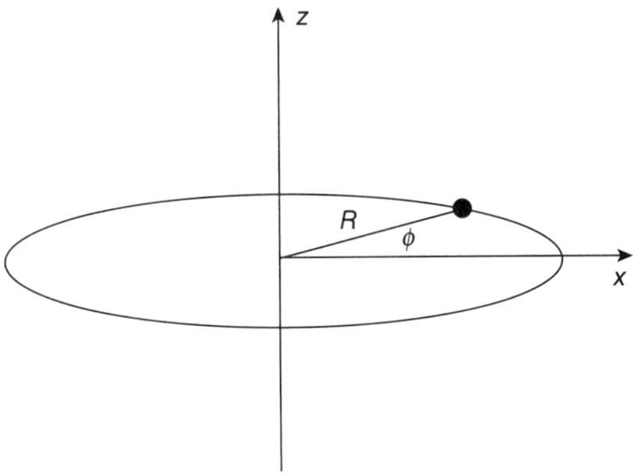

(a) Find the eigenvalues and eigenfunctions of this Hamiltonian.

(b) Consider the Hamiltonian

$$\hat{\mathbb{H}} = -\frac{\hbar^2}{2I}\frac{d^2}{d\varphi^2} + k\sin^2\varphi$$

where k is a constant. Use first-order perturbation theory to estimate the three lowest energy levels of this Hamiltonian.

16. The vibrational motion of a diatomic molecule having reduced mass μ and equilibrium bond length R_e can be well represented as a harmonic oscillator. The Hamiltonian is

$$\hat{\mathbb{H}}_a(R) = -\frac{\hbar^2}{2\mu}\frac{d^2}{dR^2} + \frac{1}{2}k(R - R_e)^2$$

The system starts out in the ground state of this harmonic oscillator potential, $\psi_0(R)$.

(a) The potential function is changed suddenly by adding a new term, $\mathbb{V}'(R) = -\gamma R$ where γ is a constant. Show that this new Hamiltonian, $\hat{\mathbb{H}}_b(R) = \hat{\mathbb{H}}_a(R) + \mathbb{V}'(R)$, still represents a harmonic oscillator with an added constant potential and a different equilibrium bond length.

(b) What is the probability that the molecule initially in $\psi_0(R)$ will be found in the ground state of this new harmonic oscillator potential?

CHAPTER 2

ELECTROMAGNETIC RADIATION

Spectroscopy involves the interaction of matter with electromagnetic radiation. Electromagnetic radiation may be described either classically or quantum mechanically. It is most often treated as a classical wave phenomenon involving oscillating electric and magnetic fields that propagate (in a vacuum) at the speed of light, c. For many spectroscopic applications, the classical description is fully adequate. For certain applications, however, it becomes important to recognize that the electromagnetic field is also quantized and consists of individual "particles" of energy known as photons. This chapter outlines the classical and quantum mechanical treatments of electromagnetic radiation, summarizes some of its properties, and describes briefly the characteristics of radiation sources used in spectroscopy.

2.1. CLASSICAL DESCRIPTION OF LIGHT

An electromagnetic field is described classically by the electric field $\mathbf{E}(\mathbf{r},t)$ and the magnetic field $\mathbf{H}(\mathbf{r},t)$. These quantities are vectors, meaning that they have both a magnitude and a direction that depend on time t and three-dimensional position $\mathbf{r}$. To include the effect of the field on matter, two more vectors are defined, the electric displacement $\mathbf{D}(\mathbf{r},t)$ and the magnetic induction $\mathbf{B}(\mathbf{r},t)$. These vectors are related through Maxwell's equations:

Condensed-Phase Molecular Spectroscopy and Photophysics, First Edition. Anne Myers Kelley.

$$\nabla \cdot \mathbf{D} = \rho \tag{2.1a}$$

$$\nabla \times \mathbf{E} = -\partial \mathbf{B}/\partial t \tag{2.1b}$$

$$\nabla \cdot \mathbf{B} = 0 \tag{2.1c}$$

$$\nabla \times \mathbf{H} - (\partial \mathbf{D}/\partial t) = \mathbf{J}, \tag{2.1d}$$

where ρ is the electric charge density (coulombs per cubic meter), and **J** is the electric current density (amperes per square meter). The vector operators appearing in Equation (2.1) are defined in Appendix B.

The vectors **D** and **E**, and **B** and **H**, are related through properties of the material in which the radiation propagates:

$$\mathbf{D} = \varepsilon \mathbf{E} = \varepsilon_0 \mathbf{E} + \mathbf{P} \tag{2.2a}$$

$$\mathbf{B} = \mu \mathbf{H} = \mu_0 \mathbf{H} + \mathbf{M}. \tag{2.2b}$$

Here, ε_0 and μ_0 are constants known as the permittivity and permeability of free space, respectively. **P** and **M** are the electric and magnetic polarization, respectively; these are properties of the medium. In the most general case, for example, in ordered materials such as crystals, ε and μ are second-rank tensors, and the vectors **D** and **E** (or **B** and **H**) are not necessarily parallel. However, these quantities are scalars in isotropic media and we will normally ignore their tensor properties. In addition, for most common materials at frequencies in the infrared to ultraviolet (corresponding to electronic and vibrational transitions), μ differs little from μ_0, and in the following treatment, we will normally assume $\mu = \mu_0$. On the other hand, for most condensed-phase materials, ε and ε_0 are significantly different at these frequencies. The quantity ε is known as the dielectric function or dielectric constant of the material, although it depends on the frequency of the radiation so it is not really a "constant" despite the common use of that term.

The quantities ρ and **J** act as sources of electromagnetic radiation. For spectroscopic applications, we are usually interested in the behavior of fields that are generated external to the spectroscopically active sample, for example, in a laser or a lamp. Thus, for the remainder of this discussion, we will set ρ and **J** to zero. Under these restrictions, one set of solutions to Maxwell's equations are the well-known monochromatic plane waves:

$$\mathbf{E}(\mathbf{r}, t) = (\mathbf{e}_1 E_0/2)\{\exp[i(\omega t - \mathbf{k} \cdot \mathbf{r})] + \exp[-i(\omega t - \mathbf{k} \cdot \mathbf{r})]\} \tag{2.3a}$$

$$\mathbf{H}(\mathbf{r}, t) = (\mathbf{e}_2 H_0/2)\{\exp[i(\omega t - \mathbf{k} \cdot \mathbf{r})] + \exp[-i(\omega t - \mathbf{k} \cdot \mathbf{r})]\}. \tag{2.3b}$$

Here, $\mathbf{e}_1$ and $\mathbf{e}_2$ are constant unit vectors, E_0 and H_0 are constant amplitudes assumed here to be real, and ω is the frequency in units of radians per second or angular frequency; $\omega = 2\pi\nu$, where ν is the frequency in cycles per second. E_0 and H_0 are the maximum amplitudes of **E** and **H**, respectively, and with $\rho = 0$ and $\mathbf{J} = 0$,

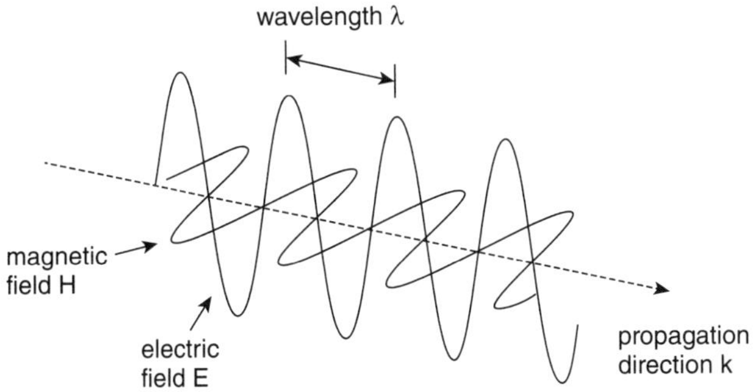

Figure 2.1. The transverse electric and magnetic fields associated with an electromagnetic plane wave.

$$H_0 = \sqrt{\frac{\varepsilon}{\mu}} E_0 \tag{2.4}$$

The quantity **k** is known as the wavevector, and its magnitude is $k = \omega\sqrt{\mu\varepsilon}$. In the absence of charges, Maxwell's equations require that $\nabla \cdot \mathbf{E} = 0$ and $\nabla \cdot \mathbf{H} = 0$, resulting in the requirement $\mathbf{e}_1 \cdot \mathbf{k} = \mathbf{e}_2 \cdot \mathbf{k} = 0$. This means that the **E** and **H** vectors are perpendicular to each other [by Eq. (2.1b)], and both are perpendicular to **k**, which we show later is the direction of propagation of the wave. **E** and **H** are transverse waves, as illustrated in Figure 2.1.

The condition $\omega t - \mathbf{k} \cdot \mathbf{r}$ = constant defines a surface of constant field strength, which is a plane normal to **k**; this surface travels in the direction of **k** with velocity $v = \omega/k$. This velocity is known as the phase velocity. At any instant in time, the distance between two adjacent peaks in **E**, the definition of the wavelength, is given by $\lambda = 2\pi/k$. Substituting the value of k into the phase velocity expression yields $v = 1/\sqrt{\mu\varepsilon}$. In a vacuum, $\varepsilon = \varepsilon_0$ and $\mu = \mu_0$, and the phase velocity defines the speed of light:

$$v = 1/\sqrt{\mu_0\varepsilon_0} = c. \tag{2.5a}$$

In any other medium, the phase velocity has the value

$$v = c/n, \tag{2.5b}$$

where $n = \mathrm{Re}\left(\sqrt{\mu\varepsilon/\mu_0\varepsilon_0}\right)$ is the refractive index of the medium, and Re indicates the real part. Since in most media $\mu \approx \mu_0$, this can usually be written as $n = \mathrm{Re}\left(\sqrt{\varepsilon/\varepsilon_0}\right) = \mathrm{Re}\left(\sqrt{\varepsilon_r}\right)$ where $\varepsilon_r = \varepsilon/\varepsilon_0$ is the "relative" refractive index. Because ε_r is a function of the frequency, the quantity $\sqrt{\varepsilon_r}$ varies with frequency, and is a complex-valued quantity at frequencies where the medium absorbs. In general,

$$\sqrt{\varepsilon_r} = n + i\kappa, \tag{2.6}$$

where the imaginary part, κ, is related to the absorption coefficient. In common usage, the term "refractive index" is usually applied to only the real part of this quantity (n) and is often assumed to be measured at optical frequencies, usually the sodium D line (589 nm).

The energy density u (energy per unit volume) of the electromagnetic field is given in general by

$$u = \frac{1}{2}(\boldsymbol{H} \cdot \boldsymbol{B} + \boldsymbol{E} \cdot \boldsymbol{D}). \tag{2.7}$$

Using Equations (2.2)–(2.4) and doing a little algebra, this becomes

$$u = \frac{\varepsilon E_0^2}{4}[2\cos(2\omega t - 2\boldsymbol{k} \cdot \boldsymbol{r}) + 2], \tag{2.8}$$

and when averaged over multiple cycles of the field, it becomes

$$u_{\text{avg}} = \frac{\varepsilon E_0^2}{2}. \tag{2.9}$$

Then setting $\varepsilon = \varepsilon_r \varepsilon_0$ and assuming $\varepsilon_r = n^2$, we have

$$u_{\text{avg}} = \frac{\varepsilon_0 n^2 E_0^2}{2}. \tag{2.10}$$

The electric field E_0 has units of $\text{J m}^{-1}\,\text{C}^{-1}$, while ε_0, the permittivity of free space, has units of $\text{C}^2\,\text{J}^{-1}\,\text{m}^{-1}$, so u_{avg} has units of J m^{-3}, or energy per unit volume.

Finally, for later reference, we need the expression for the energy flux across a surface perpendicular to **k**. This is referred to as the Poynting vector, **S**. Its magnitude, which we will refer to as the intensity of the radiation (energy per unit time per unit area), is simply the energy density times the speed of light in the medium:

$$I = u_{\text{avg}} \cdot v = \frac{\varepsilon_0 n^2 E_0^2}{2}\frac{c}{n} = \frac{\varepsilon_0 n c E_0^2}{2}. \tag{2.11}$$

The fields **E** and **B** are derivable from a vector potential **A** and a scalar potential Φ:

$$\mathbf{E}(\mathbf{r}, t) = -\{\partial \mathbf{A}(\mathbf{r}, t)/\partial t\} - \nabla \Phi(\mathbf{r}, t) \tag{2.12a}$$

$$\mathbf{B}(\mathbf{r}, t) = \nabla \times \mathbf{A}(\mathbf{r}, t). \tag{2.12b}$$

E and **B** defined in this way obey Maxwell's equations. It turns out that the same **E** and **B** can be generated by an infinite number of pairs of **A** and Φ related to one another by gauge transformations. We are free to choose the most convenient one. For spectroscopic applications, one typically chooses the

Coulomb gauge, in which $\nabla \cdot A(r,t) = 0$, and, when no charges are present, $\Phi(\mathbf{r},t) = 0$; then,

$$\mathbf{E}(\mathbf{r}, t) = -\partial \mathbf{A}(\mathbf{r}, t)/\partial t. \tag{2.13a}$$

$$\mathbf{B}(\mathbf{r}, t) = \nabla \times \mathbf{A}(\mathbf{r}, t). \tag{2.13b}$$

In order to generate the transverse plane wave fields given previously, the vector potential must have the form

$$\mathbf{A}(\mathbf{r}, t) = (iE_0/\omega)\mathbf{e}_1\{\exp[i(\omega t - \mathbf{k}\cdot\mathbf{r})] - \exp[-i(\omega t - \mathbf{k}\cdot\mathbf{r})]\}/2. \tag{2.14}$$

We will use both the electric field form and the vector and scalar potential form to describe spectroscopy in later chapters.

2.2. QUANTUM MECHANICAL DESCRIPTION OF LIGHT

For sufficiently strong fields, which practically means most light sources used in the laboratory, it is usually adequate to treat the electromagnetic field as a classical time-dependent perturbation. But in reality, radiation is quantized too, and this becomes important when the fields are weak, particularly when an excited molecule emits radiation at a frequency where there is no deliberately applied field (spontaneous emission).

Here we develop the quantum mechanical description of light under the assumption that the medium is nondissipative and that the dispersion (variation in the optical properties with wavelength) is small enough to be neglected. These somewhat restrictive conditions are chosen to keep the development reasonably simple. This section is based on the development of Nienhuis and Alkemade (1976) (see references at end of chapter), who additionally consider the case of a weakly dispersive medium. We again assume $\rho = 0$ and $\mathbf{J} = 0$, and work in the Coulomb gauge.

We first consider the radiation field to be confined to a three-dimensional box (like the particle in a box problem), where stationary, standing-wave solutions must have nodes at the ends of the box. The wavevector and the frequency are related by $k = \omega_k/(\varepsilon\mu)^{1/2}$ as introduced in Section 2.1. If the box has length L on each side, then to have standing waves the components of the wavevector must satisfy $k_x = 2\pi n_x/L$, where n_x is an integer, and similarly for k_y and k_z. The classical vector potential $\mathbf{A}$, subject to these box quantization conditions and Maxwell's equations, may be written as a sum of amplitudes for each of the allowed radiation modes ω_k and polarization directions α:

$$\mathbf{A}(\boldsymbol{r}, t) = \sum_k \sum_\alpha \left[c_{k\alpha}(t)u_{k\alpha}(\boldsymbol{r}) + c^*_{k\alpha}(t)u^*_{k\alpha}(\boldsymbol{r})\right], \tag{2.15}$$

where $u_{k\alpha}(\boldsymbol{r}) = \mathbf{e}_{k\alpha}e^{i\boldsymbol{k}\cdot\boldsymbol{r}}$, with $\mathbf{e}_\alpha$ being the unit polarization vector, and $c_{k\alpha}(t) = c_{k\alpha}e^{-i\omega t}$, with $c_{k\alpha}$ being a (complex) amplitude for the wave with wavevector $\mathbf{k}$

and polarization vector α. Since we have not specified the values of these amplitudes, this result is completely general.

The total energy of the pure radiation field in a vacuum is given classically by integrating Equation (2.7) over the volume of the box containing the field:

$$H_{\mathrm{rad}} = \frac{1}{2}\int_V dv(\boldsymbol{H}\cdot\boldsymbol{B}+\boldsymbol{E}\cdot\boldsymbol{D}). \tag{2.16a}$$

Making use of Equation (2.2), we can also write this as

$$\mathrm{H}_{\mathrm{rad}} = \frac{1}{2}\int_V dv\left(\frac{1}{\mu}|\boldsymbol{B}|^2+\varepsilon|\boldsymbol{E}|^2\right). \tag{2.16b}$$

Using this along with Equations (2.13)–(2.15) and performing the integral gives, after some manipulation,

$$H_{\mathrm{rad}} = V\sum_k\sum_\alpha \varepsilon\omega_k^2\left(c_{k\alpha}c_{k\alpha}^* + c_{k\alpha}^*c_{k\alpha}\right), \tag{2.17}$$

where V is the volume of the box. The sum over k runs over all allowed values of k_x, k_y, and k_z as given previously, and the sum over α runs over the two polarization directions $\mathbf{e}_\alpha$ allowed for each wavevector (two mutually perpendicular directions each perpendicular to $\mathbf{k}$). We next define the quantity

$$a_{k\alpha}(t) = \left(\frac{2V\varepsilon\omega_k}{\hbar}\right)^{\frac{1}{2}} c_{k\alpha}(t). \tag{2.18}$$

Inserting this into Equation (2.17) gives

$$H_{\mathrm{rad}} = \frac{1}{2}\sum_k\sum_\alpha \hbar\omega_k\left(a_{k\alpha}a_{k\alpha}^* + a_{k\alpha}^*a_{k\alpha}\right). \tag{2.19}$$

Now H_{rad} looks just like the Hamiltonian for a collection of quantum mechanical harmonic oscillators if we replace the classical quantities $a_{k\alpha}(t)$ and $a_{k\alpha}^*(t)$ with the corresponding quantum mechanical operators $\mathbb{a}_{k\alpha}(t)$ and $\mathbb{a}_{k\alpha}^\dagger(t)$, which are assumed to follow the same commutation relations as for the quantum harmonic oscillator:

$$[\mathbb{a}_{k\alpha}, \mathbb{a}_{k'\alpha'}^\dagger] = \delta_{\alpha\alpha'}\delta_{kk'} \tag{2.20a}$$

$$[\mathbb{a}_{k\alpha}, \mathbb{a}_{k'\alpha'}] = [\mathbb{a}_{k\alpha}^\dagger, \mathbb{a}_{k'\alpha'}^\dagger] = 0. \tag{2.20b}$$

Equation (2.19) then becomes

$$H_{\mathrm{rad}} = \sum_k\sum_\alpha \hbar\omega_k\left(\mathbb{a}_{k\alpha}^\dagger\mathbb{a}_{k\alpha} + \frac{1}{2}\right) = \sum_k\sum_\alpha \hbar\omega_k\left(\mathbb{N}_{k\alpha} + \frac{1}{2}\right), \tag{2.21}$$

where N_{ka} is the number operator. The actions of the raising and lowering operators for radiation are the same as for the harmonic oscillator:

$$\mathbb{a}_{k\alpha}\left|n_{k\alpha}\right\rangle = n_{k\alpha}^{1/2}\left|(n-1)_{k\alpha}\right\rangle \text{ and } \mathbb{a}_{k\alpha}^{\dagger}\left|n_{k\alpha}\right\rangle = (n_{k\alpha}+1)^{1/2}\left|(n+1)_{k\alpha}\right\rangle, \quad (2.22a)$$

and

$$\mathbb{a}_{k\alpha}\left|0_{k\alpha}\right\rangle = 0. \quad (2.22b)$$

Although the mathematics is analogous, the interpretation of the state $|n_{k1\alpha1}n_{k2\alpha2}\ldots\rangle$ is different for radiation than for matter. For matter, it would correspond to a state in which the physical oscillator described by index $(k1\alpha1)$ is excited to the $n_{k1\alpha1}$ quantum level, the oscillator given by mode $(k2\alpha2)$ is excited to level $n_{k2\alpha2}$, and so on. For radiation, the interpretation is that there are $n_{k1\alpha1}$ photons in the radiation mode defined by wavevector $k1$ and polarization direction $\alpha1$, $n_{k2\alpha2}$ photons in the mode described by $k2$ and $\alpha2$, and so on. Photons are massless particles that behave like quantum mechanical excitations of the radiation field. The operators $\mathbb{a}^{\dagger}$ and $\mathbb{a}$ for radiation are more often called creation and annihilation operators because they increase and decrease, respectively, the number of photons in the radiation field by one. The state $|0_{k1\alpha1}0_{k2\alpha2}\ldots\rangle = |0\rangle$, having no photons in any mode, is known as the vacuum state.

Photons are bosons, that is, they obey Bose–Einstein statistics; any number of photons may occupy the same mode of the radiation field. If this were not the case, lasers could not exist.

The number operator, $\mathbb{N}_{k\alpha} = \mathbb{a}_{k\alpha}^{\dagger}\mathbb{a}_{k\alpha}$, behaves the same as for material oscillators:

$$\mathbb{N}_{k\alpha}\left|\ldots n_{k\alpha}n_{k'\alpha'}\ldots\right\rangle = n_{k\alpha}\left|\ldots n_{k\alpha}n_{k'\alpha'}\ldots\right\rangle. \quad (2.23)$$

That is, it simply measures the number of photons in radiation mode $k\alpha$.

Equation (2.15) gave an expression for the classical vector potential in terms of radiation modes confined to a box of volume V. The corresponding quantum-mechanical operator, obtained by using Equation (2.18) and treating $a_{k\alpha}$ as a quantum mechanical operator, is

$$\mathbb{A}(\mathbf{r},t) = \frac{1}{\sqrt{\varepsilon V}}\sum\nolimits_k \sum\nolimits_\alpha \sqrt{\frac{\hbar}{2\omega_k}}\,\mathbf{e}_\alpha\left[\mathbb{a}_{k\alpha}e^{-i\omega_k t}e^{i\boldsymbol{k}\cdot\boldsymbol{r}} + \mathbb{a}_{k\alpha}^{\dagger}e^{i\omega_k t}e^{-i\boldsymbol{k}\cdot\boldsymbol{r}}\right], \quad (2.24)$$

and its time derivative gives the electric field in operator form,

$$\mathbb{E}(\mathbf{r},t) = \frac{i}{\sqrt{\varepsilon V}}\sum\nolimits_k \sum\nolimits_\alpha \sqrt{\frac{\hbar\omega_k}{2}}\,\mathbf{e}_\alpha\left[\mathbb{a}_{k\alpha}e^{-i\omega_k t}e^{i\boldsymbol{k}\cdot\boldsymbol{r}} - \mathbb{a}_{k\alpha}^{\dagger}e^{i\omega_k t}e^{-i\boldsymbol{k}\cdot\boldsymbol{r}}\right]. \quad (2.25)$$

From the standpoint of the radiation field, $\mathbb{a}_{k\alpha}$ and $\mathbb{a}_{k\alpha}^{\dagger}$ are operators, while t and $\mathbf{r}$ are just parameters. The position $\mathbf{r}$ is an operator on the states of matter but not on the states of radiation.

The quantum mechanical form for the radiation field has to be used in order to properly develop the theory of spontaneous emission or multiphoton processes that involve a spontaneous emission step, such as Raman scattering. It can also be used to describe processes, such as absorption, which are equally well handled by treating the electromagnetic field classically.

2.3. FOURIER TRANSFORM RELATIONSHIPS BETWEEN TIME AND FREQUENCY

Most spectroscopy experiments involve exciting a sample with some source of electromagnetic radiation. Specific types of sources are discussed in the final section in this chapter. Here we discuss some of the general considerations that govern the resulting electromagnetic fields.

Consider a monochromatic electromagnetic field. The electric field has the form given previously,

$$\mathbf{E}(\mathbf{r},t) = \mathbf{e}\mathrm{E}_0\{\exp[i(\omega t - \mathbf{k}\cdot\mathbf{r})] + \mathrm{c.c.}\}/2 = \mathbf{e}E_0\cos(\omega t - \mathbf{k}\cdot\mathbf{r}), \tag{2.26}$$

where "c.c." refers to complex conjugate. To further simplify, let us focus on the behavior of the field as a function of time at a given point in space, which we take to be the origin; then $\mathbf{E}(t) = \mathbf{e}E_0\{e^{i\omega t} + e^{-i\omega t}\}/2 = \mathbf{e}E_0\cos(\omega t)$.

Most light detectors used in the laboratory are so-called "square law" detectors; they produce a signal that is proportional to the intensity of the radiation, energy per unit area per unit time. The intensity is proportional to the square of the electric field, $|E(t)|^2$. At any selected point in space, the intensity varies with time as $\cos^2(\omega t)$. However, this oscillation is almost never observed directly because most detectors used in the laboratory have a finite response time that is much longer than the oscillation frequency of the field (several cycles per femtosecond for visible light). The signal reported by the detector is then obtained by averaging $|E(t)|^2$ over many optical cycles (i.e., over a time long compared with $1/\omega$), and the time-averaged value of $\cos^2(\omega t)$ is ½. The measured intensity of the monochromatic field defined in Equation (2.26) is therefore constant in time and equal to $\varepsilon_0 ncE_0^2/2$ as given in Equation (2.11).

Most real light sources consist of more than one frequency and/or polarization component, and therefore need to be expressed more generally as

$$\mathbf{E}(\mathbf{r},t) = \sum_k \sum_\alpha \mathbf{e}_\alpha E_{k\alpha}\{\exp[i(\omega_k t - \mathbf{k}\cdot\mathbf{r})] + \mathrm{c.c.}\}/2. \tag{2.27}$$

The coefficient $E_{k\alpha}$ gives the amplitude of the wave having wavevector $\mathbf{k}$ and polarization α. Since we wish to focus on the time dependence of the field, we will again assume that all components have the same polarization direction and drop the $\mathbf{e}_\alpha$. If many waves with similar frequencies are superimposed, the sum over $\mathbf{k}$ may be replaced by an integral over frequencies ω, as

$$E(t) = \frac{1}{2}\int d\omega\{E(\omega)\exp(i\omega t) + \text{c.c.}\}. \tag{2.28}$$

Now let us consider some special cases. If only one frequency ω_0 is present, then $E(\omega) = E\delta(\omega - \omega_0)$. Performing the integral gives

$$E(t) = \frac{1}{2}\{E\exp(i\omega_0 t) + \text{c.c.}\},$$

a monochromatic wave as in Equation (2.26); as discussed previously, the intensity of this field averaged over multiple cycles is constant in time. In the opposite limit where all frequencies are present with equal amplitudes, $E(\omega) = E$, we have

$$E(t) = \frac{1}{2}E\int d\omega\exp(i\omega t) = \pi E\delta(t).$$

That is, the result of superimposing plane waves with all possible frequencies and equal amplitudes is a disturbance that is infinitely sharply peaked at one time, here defined as $t = 0$ at the origin. Notice that when the frequency of the wave is completely defined (monochromatic wave), there is no localization in time; a molecule sitting at a particular point in space will see a constant light intensity at all times. If the frequency of the wave is completely undefined (all frequencies are equally weighted), then there is perfect localization in time; a molecule sitting at the origin will see a zero field intensity for times before or after zero, and a sharp spike right at $t = 0$.

These are two extreme examples, neither of which is ever exactly realized in the laboratory. A more realistic description is a distribution given by $E(\omega) = (2\pi\sigma^2)^{-1/2}\exp[-(\omega-\omega_0)^2/2\sigma^2]$. This is a Gaussian distribution in frequency, centered around ω_0 with a standard deviation of σ; the pre-exponential factor normalizes the distribution. The temporal distribution of the field is given by

$$\begin{aligned} E(t) &= \frac{1}{2}(2\pi\sigma^2)^{-1/2}\int d\omega\exp[-(\omega-\omega_0)^2/2\sigma^2]\{\exp(i\omega t) + \text{c.c.}\} \\ &= \frac{1}{2}\exp(-\sigma^2 t^2/2)\{\exp(i\omega_0 t) + \text{c.c.}\}, \end{aligned}$$

and the cycle-averaged intensity as a function of time is $I(t) \sim \exp(-\sigma^2 t^2)$. This is a Gaussian function of time with a standard deviation of $1/\sigma$. That is, as the field becomes more sharply peaked in frequency (smaller σ), it becomes broader in time (Fig. 2.2). This is often referred to as the "frequency-time uncertainty principle," but it has a fundamentally different origin than do the quantum-mechanical uncertainty relationships, such as $\Delta x\Delta p_x \geq \hbar/2$. The quantum-mechanical relationships derive from the commutation relations among operators, while the frequency–time relation is a purely classical result of the mathematics involved in superimposing waves.

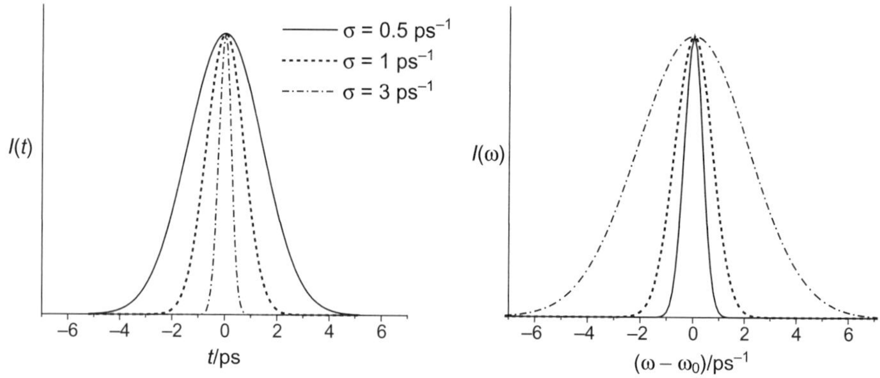

Figure 2.2. Intensity as a function of time (left) and of frequency (right) for Gaussian envelope pulses for different values of the width parameter σ.

Equation (2.28) fits the mathematical definition of a Fourier transform (see Appendix E). If $F(x)$ is some arbitrary function of x, then its Fourier transform $f(\alpha)$ is a function of another variable, α, and is defined by

$$f(\alpha) = \frac{1}{\sqrt{2\pi}} \int_{-\infty}^{\infty} dx F(x) e^{i\alpha x} \tag{2.29}$$

(different definitions sometimes differ in the $1/\sqrt{2\pi}$ prefactor). Many functions do not have analytic Fourier transforms, but these cases can be solved numerically on a computer very quickly and accurately. However, it is generally true that as $F(x)$ becomes a more sharply peaked function of x, $f(\alpha)$ becomes a less sharply peaked function of α. In our case, the variables x and α correspond to frequency and time, respectively. Whether $E(\omega)$ is Gaussian or some other reasonably smooth, well-behaved distribution, narrowing the frequency distribution will make the temporal distribution broader, and vice versa.

These considerations are important when planning or analyzing experiments carried out with pulsed lasers. As discussed at the end of this chapter, it is now possible to create very short light pulses (tens to hundreds of femtoseconds, where $1\,\text{fs} = 10^{-15}$ seconds) with commercial laser sources, and such short pulses are often desirable for a number of reasons. However, light pulses that are very short in time contain a very broad range of frequencies, and that can be disadvantageous for spectroscopy.

2.4. BLACKBODY RADIATION

Objects that are hot can transfer energy to their environment both by contact and through emission of radiation. Objects that are only moderately hot emit mainly infrared and longer wavelengths, while very hot objects also emit significant amounts of visible and even ultraviolet radiation. The sun is one example of an object at a high temperature that emits infrared, visible, and some ultraviolet light. An idealized description of such an emitter is a body that is at thermal equilibrium with its environment and can absorb and emit radiation at all frequencies. Such a body is referred to as a blackbody, and its emission spectrum as blackbody radiation or blackbody emission.

We will not derive the blackbody emission spectrum here, but merely give the result. Blackbody radiation at temperature T has a spectrum $\rho_{rad}(\nu)$ (energy per unit volume per unit frequency) given by (Yariv, 1975)

$$\rho_{\text{rad}}(\nu) = \frac{8\pi h n^3 \nu^3}{c^3} \frac{1}{\exp(h\nu / k_B T) - 1}, \tag{2.30a}$$

or alternatively in terms of angular frequency as

$$\rho_{\text{rad}}(\omega) = \frac{\hbar n^3 \omega^3}{\pi^2 c^3} \frac{1}{\exp(\hbar\omega / k_B T) - 1}, \tag{2.30b}$$

where h is Planck's constant, c is the speed of light, n is the refractive index of the medium, and k_B is Boltzmann's constant. The subscript "rad" is used to distinguish this ρ from the density of states encountered in the next chapter. Examples of blackbody emission curves for several different temperatures are shown in Figure 2.3. Note that the units of this plot, and of Equation (2.30a), are energy per unit frequency per unit volume. If plotted as energy per unit wavelength per unit volume, these curves look different because of the inverse relationship between frequency and wavelength.

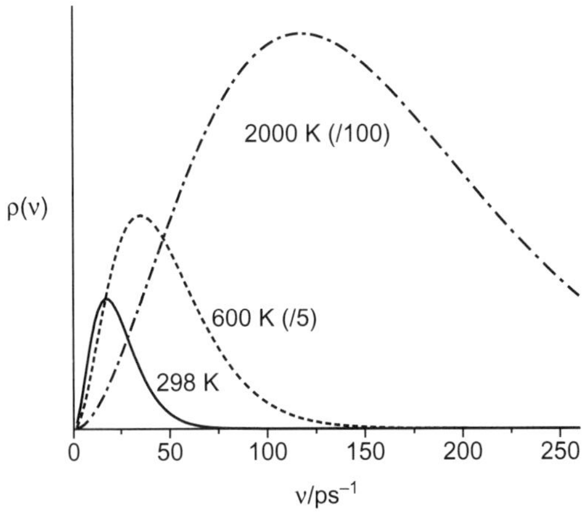

Figure 2.3. Blackbody radiation spectrum (energy per unit frequency per unit volume) for three different temperatures, scaled by the indicated factors.

While real objects are not perfect blackbody emitters, the sun and some broadband light sources used for laboratory spectroscopy are reasonably well approximated as such. This assumption is also used in the derivation of the Einstein A and B coefficients that link absorption, stimulated emission, and spontaneous emission rates (Chapter 4).

2.5. LIGHT SOURCES FOR SPECTROSCOPY

Most light sources used for laboratory spectroscopy fall into one of two broad classes: lamps and lasers.

Many commercial optical instruments, such as UV-visible spectrophotometers and fluorimeters, require relatively low intensity, continuously tunable radiation over a broad frequency range. Such instruments typically use high-pressure deuterium or tungsten lamps to generate the radiation, followed by some wavelength selective element, such as a monochromator or optical filter, to select the desired band of wavelengths from the broadband source. Fourier transform infrared spectrometers use a heated material to generate a blackbody-like emission peaked in the infrared, pass all wavelengths through the sample, and then analyze the spectrum of the transmitted light. Other spectrometers utilize low-pressure lamps based on atomic emission to generate a small number of emission lines sharply peaked around specific frequencies.

What all of these sources have in common is that the light is generated mainly through spontaneous emission and the atoms or molecules undergoing the emission are usually randomly oriented. Therefore, all directions of the wavevector and both polarization directions associated with a particular **k** are equally probable, although the packaging and optics associated with the lamp may limit emission to a narrow band of wavevector directions. Furthermore, since there is no particular relationship between the times at which different emitters emit, there is no phase relationship between the different waves. For these reasons lamps are characterized as incoherent light sources. Some lamps emit this incoherent light in a continuous manner while others are pulsed, but light pulses from incoherent sources are usually fairly long compared with most molecular time scales, hundreds of nanoseconds to milliseconds.

Laser sources are fundamentally different in that the radiation is built up by stimulated emission in a one-dimensional cavity. Therefore, the only radiation modes significantly populated are those having **k** oriented along a particular axis, say the x-axis. (More precisely, the wavevectors are distributed within some very small range of solid angles $d\Omega$ around the x-axis.) Also, most laser cavities contain a polarization selective element such that modes with only one polarization vector **e** are significantly populated. Finally, since the output of a laser is built up by oscillation in a cavity of finite length, there is a quantization condition on the allowed wavevectors just as there was when we quantized the radiation field in a three-dimensional cavity: $k_x = 2\pi n_x/L$, where n_x is an integer and L is the laser cavity length. The allowed modes of the radiation field are therefore spaced by $2\pi/L$. For visible radiation (500 nm) and a typical cavity length around 1 m, the wavelength spacing between successive modes is

$$\Delta\lambda = 2\pi\Delta k/k^2 = 2\pi\Delta k/(2\pi/\lambda)^2 = \lambda^2\Delta k/2\pi = (500\text{ nm})^2(2\pi/10^9\text{ nm})/2\pi$$
$$= 2.5\times10^{-4}\text{ nm},$$

or a wavenumber spacing of

$$\Delta(1/\lambda) = \Delta\lambda/\lambda^2 = (2.5\times10^{-4}\text{ nm})/(500\text{ nm})^2 = 10^{-9}\text{ nm}^{-1} = 0.01\text{ cm}^{-1}.$$

Note that the wavelengths used in these calculations need to be the wavelengths in the lasing medium, not the vacuum wavelengths. If the linewidth of

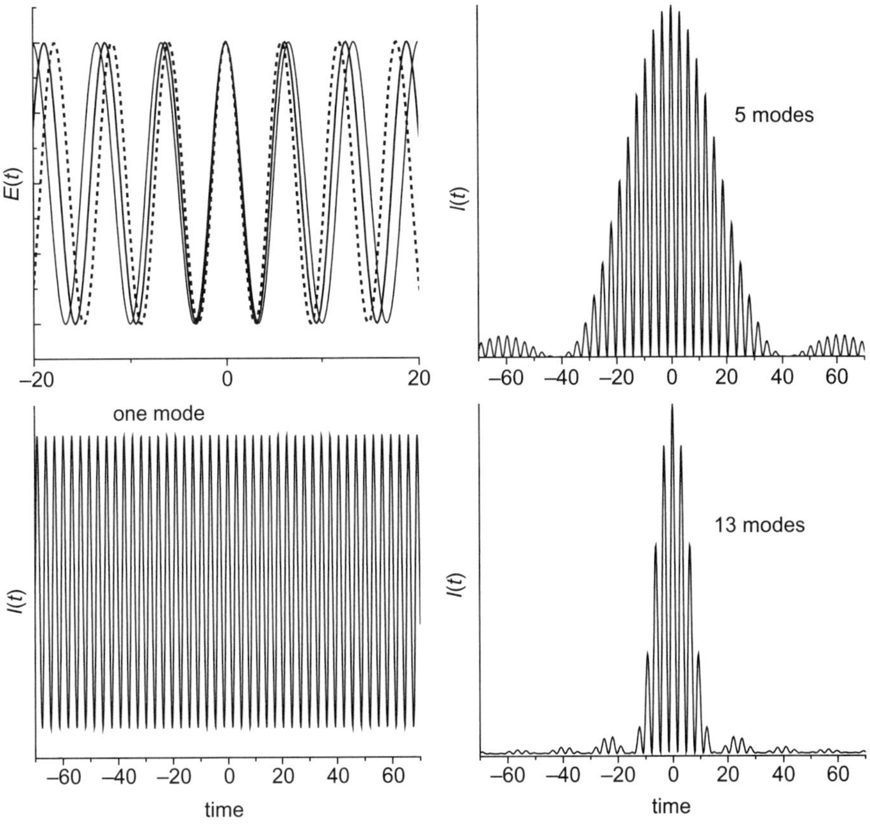

Figure 2.4. Upper left: Electric field amplitudes for three different modes of a laser cavity having slightly different frequencies. The peaks in $E(t)$ all occur at slightly different times except for the common peak at $t = 0$. Lower left: Field intensity, $I(t) \sim E^2(t)$, for a single mode. When averaged over a cycle of the radiation, the intensity is constant in time. Upper right: Field intensity for a sum of five modes of the cavity. Lower right: Field intensity for a sum of 13 modes. As the number of modes increases, the intensity peak around $t = 0$ becomes sharper, that is, the pulse becomes shorter.

the material transition to be probed is very much broader than this spacing, then the discreteness of the laser modes does not matter. If, however, the material linewidth is on the same order as or narrower than the intermode spacing, the discreteness certainly does matter. In either case, the distribution of wavevectors making up the electric field really is given by a discrete sum, not an integral, but it may often be approximated by an integral.

In an ordinary continuous-wave laser, lasing "at a particular wavelength" corresponds to significant amplitudes in a number of modes (tens to thousands) near some central wavelength. The phases of the different modes are randomly distributed, so the oscillations in the different modes interfere with one another, and the intensity of the resulting field, $I(t) \sim |E(t)|^2$, has a lot of "noise" in it when examined on short time scales. In a single-mode laser, additional frequency selective elements are introduced into the cavity to make it possible for only one single mode (one $\mathbf{k}$) to lase at a time. Now since there is no interference among modes, the intensity of the output is very stable even on the shortest time scales, and the frequency of the laser is very well defined. In a mode-locked laser, some group of modes around a central wavevector are allowed to lase, and a time-dependent loss is introduced into the cavity. The result is that the *phases* of the different modes develop a well-defined relationship such that at any point in space, they all add up constructively (*in phase*) at certain times, producing a pulse in the intensity envelope. At all other times, the phases are random, and the different frequency components interfere destructively, producing little or no net intensity. The larger the number of modes present (more frequency components), the sharper this pulse becomes in time; that is, temporally short pulses are broad in frequency and *vice versa*, as discussed previously (Fig. 2.4).

REFERENCES AND FURTHER READING

J. D. Jackson, *Classical Electrodynamics*, 2nd edition (John Wiley & Sons, New York, 1975).

G. Nienhuis and C. T. J. Alkemade, Physica **81C**, 181 (1976).

A. Yariv, *Quantum Electronics*, 2nd edition (John Wiley & Sons, New York, 1975).

PROBLEMS

1. A CO_2 laser producing light of wavelength 9.6 μm (infrared radiation) uses an electrical power of 5 kW (5000 W). If the laser is 27% efficient (i.e., 27% of the electrical power consumed is converted into optical power), how many photons per second does the laser emit?

2. Find the functional form of the cycle-averaged intensity as a function of time, $I(t)$, for electric fields having the following frequency distributions:

(a) $E(\omega) = E_0\exp[-|\omega-\omega_0|/\sigma]$
(b) $E(\omega) = E_0\sin[(\omega-\omega_0)/\sigma]/[(\omega-\omega_0)/\sigma]$.

3. A classical vector potential is given by $\boldsymbol{A}(\boldsymbol{r},t) = A_0(\hat{\boldsymbol{x}}+\hat{\boldsymbol{y}})\cos(\boldsymbol{k}\cdot\boldsymbol{r}-\omega t)$, where $\boldsymbol{k}$ is along $\hat{\boldsymbol{z}}$, and $\hat{\boldsymbol{x}}$, $\hat{\boldsymbol{y}}$, and $\hat{\boldsymbol{z}}$ are unit vectors in the three Cartesian directions. Calculate $\boldsymbol{E}(\boldsymbol{r},t)$ and $\boldsymbol{B}(\boldsymbol{r},t)$ and show that they obey Maxwell's equations in a vacuum.

4. Show that in a vacuum, the vector potential in the Coulomb gauge must satisfy the equation

$$\nabla^2 \boldsymbol{A}(\boldsymbol{r},t) = \mu_0\varepsilon_0 \frac{\partial^2}{\partial t^2}\boldsymbol{A}(\boldsymbol{r},t).$$

5. The output of a He-Ne laser (wavelength 632.8 nm) propagating through air has an intensity of $0.1\,\mathrm{W/cm^2}$. Treating it as a monochromatic plane wave,

(a) Calculate the associated electric field amplitude, E_0.

(b) Calculate the photon flux (number of photons per square meter per second).

6. A solid-state diode laser has an optical cavity length of 600 μm. The refractive index of the laser material is $n = 4$ at the lasing wavelength of 1550 nm (measured in air). The laser linewidth (full width at half maximum) is 500 GHz.

(a) What is the linewidth in nm?

(b) How many modes of the laser cavity fall within the linewidth?

7. (a) Show explicitly where the factor of 1/2 in Equation (2.21) comes from in converting the classical expression in Equation (2.19) to its quantum mechanical analog.

(b) Equation (2.21) may be rewritten as

$$\mathbb{H}_{\mathrm{rad}} = \sum_{k,\alpha} \hbar\omega_k \mathbb{N}_{k\alpha} + \sum_{k,\alpha} \frac{\hbar\omega_k}{2}.$$

What does this say about the energy associated with the field when the number of photons is zero? What happens to this energy as the size of the box used for quantization becomes infinitely large? Is this a problem?

8. In the chapter, the statement is made that lasers could not exist if photons were not bosons (particles whose spin statistics dictate that any number may occupy the same quantum mechanical state). Based on what you know about lasers, explain this statement.

CHAPTER 3

RADIATION–MATTER INTERACTIONS

Chapter 1 summarized key principles and results from time-independent quantum mechanics. This chapter begins with a summary and review of time-dependent quantum mechanics, and then addresses the interaction of time-dependent electromagnetic fields with matter to produce transitions between their quantum-mechanical stationary states.

3.1. THE TIME-DEPENDENT SCHRÖDINGER EQUATION

The evolution of a quantum mechanical state between two time intervals is entirely determined by the time-dependent Schrödinger equation:

$$i\hbar\frac{d}{dt}\Psi(t)=\mathbb{H}(t)\Psi(t). \tag{3.1}$$

This is true regardless of whether Ψ is an eigenstate of the Hamiltonian $\mathbb{H}(t)$, which in general may be time dependent. If the Hamiltonian is time independent, and the states are expressed as wavefunctions in three-dimensional space, the time-dependent Schrödinger equation is

$$i\hbar\frac{d}{dt}\Psi(\boldsymbol{r},t)=\left\{-\frac{\hbar^2}{2m}\boldsymbol{\nabla}^2+\mathrm{V}(\mathbf{r})\right\}\Psi(\boldsymbol{r},t). \tag{3.2}$$

Condensed-Phase Molecular Spectroscopy and Photophysics, First Edition. Anne Myers Kelley.

Expectation values in general change with time, and we can determine their time evolution by using the time-dependent Schrödinger equation. Let $\mathbb{O}$ be an observable; how does $<\mathbb{O}>_t = <\Psi(t)|\mathbb{O}|\Psi(t)>$ evolve with time? Differentiate the expectation value with respect to time:

$$\frac{d}{dt}\int \Psi^*(t)\mathbb{O}\Psi(t) = \int \frac{d\Psi^*(t)}{dt}\mathbb{O}\Psi(t) + \int \Psi^*(t)\mathbb{O}\frac{d\Psi(t)}{dt} + \int \Psi^*(t)\frac{d\mathbb{O}}{dt}\Psi(t). \quad (3.3)$$

The first two terms can be simplified by using the time-dependent Schrödinger equation and its Hermitian conjugate:

$$i\hbar\frac{d}{dt}\Psi(t) = \mathbb{H}(t)\Psi(t) \quad \text{and} \quad -i\hbar\frac{d}{dt}\Psi^*(t) = \Psi^*(t)\mathbb{H}(t) \quad (3.4)$$

(operators operate to the left on the complex conjugate of the state vector). So,

$$\begin{aligned}\frac{d}{dt}\langle\mathbb{O}\rangle_t &= -\frac{1}{i\hbar}\langle\Psi(t)|\mathbb{H}(t)\mathbb{O}|\Psi(t)\rangle + \frac{1}{i\hbar}\langle\Psi(t)|\mathbb{O}\mathbb{H}(t)|\Psi(t)\rangle + \left\langle\Psi(t)\left|\frac{d\mathbb{O}}{dt}\right|\Psi(t)\right\rangle \\ &= \frac{1}{i\hbar}\langle[\mathbb{O},\mathbb{H}(t)]\rangle + \left\langle\frac{d\mathbb{O}}{dt}\right\rangle.\end{aligned} \quad (3.5)$$

The last term is zero unless the operator $\mathbb{O}$ contains time *explicitly* as a parameter. This shows that if an operator commutes with the Hamiltonian and does not depend explicitly on time, its expectation values do not change with time although the wavefunction itself does change. Such operators are called constants of the motion.

An important application of this result is to position and momentum, the result of which has a special name: Ehrenfest's theorem. Let the Hamiltonian be that of a particle of mass m moving in a static, one-dimensional potential: $\mathbb{H} = (1/2m)\mathbb{P}^2 + V(\mathbb{Q})$. By the previous result,

$$\frac{d}{dt}\langle\mathbb{Q}\rangle_t = (1/i\hbar)\langle[\mathbb{Q},\mathbb{H}]\rangle = (1/2im\hbar)\langle[\mathbb{Q},\mathbb{P}^2]\rangle$$

and

$$\frac{d}{dt}\langle\mathbb{P}\rangle_t = (1/i\hbar)\langle[\mathbb{P},\mathbb{H}]\rangle = (1/i\hbar)\langle[\mathbb{P},V(\mathbb{Q})]\rangle.$$

One can show that $[\mathbb{Q}, \mathbb{P}^2] = 2i\hbar\mathbb{P}$ and $[\mathbb{P}, V(\mathbb{Q})] = -i\hbar(d/d\mathbb{Q})V(\mathbb{Q})$, giving

$$\frac{d}{dt}\langle\mathbb{Q}\rangle_t = \frac{1}{m}\langle\mathbb{P}\rangle_t \quad \text{and} \quad \frac{d}{dt}\langle\mathbb{P}\rangle_t = -\langle V'(\mathbb{Q})\rangle_t, \quad (3.6)$$

where the prime indicates differentiation with respect to $\mathbb{Q}$. Now take the second time derivative:

$$\frac{d^2}{dt^2}\langle\mathbb{Q}\rangle_t = \frac{1}{m}\frac{d}{dt}\langle\mathbb{P}\rangle_t = -\frac{1}{m}\langle \mathrm{V}'(\mathbb{Q})\rangle_t \tag{3.7}$$

or, $-\langle \mathrm{V}'(\mathbb{Q})\rangle_t = \mathrm{m}\frac{d^2}{dt^2}\langle\mathbb{Q}\rangle_t$, which is analogous to $F = ma$ in Newton's equations of motion. A system in which the Hamiltonian does not depend on time is called conservative. The state vector for such a system does, in general, depend on time even if the Hamiltonian does not. The usual way to deal with this is to write the time-dependent state as a linear combination of the time-independent eigenstates of the Hamiltonian, $\{\varphi_i\}$, with time-dependent coefficients $c_i(t)$, $\Psi(t) = \Sigma_i c_i(t)\varphi_i$. Plug this into the time-dependent Schrödinger equation, multiply by a particular state φ^*_n, and integrate to get

$$i\hbar\sum_i \langle\phi_n|\phi_i\rangle\frac{d}{dt}c_i(t) = \sum_i \langle\phi_n|\mathbb{H}|\phi_i\rangle c_i(t),$$

φ_i is an eigenstate of $\mathbb{H}$ with eigenvalue E_i, and since different eigenstates are orthogonal, only the $i = n$ terms survive, leaving

$$i\hbar\frac{d}{dt}c_n(t) = \sum_i c_i(t)E_i\langle\phi_n|\phi_i\rangle = c_n(t)E_n. \tag{3.8}$$

Solving for the coefficient gives

$$c_n(t) = c_n(t_0)\exp\left[-\frac{iE_n(t-t_0)}{\hbar}\right].$$

Thus, the time evolution of the state vector is given by

$$\Psi(t) = \sum_i c_i(t_0)\exp\left[-\frac{iE_i(t-t_0)}{\hbar}\right]\phi_i. \tag{3.9}$$

In the case where $\Psi(t_0)$ is already an eigenstate of the Hamiltonian with eigenvalue E, then only one of the coefficients is nonzero, and

$$\Psi(t) = \exp\left[-\frac{iE(t-t_0)}{\hbar}\right]\Psi(t_0). \tag{3.10}$$

The time evolution changes the state by just an overall phase factor. All expectation values of observables, probabilities, and so on become time-independent. Eigenfunctions of the Hamiltonian are therefore called stationary states.

3.2. TIME-DEPENDENT PERTURBATION THEORY

Thus far, we have dealt only with Hamiltonians that are not explicitly time dependent. Many interesting dynamics can occur even in this situation if the system starts out in a state that is not an eigenstate of that constant Hamiltonian. But spectroscopy involves *explicitly* time-dependent potentials.

In "traditional" spectroscopy, the forces on the particles due to the external time-varying field are generally rather small compared with those already present in the absence of the field. In this case, it is appropriate to treat the field as a perturbation, and that is the approach we will take in most of what follows. Note, however, that with modern lasers, it is actually quite easy to subject matter to fields that cannot be considered small in the perturbative sense, and nonperturbative descriptions of radiation–matter interactions are sometimes necessary.

Consider a Hamiltonian composed of a time-invariant "zeroth-order" part, $\mathbb{H}_0$, and a time-varying "perturbative" part, $\mathbb{W}(t)$. Let the eigenstates of $\mathbb{H}_0$ be $\{\varphi_n\}$.

$$\mathbb{H} = \mathbb{H}_0 + \mathbb{W}(t) \tag{3.11}$$

Assume that $\mathbb{W}(t)$ is turned on suddenly at time $t = 0$ and that before the perturbation is turned on the system is in one of the eigenstates of $\mathbb{H}_0$, ϕ_i. At a later time, the system will in general be in a superposition of all of the ϕ_n due to the coupling through $\mathbb{W}(t)$:

$$\Psi(t) = \sum_n c_n(t)\phi_n. \tag{3.12}$$

We seek the probability for finding the system in an arbitrary different final state, ϕ_f, at a later time t:

$$P_{if}(t) = |\langle\phi_f|\Psi(t)\rangle|^2 = |c_f(t)|^2. \tag{3.13}$$

Start with the time-dependent Schrödinger equation:

$$i\hbar\frac{d}{dt}\Psi(t) = [\mathbb{H}_0 + \mathbb{W}(t)]\Psi(t).$$

Write $\Psi(t)$ as a superposition of $\mathbb{H}_0$ eigenstates:

$$i\hbar\sum_n \phi_n \frac{d}{dt}c_n(t) = \sum_n c_n(t)[\mathbb{H}_0 + \mathbb{W}(t)]\phi_n$$

Take the product of both sides with ϕ_k^* and integrate:

$$i\hbar\sum_n \langle\phi_k|\phi_n\rangle\frac{d}{dt}c_n(t)=\sum_n c_n(t)[\langle\phi_k|\mathbb{H}_0|\phi_n\rangle+\langle\phi_k|\mathbb{W}(t)|\phi_n\rangle]$$

$$i\hbar\sum_n \delta_{kn}\frac{d}{dt}c_n(t)=\sum_n c_n(t)[E_n\delta_{kn}+W_{kn}(t)]$$

$$i\hbar\frac{d}{dt}c_k(t)=c_k(t)E_k+\sum_n c_n(t)W_{kn}(t). \tag{3.14}$$

Now if there were no perturbation, we would have

$$c_k(t)=c_k(0)\exp\left[-\frac{iE_kt}{\hbar}\right].$$

We will assume that in the presence of the perturbation,

$$c_k(t)=b_k(t)\exp\left[-\frac{iE_kt}{\hbar}\right],$$

where we expect that $b_k(t)$ will vary slowly in time compared with $\exp(-iE_kt/\hbar)$. But we have made no approximation yet, since $b_k(t)$ is still completely general. Notice that with this definition, $|c_k(t)|^2 = |b_k(t)|^2$. Plug this in to get

$$i\hbar\exp(-\mathrm{i}\mathrm{E}_k\mathrm{t}/\hbar)\frac{d}{dt}b_k(t)+E_kb_k(t)\exp(-iE_kt/\hbar)$$
$$=b_k(t)E_k\exp(-iE_kt/\hbar)+\sum_n b_n(t)W_{kn}(t)\exp(-iE_nt/\hbar),$$

which simplifies to

$$\mathrm{i}\hbar\frac{d}{dt}b_k(t)=\sum_n b_n(t)W_{kn}(t)\exp\{i(E_k-E_n)t/\hbar\}. \tag{3.15}$$

These are a set of coupled differential equations for the coefficients $b_k(t)$, with no assumptions yet made about the relative magnitudes of $\mathbb{H}_0$ and $\mathbb{W}(t)$.

We now assume that the perturbation is weak enough, and/or the time is short enough, that the $b_k(t)$ differ only slightly from what they were at $t = 0$. Thus we replace $b_n(t)$ on the right-hand side by $b_n(0)$, and since we assumed that the system started out in state ϕ_i, $b_n(0) = \delta_{ni}$. This leaves

$$\mathrm{i}\hbar\frac{d}{dt}b_k^{(1)}(t)=W_{ki}(t)\exp\{i(E_k-E_i)t/\hbar\}=W_{ki}(t)\exp(i\omega_{ki}t), \tag{3.16}$$

where we define $\omega_{ki} = (E_k - E_i)/\hbar$. The $^{(1)}$ superscript signifies first-order approximation. Integration yields

$$b_k^{(1)}(t)=\frac{1}{i\hbar}\int_0^t d\tau\exp(i\omega_{ki}\tau)W_{ki}(\tau), \tag{3.17}$$

and, choosing the particular state ϕ_f ("f" standing for "final"),

$$P_{if}^{(1)}(t) = |b_f^{(1)}(t)|^2 = \left(\frac{1}{\hbar}\right)^2 \left|\int_0^t d\tau \exp(i\omega_{fi}\tau) W_{fi}(\tau)\right|^2. \tag{3.18}$$

This is the 1st order probability for finding the system in state ϕ_f at time t, assuming it started in ϕ_i at time zero.

We now need to specify the nature of $\mathbb{W}(t)$. The most common type of time-dependent perturbation is the oscillatory one originating from electromagnetic radiation. Let us assume

$$\mathbb{W}(t) = \mathbb{W}\{\exp(i\omega t) + \exp(-i\omega t)\} \tag{3.19}$$

where $\mathbb{W}$ is a time-independent operator and ω is a real, positive constant frequency. Substituting into the previous general formula gives

$$\begin{aligned} P_{if}^{(1)}(t) &= \left(\frac{|W_{fi}|}{\hbar}\right)^2 \left|\int_0^t d\tau \{\exp[i(\omega_{fi}+\omega)\tau] + \exp[i(\omega_{fi}-\omega)\tau]\}\right|^2 \\ &= \left(\frac{|W_{fi}|}{\hbar}\right)^2 \left|\frac{\exp[i(\omega_{fi}+\omega)t]-1}{i(\omega_{fi}+\omega)} + \frac{\exp[i(\omega_{fi}-\omega)t]-1}{i(\omega_{fi}-\omega)}\right|^2. \end{aligned} \tag{3.20}$$

where W_{fi} is the matrix element of the time-independent part of the perturbation.

Now for positive ω_{fi} (i.e., the final state is higher in energy than the initial state), the second term will always be larger than the first, and will be much larger for a near-resonant perturbation. So neglect the first term and rewrite the second to get

$$\begin{aligned} P_{if}^{(1)}(t) &= \left(\frac{|W_{fi}|}{\hbar}\right)^2 \left|\exp\left[\frac{i(\omega_{fi}-\omega)t}{2}\right]\left\{\frac{\exp[i(\omega_{fi}-\omega)t/2] - \exp[-i(\omega_{fi}-\omega)t/2]}{i(\omega_{fi}-\omega)}\right\}\right|^2 \\ &= \left(\frac{|W_{fi}|}{\hbar}\right)^2 \left\{\frac{\sin[(\omega_{fi}-\omega)t/2]}{(\omega_{fi}-\omega)/2}\right\}^2. \end{aligned} \tag{3.21}$$

This is an oscillatory function peaked around $\omega = \omega_{fi}$; as t gets longer, the peak gets sharper (Fig. 3.1).

If the oscillatory perturbation goes on for a very long time, the transition probability becomes

$$P_{if}^{(1)}(t \to \infty) = \left(\frac{|W_{fi}|}{\hbar}\right)^2 \lim_{t\to\infty} \left\{\frac{\sin[(\omega_{fi}-\omega)t/2]}{(\omega_{fi}-\omega)/2}\right\}^2. \tag{3.22}$$

This is *very* sharply peaked around $\omega = \omega_{fi}$; in fact, it is one of the functions that approaches a delta function as a limit:

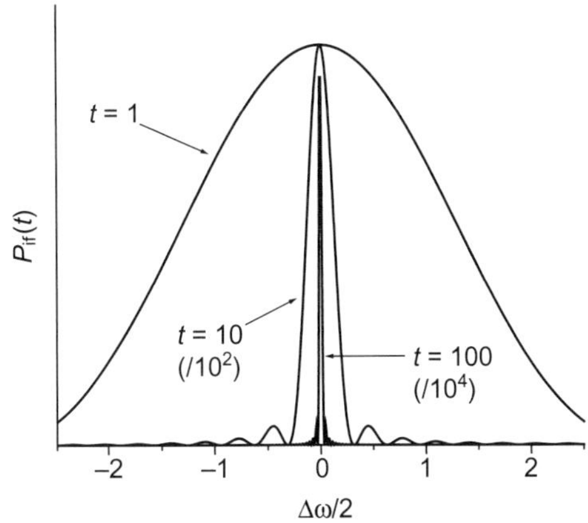

Figure 3.1. Plot of the transition probability in Equation (3.21) as a function of $(\omega_{fi} - \omega)/2$, for three different values of the time. The $t = 10$ and $t = 100$ plots are scaled by factors of 100 and 10,000, respectively.

$$\delta(x) = \lim_{\alpha\to\infty}\left\{\frac{1}{\pi}\frac{\sin^2(\alpha x)}{\alpha x^2}\right\}.$$

Therefore,

$$\lim_{t\to\infty}\left\{\frac{\sin[(\omega_{fi}-\omega)t/2]}{(\omega_{fi}-\omega)/2}\right\}^2 = \pi t \lim_{t\to\infty}\left\{\frac{1}{\pi}\frac{\sin^2[(\omega_{fi}-\omega)t/2]}{[(\omega_{fi}-\omega)/2]^2 t}\right\}$$
$$= \pi t\delta\left(\frac{\omega_{fi}-\omega}{2}\right) = 2\pi t\delta(\omega_{fi}-\omega).$$

So,

$$P_{if}^{(1)}(t\to\infty) = \left(\frac{|W_{fi}|}{\hbar}\right)^2 2\pi t\delta(\omega_{fi}-\omega). \tag{3.23}$$

This says that under the influence of a weak, long-lasting oscillatory perturbation, the system can be found in the upper state only if the frequency of the perturbation exactly matches the frequency difference between the states, and the probability that a transition has been made increases linearly with time. The *rate* of making transitions per unit time is then

$$R_{if}^{(1)}(\omega) = \frac{d}{dt}P_{if}^{(1)}(t\to\infty) = \frac{2\pi}{\hbar^2}|W_{fi}|^2\,\delta(\omega_{fi}-\omega), \tag{3.24}$$

to first order in the perturbation.

This approach can be carried through to higher orders in perturbation theory as well. To do that, go back to the original Equation (3.15) for the time-dependent coefficients,

$$i\hbar \frac{d}{dt} b_k(t) = \sum_n b_n(t) W_{kn}(t) \exp\{i\omega_{kn} t\},$$

and plug in $b_n(t) = b_n^{(1)}(t)$ to solve for $b_k^{(2)}(t)$, and so on. Higher orders of perturbation theory are needed to describe spectroscopies that involve more than one interaction with an electromagnetic field, including Raman scattering and all nonlinear processes.

Equation (3.24) is a very simple result but it often needs some modifications to be applicable to real situations. For example, we often detect transitions not to a single final state but to a group of closely spaced final states. If these states are discrete, we can write the total rate simply as $R_{\text{total}}(\omega) = \Sigma_f R_{\text{if}}(\omega)$, but if they are continuous, we must integrate:

$$R_{\text{total}}(\omega) = \int d\omega_{\text{fi}} R_{\text{if}}(\omega) \rho(\omega_{\text{fi}}), \tag{3.25}$$

where $\rho(\omega) = dN/d\omega$ is the density of states, the number of states per unit frequency. Substituting in Equation (3.24) for the state-to-state rate and doing the integral gives

$$R_{\text{total}}(\omega) = (2\pi/\hbar^2) \int d\omega_{\text{fi}} |W_{\text{fi}}|^2 \rho(\omega_{fi}) \delta(\omega_{\text{fi}} - \omega) = (2\pi/\hbar^2) |W_{\text{avg}}(\omega)|^2 \rho(\omega). \tag{3.26}$$

where it is assumed that all of the states of about the same frequency also have about the same matrix element, $W_{\text{avg}}(\omega)$. Note the units: $\rho(\omega)$ has units of ω^{-1} = time, $\hbar$ is in energy·time, W_{fi} is in energy, so $R_{\text{total}}(\omega)$ has units of time^{-1}, which is right for a rate.

Equation (3.26) is often called the Fermi golden rule. It says that a transition can occur only to those states for which the frequency difference between initial and final states equals the perturbation frequency, and the total rate for making a transition is given by the square of the matrix element for each state multiplied by the density of states at the perturbation frequency. Fermi's golden rule is perhaps more often applied to time-dependent processes that are induced by a constant rather than a time-dependent perturbation, and these will be discussed in Chapter 9.

3.3. INTERACTION OF MATTER WITH THE CLASSICAL RADIATION FIELD

In spectroscopy, the relevant time-dependent perturbation is usually an electromagnetic field. This section will concentrate on the electric field interactions that give rise to most spectroscopic transitions. This will be done at three different levels of approximation. We first consider interactions of matter with the classical electric field only. We then take a more complete approach that starts from the classical scalar and vector potentials from which the electric and magnetic fields arise. Finally, we treat both the matter and the radiation quantum mechanically. This is necessary for a correct description of processes involving spontaneous emission, such as fluorescence, phosphorescence, and Raman scattering.

The simplest approach considers only the electric field associated with a stationary, monochromatic electromagnetic wave as in Equation (2.3a), $\mathbf{E}(\mathbf{r},t) = (\mathbf{e}E_0/2)\{\exp[i(\omega t - \mathbf{k}\cdot\mathbf{r})] + \exp[-i(\omega t - \mathbf{k}\cdot\mathbf{r})]\}$. A molecule or atom in the presence of this field feels a potential energy due to the interaction of its charge distribution with the electric field. It is conventional to write the molecular charge distribution as a multipole expansion. For a molecule with zero net charge (i.e., not an ion), the leading term in this multipole expansion is the dipole interaction,

$$\mathbb{W}(t) = \mathbf{E}(\mathbf{r}, \mathrm{t})\cdot\boldsymbol{\mu}, \tag{3.27}$$

where the dipole moment operator is defined as $\boldsymbol{\mu} = -\Sigma_i q_i r_i$. The sum is over all the charged particles (nuclei and electrons) in the molecule; q_i is the charge on the ith particle and r_i is its position. For a molecule having no net charge, it does not matter where we choose the origin in calculating the dipole moment. It is convenient to choose the origin to be at the center of positive charge (which usually is almost equal to the center of mass), because then we only have to sum over the positions of the electrons and can ignore the nuclei.

Let us choose for concreteness $\mathbf{e}$ along $\hat{z}$ and $\mathbf{k}$ along $\hat{x}$. Then,

$$\mathbb{W}(\mathrm{t}) = \frac{1}{2}E_0\{\exp[i(\omega t - kx) + \exp[-i(\omega t - kx)\}\hat{z}\cdot\boldsymbol{\mu}. \tag{3.28}$$

Now choose the coordinate system to have its origin at the center of the molecule of interest, and use the series expansion of the exponential (Appendix B):

$$\exp(\pm ikx) \approx 1 \pm ikx - k^2x^2/2 \ldots. \tag{3.29}$$

Keeping only the first term, a good approximation when $kx << 1$, gives

$$\mathbb{W}(t) = \frac{1}{2}E_0\{\exp(i\omega t) + \exp(-i\omega t)\}\mu_z, \tag{3.30}$$

where $\mu_z = \hat{z}\cdot\boldsymbol{\mu}$ is the z component of the dipole moment operator. This is known as the electric dipole approximation. It amounts to assuming that the field strength hardly varies over the spatial dimensions of the molecule; that is, $kx << 1$ over the range of x covered by a single molecule. For visible light $k = 2\pi/\lambda \approx 2\pi/5000\,\text{Å} \approx 0.001\,\text{Å}$, while most molecules are only a few angstroms in size, so indeed $kx << 1$ is satisfied. However, this approximation may need to be reconsidered when the material system of interest is a metal (Chapter 14) or semiconductor nanoparticle (Chapter 16) or a conjugated polymer (Chapter 13).

The time-independent part of the matrix element that goes into Equation (3.24) is then

$$W_{\text{fi}} = \frac{E_0}{2}\langle \phi_f | \mu_z | \phi_i \rangle$$

However, most realistic laboratory applications do not involve an absolutely monochromatic electromagnetic field. The field is instead composed of a linear combination of waves with slightly different frequencies, as discussed in Section 2.3. In order to account for this, we need to multiply Equation (3.24) by the frequency distribution for the incident field and integrate over all frequencies present:

$$\begin{aligned} R_{\text{if}}^{(1)}(\omega) &= \frac{2\pi}{\hbar^2}\int d\omega \rho(\omega) |W_{\text{fi}}(\omega)|^2 \,\delta(\omega_{\text{fi}} - \omega) \\ &= \frac{2\pi}{\hbar^2}\left(\frac{E_0}{2}\right)^2 |\langle \phi_{\text{f}} | \mu_z | \phi_{\text{i}} \rangle|^2 \rho(\omega_{\text{fi}}). \end{aligned} \tag{3.31}$$

where $\rho(\omega)$ is the number of plane waves per unit frequency having average amplitude E_0. Finishing the problem requires finding the time-independent matrix elements of the dipole operator.

We will return to this matrix element after first taking a more rigorous approach starting from the vector potential derived from Maxwell's equations [Eq. (2.14)]. The Hamiltonian of a particle of mass m and charge e in the presence of a vector potential $\mathbf{A}(\mathbf{r},t)$ can be shown to be (McHale, 1999)

$$\mathbb{H} = (1/2m)(\mathbb{p} - e\mathbb{A})^2 + \mathbb{V}(\mathbf{r}) \tag{3.32}$$

The quantity $(1/\text{m})(\mathbb{p} - e\mathbb{A})$ corresponds to the classical velocity, which is different from $\mathbb{p}/m$ where $\mathbb{p}$ is the quantum mechanical momentum operator. Expanding the square gives

$$\mathbb{H} = (1/2m)(\mathbb{p}\cdot\mathbb{p} - e\mathbb{p}\cdot\mathbb{A} - e\mathbb{A}\cdot\mathbb{p} + e^2\mathbb{A}\cdot\mathbb{A}) + \mathbb{V}(\mathbf{r}) = \mathbb{H}_0 + \mathbb{W}(\mathbf{r}, t).$$

where $\mathbb{H}_0 = \mathbb{p}^2/2m + \mathbb{V}(\mathbf{r})$, the usual Hamiltonian in the absence of the applied field, and

$$\mathbb{W}(\mathbf{r}, t) = -(e/2m)(\mathbb{p}\cdot\mathbb{A} + \mathbb{A}\cdot\mathbb{p}) + (e^2/2m)\mathbb{A}\cdot\mathbb{A} \tag{3.33}$$

Note that since A contains the position coordinate $\mathbf{r}$, and position and momentum do not commute, we cannot generally assume $\mathbb{p}\cdot\mathbb{A} = \mathbb{A}\cdot\mathbb{p}$ in Equation

(3.33). However, once we make the dipole approximation [$\exp(\pm i\mathbf{k}\cdot\mathbf{r}) \approx 1$], then the $\mathbf{r}$ dependence of A goes away and we may write

$$\mathbb{W}(\mathbf{r},t) = -(e/m)\mathbb{A}\cdot\mathbb{p} + (e^2/2m)\mathbb{A}\cdot\mathbb{A} \tag{3.34}$$

the form we will normally use.

For most conventional light sources, although not always for intense pulsed lasers, the $\mathbb{A}\cdot\mathbb{A}$ term (quadratic in the fields) can be neglected relative to the terms linear in $\mathbb{A}$. The vector potential that corresponds to an electric field polarized along z and propagating along x is

$$\mathbf{A}(\mathbf{r},t) = (iE_0/2\omega)\{\exp[i(\omega t - kx)] - \exp[-i(\omega t - kx)]\}\hat{z}.$$

As the vector potential has only a z-component, and it depends only on the x-coordinate, and x-commutes with p_z, Equation (3.34) becomes

$$\mathbb{W}(\mathbf{r},t) = -(e/m)\mathbb{p}_z\mathbb{A}_z = -(ieE_0/2m\omega)\mathbb{p}_z\{\exp[\mathrm{i}(\omega t - kx)] - \exp[-i(\omega t - kx)]\}.$$

Making the approximation we did previously that $\exp(\pm ikx) \approx 1$ leaves

$$\mathbb{W}^{\mathrm{DE}}(t) = -(ieE_0/2m\omega)\mathbb{p}_z\{\exp(i\omega t) - \exp(-i\omega t)\}, \tag{3.35}$$

where DE stands for electric dipole Hamiltonian. The matrix elements of its time-independent part are

$$\langle\phi_\mathrm{f}|\mathbb{W}^{\mathrm{DE}}|\phi_\mathrm{i}\rangle = -(ieE_0/2m\omega)\langle\phi_\mathrm{f}|\mathbb{p}_z|\phi_\mathrm{i}\rangle. \tag{3.36a}$$

This at first seems to be a very different result from the one obtained starting by treating the electric field as a perturbing potential [Eq. (3.31)]. That resulted in matrix elements of position, the dipole moment operator being proportional to position. Here, starting with the effect of the vector potential on the Hamiltonian results in matrix elements of momentum. It turns out there is a trick that converts one to the other. Notice that

$$[\mathbb{z},\mathbb{H}_0] = [\mathbb{z}, \mathbb{p}^2/2m + \mathbb{V}(\mathbf{r})] = (1/2m)[\mathbb{z},\mathbb{p}^2] = (\mathrm{i}\hbar/m)\mathbb{p}_z \text{ since } [\mathbb{z},\mathbb{p}^2] = 2\mathrm{i}\hbar\mathbb{p}_z$$

Therefore,

$$\begin{aligned}\langle\phi_\mathrm{f}|\mathbb{p}_z|\phi_\mathrm{i}\rangle &= -(im/\hbar)\langle\phi_\mathrm{f}|[\mathbb{z},\mathbb{H}_0]|\phi_\mathrm{i}\rangle = -(im/\hbar)\langle\phi_\mathrm{f}|\mathbb{z}\mathbb{H}_0 - \mathbb{H}_0\mathbb{z}|\phi_\mathrm{i}\rangle\\ &= -(im/\hbar)(E_i - E_f)\langle\phi_\mathrm{f}|\mathbb{z}|\phi_\mathrm{i}\rangle = im[(E_f - E_i)/\hbar)]\langle\phi_\mathrm{f}|\mathbb{z}|\phi_\mathrm{i}\rangle\\ &= im\omega_{\mathrm{fi}}\langle\phi_\mathrm{f}|\mathbb{z}|\phi_\mathrm{i}\rangle.\end{aligned}$$

The time-independent part of the matrix element thus becomes

$$\begin{aligned}\langle\phi_\mathrm{f}|\mathbb{W}^{\mathrm{DE}}|\phi_\mathrm{i}\rangle &= (eE_0\omega_{\mathrm{fi}}/2\omega)\langle\phi_\mathrm{f}|\mathbb{z}|\phi_\mathrm{i}\rangle\\ &= (E_0\omega_{\mathrm{fi}}/2\omega)\langle\phi_\mathrm{f}|\mathbb{\mu}_z|\phi_\mathrm{i}\rangle.\end{aligned} \tag{3.36b}$$

and the transition rate is derived as for Equation (3.31),

$$\begin{aligned} R_{\text{if}}^{(1)}(\omega) &= \frac{2\pi}{\hbar^2}\int d\omega\rho(\omega)|W_{\text{fi}}(\omega)|^2\,\delta(\omega_{\text{fi}}-\omega) \\ &= \frac{2\pi}{\hbar^2}\left(\frac{E_0}{2}\right)^2 |\langle\phi_{\text{f}}|\mu_z|\phi_{\text{i}}\rangle|^2 \int d\omega\left(\frac{\omega_{\text{fi}}}{\omega}\right)^2 \rho(\omega)\delta(\omega_{\text{fi}}-\omega) \\ &= \frac{2\pi}{\hbar^2}\left(\frac{E_0}{2}\right)^2 |\langle\phi_{\text{f}}|\mu_z|\phi_{\text{i}}\rangle|^2 \rho(\omega_{\text{fi}}). \end{aligned} \tag{3.37}$$

This is the same result obtained in Equation (3.31).

Finally, we wish to relate this rate to a property of the applied field that is more easily measured. Equation (2.10) gives the cycle-averaged energy density in terms of E_0 and the refractive index of the medium for a single monochromatic wave, $u_{\text{avg}} = (\varepsilon_0 n^2 E_0^2)/2$. Substituting that result into Equation (3.37) gives the absorption rate in terms of the energy density,

$$R_{\text{if}}^{(1)}(\omega) = \frac{\pi}{\hbar^2 n^2 \varepsilon_0}|\langle\phi_{\text{f}}|\mu_z|\phi_{\text{i}}\rangle|^2 u_{\text{avg}}\rho(\omega_{\text{fi}}) = \frac{\pi}{\hbar^2 n^2 \varepsilon_0}|\langle\phi_{\text{f}}|\mu_z|\phi_{\text{i}}\rangle|^2 \rho_{\text{rad}}(\omega_{\text{fi}}), \tag{3.38}$$

where $\rho_{\text{rad}}(\omega_{\text{fi}})$ is the energy density of the radiation per unit frequency in units of $\text{J}\cdot\text{m}^{-3}/\text{s}^{-1}$ as defined for blackbody radiation in Equation (2.30b). This gives $R_{\text{if}}^{(1)}(\omega)$ the correct units for a rate, s^{-1}.

If we relax the approximation $\exp(\pm ikx) \approx 1$ and instead keep two terms in the Taylor series expansion, then $\exp(\pm ikx) \approx 1 \pm ikx$. This leaves for the total perturbation

$$\begin{aligned} \mathbb{W}(\mathbf{r},t) &= -(ieE_0/2m\omega)\mathbb{p}_z\{\exp(i\omega t)(1-ik\mathbb{x}) - \exp(-i\omega t)(1+ik\mathbb{x})\} \\ &= \mathbb{W}^{\text{DE}}(t) - (keE_0/2m\omega)\mathbb{p}_z\mathbb{x}\{\exp(i\omega t)+\exp(-i\omega t)\}. \end{aligned}$$

The second term can be rewritten using

$$\mathbb{p}_z\mathbb{x} = (\mathbb{p}_z\mathbb{x} - \mathbb{z}\mathbb{p}_x)/2 + (\mathbb{z}\mathbb{p}_x + \mathbb{x}\mathbb{p}_z)/2$$

(note $\mathbb{p}_z$ and $\mathbb{x}$ commute) and noting that the y component of angular momentum is $\mathbb{L}_y = \mathbb{z}\mathbb{p}_x - \mathbb{p}_z\mathbb{x}$, we have

$$\mathbb{p}_z\mathbb{x} = -\mathbb{L}_y/2 + (\mathbb{z}\mathbb{p}_x + \mathbb{x}\mathbb{p}_z)/2.$$

The second term can be rewritten as one involving only components of the position in the same way as was used to convert matrix elements of momentum to those of position in $\mathbb{W}^{\text{DE}}(t)$:

$$\begin{aligned} [\mathbb{z}\mathbb{x}, \mathbb{H}_0] &= (1/2m)[\mathbb{z}\mathbb{x}, \mathbb{p}^2] = (1/2m)[\mathbb{z}\mathbb{x}, \mathbb{p}_x^2 + \mathbb{p}_z^2] = (1/2m)(\mathbb{z}[\mathbb{x}, \mathbb{p}_x^2] + \mathbb{x}[\mathbb{z}, \mathbb{p}_z^2]) \\ &= (1/2m)(2i\hbar\mathbb{z}\mathbb{p}_x + 2i\hbar\mathbb{x}\mathbb{p}_z) = (i\hbar/m)(\mathbb{z}\mathbb{p}_x + \mathbb{x}\mathbb{p}_z). \end{aligned}$$

Therefore,

$$\mathbb{W}(\mathbf{r},t)=\mathbb{W}^{\mathrm{DE}}(t)+(keE_0/2m\omega)\{\exp(i\omega t)+\exp(-i\omega t)\}\{\mathbb{L}_y+(i\hbar/m)[\mathbb{z}\mathbb{x},\mathbb{H}_0]\},$$

and the matrix elements of the time-independent part of the perturbation are

$$\langle\phi_{\mathrm{f}}|\mathbb{W}|\phi_{\mathrm{i}}\rangle=\mathbb{W}_{\mathrm{fi}}^{\mathrm{DE}}+\mathbb{W}_{\mathrm{fi}}^{\mathrm{DM}}+\mathbb{W}_{\mathrm{fi}}^{\mathrm{QE}}, \tag{3.39}$$

where the DE (electric dipole), DM (magnetic dipole), and QE (electric quadrupole) terms are given by

$$\mathbb{W}_{\mathrm{fi}}^{\mathrm{DE}}=(eE_0\omega_{\mathrm{fi}}/2\omega)\langle\phi_{\mathrm{f}}|\mathbb{z}|\phi_{\mathrm{i}}\rangle \tag{3.40a}$$

$$\mathbb{W}_{\mathrm{fi}}^{\mathrm{DM}}=(ekE_0/4\mathrm{m}\omega)\langle\phi_{\mathrm{f}}|\mathbb{L}_{\mathrm{y}}|\phi_{\mathrm{i}}\rangle \tag{3.40b}$$

$$\mathbb{W}_{\mathrm{fi}}^{\mathrm{QE}}=-(iekE_0\omega_{\mathrm{fi}}/4\omega)\langle\phi_{\mathrm{f}}|\mathbb{z}\mathbb{x}|\phi_{\mathrm{i}}\rangle. \tag{3.40c}$$

The magnetic dipole and electric quadrupole terms are included in Equation (3.40) for completeness, but for most spectroscopic applications, the electric dipole term is all that is needed. These equations are for a single charged particle of charge e. For a collection of charged particles (e.g., a molecule) we must sum over all the nuclei and electrons; for example, the electric dipole matrix element becomes

$$\mathbb{W}_{\mathrm{fi}}^{\mathrm{DE}}=(E_0\omega_{\mathrm{fi}}/2\omega)\left\langle\phi_{\mathrm{f}}\left|\sum\nolimits_j q_j\mathbb{z}_j\right|\phi_{\mathrm{i}}\right\rangle=(E_0\omega_{\mathrm{fi}}/2\omega)\langle\phi_{\mathrm{f}}|\mu_{\mathrm{z}}|\phi_{\mathrm{i}}\rangle \tag{3.41}$$

where q_j is the charge on particle j and $\mu=\Sigma_j q_j\mathbb{r}_j$ is the total dipole moment operator, μ_z being its z-component.

3.4. INTERACTION OF MATTER WITH THE QUANTIZED RADIATION FIELD

We now address the treatment of radiation–matter interactions using the quantum mechanical form for the radiation field. We showed in Chapter 2 that the quantum-mechanical operator corresponding to the classical vector potential **A** is given by Equation (2.24),

$$\mathbb{A}(\mathbf{r},t)=\frac{1}{\sqrt{\varepsilon V}}\sum\nolimits_k\sum\nolimits_\alpha\sqrt{\frac{\hbar}{2\omega_k}}\mathbf{e}_\alpha\left[\mathbb{a}_{k\alpha}e^{-i\omega_k t}+\mathbb{a}_{k\alpha}^{\dagger}e^{i\omega_k t}\right].$$

Here, V is the volume of the box in which the radiation is quantized (eventually we will make this artifice go away), and we have made the dipole approximation. From the standpoint of the radiation field, $\mathbb{a}_{k\alpha}$ and $\mathbb{a}_{k\alpha}^{\dagger}$ are operators, while t and **r** are just parameters; the position **r**, however, is an operator on the states of matter.

As stated previously, the effect of a classical electromagnetic field on a charged particle is to add an interaction term to the field-free Hamiltonian of the form [Eq. (3.34)]

$$\mathbb{W}(\mathbf{r},t) = -(e/m)\mathbb{A}\cdot\mathbb{p} + (e^2/2m)\mathbb{A}\cdot\mathbb{A}.$$

When we quantize the radiation field, the interaction still has this form. The only differences are that the initial and final states involved are now states of not only the matter but also the radiation, and the quantity $\mathbb{A}(\mathbf{r},t)$ is now an operator on the states of both matter and radiation.

Neglecting as before the $\mathbb{A}\cdot\mathbb{A}$ term relative to $\mathbb{A}\cdot\mathbb{p}$ leaves

$$\mathbb{W}(t) = \frac{-e}{m}\sqrt{\frac{\hbar}{2\varepsilon V}}\sum_k\sum_\alpha\sqrt{\frac{1}{\omega_k}}\left[\mathbb{a}_{k\alpha}e^{-i\omega_k t} + \mathbb{a}_{k\alpha}^\dagger e^{i\omega_k t}\right]\mathbf{e}_\alpha\cdot\mathbb{p} \tag{3.42}$$

Note that since the interaction is linear in the photon creation and annihilation operators, the $\mathbb{A}\cdot\mathbb{p}$ term to first order can only add or subtract *one* photon from the field. This term to first order therefore describes one-photon absorption and emission processes.

Absorption of a photon of type (k,α) involves a transition from material state $|i>$ to $|f>$ and from radiation state $|\ldots n_{k\alpha}\ldots>$ to $|\ldots(n-1)_{k\alpha}\ldots>$. The total matrix element is then

$$\begin{aligned}W_{\text{fi}}(t) &= \langle \text{f};\ldots(n-1)_{k\alpha}\ldots|\mathbb{W}(t)|\text{i};\ldots n_{k\alpha}\ldots\rangle \\ &= \frac{-e}{m}\sqrt{\frac{\hbar}{2\varepsilon V}}\sum_{k'\alpha'}\sqrt{\frac{1}{\omega_{k'}}}\left[\langle\ldots(n-1)_{k\alpha}\ldots|\mathbb{a}_{k'\alpha'}e^{-i\omega_{k'}t} + \mathbb{a}_{k'\alpha'}^\dagger e^{i\omega_{k'}t}|\ldots n_{k\alpha}\ldots\rangle\right]\mathbf{e}_{\alpha'} \\ &\quad\cdot\langle \text{f}|\mathbb{p}|\text{i}\rangle.\end{aligned} \tag{3.43}$$

Of the photon operators, only $\mathbb{a}_{k\alpha}$ gives a nonzero matrix element, since

$$\mathbb{a}_{k\alpha}|\ldots n_{k\alpha}\ldots\rangle = n_{k\alpha}^{1/2}|\ldots(\text{n}-1)_{k\alpha}\ldots\rangle$$

and the different photon states are orthonormal just as for material harmonic oscillator states. This leaves, dropping the subscript k on the frequency,

$$W_{\text{fi}}(t) = -(e/m)(\hbar n_{k\alpha}/2\omega\varepsilon V)^{1/2}\exp(-i\omega t)\mathbf{e}_\alpha\cdot\langle \text{f}|\mathbb{p}|\text{i}\rangle. \tag{3.44}$$

and using the result derived previously to convert matrix elements of momentum to matrix elements of position, the time-independent part of the matrix element that goes into the Golden Rule expression becomes

$$W_{\text{fi}} = -ie\omega_{\text{fi}}(\hbar n_{k\alpha}/2\omega\varepsilon V)^{1/2}\mathbf{e}_\alpha\cdot\langle \text{f}|\mathbb{r}|\text{i}\rangle. \tag{3.45}$$

We now wish to calculate the rate of absorption transitions. The general first-order Golden Rule result found before for the rate of transitions due to a sinusoidal perturbation that is not absolutely monochromatic [refer to Eq. (3.31)] is

$$R_{\text{if}}^{(1)}(\omega) = \frac{2\pi}{\hbar^2}\int d\omega\rho(\omega)|W_{\text{fi}}(\omega)|^2\,\delta(\omega_{\text{fi}}-\omega),$$

and plugging in the result for the matrix element in Equation (3.45) gives

$$\begin{aligned} R_{\text{if}}^{(1)}(\omega) &= \frac{2\pi}{\hbar^2}\left(\frac{e^2\hbar n_{\text{avg}}}{2\varepsilon V}\right)|\boldsymbol{e}_\alpha\cdot\langle \text{f}|\mathbb{r}|\text{i}\rangle|^2\int d\omega\rho(\omega)\frac{\omega_{\text{fi}}^2}{\omega}\delta(\omega_{\text{fi}}-\omega) \\ &= \frac{2\pi}{\hbar^2}\left(\frac{e^2\hbar\omega_{\text{fi}} n_{\text{avg}}}{2\varepsilon V}\right)|\boldsymbol{e}_\alpha\cdot\langle \text{f}|\mathbb{r}|\text{i}\rangle|^2\rho(\omega_{\text{fi}}), \end{aligned} \tag{3.46}$$

where n_{avg} is the average number of photons in the radiation modes having frequencies around ω_{fi}.

The average number of photons is not a very useful quantity to work with. The energy density of the incident radiation, $\rho_{\text{rad}}(\omega)$, in units of energy per unit volume per unit frequency, is given by the average number of photons per state (n_{avg}) times the energy per photon ($\hbar\omega$) times the density of states [$\rho(\omega)$] divided by the volume (V),

$$\rho_{\text{rad}}(\omega) = (\hbar\omega \text{n}_{\text{avg}}/V)\rho(\omega), \tag{3.47}$$

so substituting $\rho(\omega_{\text{fi}}) = \rho_{\text{rad}}(\omega_{\text{fi}})(V/\hbar\omega_{\text{fi}}n_{\text{avg}})$ into Equation (3.46) and simplifying gives the transition rate in terms of energy density as

$$R_{\text{if}}^{(1)}(\omega) = \frac{\pi e^2}{\varepsilon\hbar^2}|\boldsymbol{e}_\alpha\cdot\mathbf{M}_{\text{fi}}|^2\rho_{\text{rad}}(\omega_{\text{fi}}) = \frac{\pi}{\varepsilon_0 n^2\hbar^2}|\boldsymbol{e}_\alpha\cdot\boldsymbol{\mu}_{\text{fi}}|^2\rho_{\text{rad}}(\omega_{\text{fi}}), \tag{3.48}$$

where we define the matrix element of the transition length as $\mathbf{M}_{\text{fi}} = \langle \text{f}|\mathbb{r}|i\rangle$. In the final equality, we have replaced ε by $\varepsilon_r\varepsilon_0$ and have set $\varepsilon_r = n^2$, and have also replaced the transition length $\mathbf{M}_{\text{fi}}$ by the transition dipole moment $\boldsymbol{\mu}_{\text{fi}} = \text{e}\mathbf{M}_{\text{fi}}$. Equation (3.48) is identical to the classically derived result, Equation (3.38). Alternatively, using $\rho_{\text{rad}}(\hbar\omega) = (1/\hbar)\rho_{\text{rad}}(\omega)$, we can write the rate in terms of the energy density per unit photon energy as

$$R_{\text{if}}^{(1)}(\hbar\omega) = \frac{\pi}{\varepsilon_0 n^2\hbar}|\boldsymbol{e}_\alpha\cdot\boldsymbol{\mu}_{\text{fi}}|^2\rho_{\text{rad}}(\hbar\omega_{\text{fi}}). \tag{3.49}$$

Finally, we specialize to the common case where the polarization vector of the light is randomly oriented relative to the absorbers. This may occur either because the light is randomly polarized in space, or because the material is a gas, liquid, or amorphous solid in which the absorbers are randomly oriented in space. When this is the case, we can set $|\mathbf{e}_\alpha \cdot \boldsymbol{\mu}_{fi}|^2 = |\boldsymbol{\mu}_{fi}|^2 \cos^2\theta$ and average over a random distribution of angles θ:

$$\left(\cos^2\theta\right)_{avg} = \frac{1}{4\pi}\int_0^{2\pi} d\varphi \int_0^{\pi} \sin\theta d\theta \cos^2\theta = \frac{1}{3}, \tag{3.50}$$

leaving the result for the transition rate in a randomly oriented sample,

$$R_{if}^{(1)}(\omega) = \frac{\pi}{3\varepsilon_0 n^2 \hbar^2} |\boldsymbol{\mu}_{fi}|^2 \rho_{rad}(\omega_{fi}). \tag{3.51}$$

Emission is just like absorption except that the radiation field gains a photon rather than losing one, making a transition from $|\ldots n_{k\alpha}\ldots\rangle$ to $|\ldots(n+1)_{k\alpha}\ldots\rangle$. Now it is only the term involving $\mathbb{a}_{k\alpha}^\dagger$ that will give a nonzero matrix element, since

$$\mathbb{a}_{k\alpha}^\dagger |\ldots n_{k\alpha}\ldots\rangle = (n_{k\alpha}+1)^{1/2} |\ldots (n+1)_{k\alpha}\ldots\rangle,$$

and the time-independent part of the matrix element becomes, instead of Equation (3.45),

$$W_{fi} = -ie\omega_{fi}[\hbar(n_{k\alpha}+1)/2\omega_k \varepsilon V]^{1/2} \mathbf{e}_\alpha \cdot \langle f|\mathbb{r}|i\rangle. \tag{3.52}$$

Unlike the matrix element for absorption, this one can be nonzero even if $n_{k\alpha} = 0$—that is, no photons are initially present in the field at the emitted wavevector and polarization. This describes spontaneous emission. The analog to Equation (3.46) for spontaneous emission is then

$$R_{if}^{(1)}(\omega) = \left(\frac{\pi e^2 \omega_{fi}}{3\hbar\varepsilon V}\right) |\mathbf{M}_{fi}|^2 \rho(\omega_{fi}), \tag{3.53}$$

where we have used the result from Equation (3.50) to average over all orientations. For spontaneous emission, there is no incident field at the emitted frequency, so we have to directly calculate the density of states for the radiation. Since the magnitude of the wavevector is given by $k = (k_x^2 + k_y^2 + k_z^2)^{1/2}$, we can draw a sphere of radius k whose surface area is $4\pi k^2$. The volume in k-space falling within wavevector element dk is $4\pi k^2 dk$. Now we need the volume in k-space occupied by each allowed state of the field. Recall from Chapter 2 that the quantization condition in the box of length L requires $k_x = 2\pi n_x/L$, where n_x is an integer, and similarly for k_y and k_z. We can list the allowed combinations (k_x, k_y, k_z) by their quantum numbers (n_x, n_y, n_z) and plot them as points in k-space, and assign each one as the center of a cubic volume that is $(2\pi/L)^3 = (2\pi)^3/V$. So the number of states within dk is the volume within dk divided by the volume per state. We also need to multiply by 2 to account for the two independent polarizations α associated with each wavevector **k**. The result is

$$\text{No. of states in } dk = 2\cdot 4\pi k^2 dk/[(2\pi)^3/V] = k^2 dkV/\pi^2.$$

The density of states is obtained by dividing the number of states by the frequency increment $d\omega$:

$$\rho(\omega) = (k^2 dkV/\pi^2)/d\omega. \tag{3.54}$$

Then, since $k = \omega\sqrt{\mu\varepsilon}$, $dk = \sqrt{\mu\varepsilon}d\omega$, and $k^2 = \mu\varepsilon\omega^2$. Putting this in gives

$$\rho(\omega) = (\mu\varepsilon)^{3/2}\omega^2 V/\pi^2. \tag{3.55}$$

This is the density of states (states per unit frequency) of the radiation field around ω. Inserting this into Equation (3.53) gives the rate of spontaneous transitions from |i> to |f>:

$$R_{\text{if}}^{(1)}(\omega) = \left(\frac{\pi e^2\omega_{\text{fi}}}{3\hbar\varepsilon V}\right)|\mathbf{M}_{\text{fi}}|^2\left[\frac{(\mu\varepsilon)^{3/2}\omega_{\text{fi}}^2 V}{\pi^2}\right] = \frac{e^2\omega_{\text{fi}}^3(\mu_0\varepsilon_0\varepsilon_r)^{3/2}}{3\hbar\varepsilon_0\varepsilon_r\pi}|\mathbf{M}_{\text{fi}}|^2 = \frac{e^2\omega_{\text{fi}}^3\varepsilon_r^{1/2}}{3\pi\hbar c^3\varepsilon_0}|\mathbf{M}_{\text{fi}}|^2$$
$$R_{\text{if}}^{(1)}(\omega) = \frac{n\omega_{\text{fi}}^3}{3\pi\hbar c^3\varepsilon_0}|\mu_{\text{fi}}|^2. \tag{3.56}$$

where we have assumed $\mu = \mu_0$ and $\varepsilon_r = n^2$, and have used $(\mu_0\varepsilon_0)^{3/2} = 1/c^3$. We have also converted from transition length, $\mathbf{M}_{\text{fi}}$, to transition dipole moment, $\mu_{fi} = e\mathbf{M}$.

Note the dependence on ω^3. This explains why spontaneous emission is often an important process at optical frequencies but much less so at infrared frequencies even though typical matrix elements μ_{fi} for electronic and vibrational transitions are of the same magnitude. The ratio of ω^3 for a typical electronic transition ($\omega = 20{,}000$–$50{,}000\,\text{cm}^{-1}$) versus a typical vibrational transition ($\omega = 500$–$3000\,\text{cm}^{-1}$) is 10^3–10^6.

Again, it is a good idea to check units. With ω in s^{-1}, ε_0 in $\text{C}^2\cdot\text{J}^{-1}\cdot\text{m}^{-1}$, $\hbar$ in J·s, c in $\text{m}\cdot\text{s}^{-1}$, and μ_{fi} in C·m, R_{fi} has units of s^{-1}, the expected units for a rate.

REFERENCES AND FURTHER READING

J. D. Jackson, *Classical Electrodynamics*, 2nd edition (John Wiley & Sons, New York, 1975).

J. L. McHale, *Molecular Spectroscopy* (Prentice Hall, Upper Saddle River, NJ, 1999).

G. Nienhuis and C. T. J. Alkemade, Physica **81C**, 181 (1976).

J. J. Sakurai, *Advanced Quantum Mechanics* (Addison-Wesley, Reading, MA, 1967).

PROBLEMS

1. States ϕ_i and ϕ_f are coupled by a time-dependent perturbation that has the form of a delta function in time:

$$W_{if}(t) = V_{if}\delta(t - t_0),$$

where V_{if} is a time-independent matrix element and $t_0 > 0$. Use the general form of first-order time-dependent perturbation theory, Equation (3.18), to calculate the probability that a system initially in state ϕ_i will be found in state ϕ_f at some time $t > t_0$. Why is there no dependence of this probability on the energy difference between the two states?

2. A heteronuclear diatomic molecule whose vibrations can be modeled as a harmonic oscillator is in its ground state at time $t = 0$. It is subjected to a laser pulse whose electric field has the form of a carrier frequency ω_0 multiplied by a single-sided exponential pulse envelope:

$$E(t) = \left(\frac{K}{2}\right)\{\exp(i\omega_0 t) + \exp(-i\omega_0 t)\}\exp(-t/2\tau) \text{ for } t \geq 0$$
$$= 0 \text{ for } t < 0,$$

where K is a real amplitude to be found in part (a). Assume that the perturbation frequency ω_0 is near but not necessarily equal to ω_{vib}, and that the pulse duration is long enough that $\omega_0 >> 1/\tau$. The time-dependent perturbation of the oscillator due to this electric field is given by

$$W_{fi}(t) \sim E(t)\langle\phi_f|x|\phi_i\rangle,$$

where ϕ_i and ϕ_f are the initial and final harmonic oscillator wavefunctions, respectively.

(a) Find the expression for K as a function of τ that makes the time-integrated perturbation strength a constant in some arbitrary units; that is,

$$\int dt|E(t)|^2 = 1.$$

(b) Calculate the probability, to first order in perturbation theory, that the oscillator will have made a transition from $v = 0$ to $v = 1$ at long times ($t >> \tau$).

(c) Assume now that you are free to vary the carrier frequency ω_0 and the pulse duration τ at will within the constraint $\omega_0 >> 1/\tau$, keeping the integrated intensity constant by using the K that you found in part (a). Determine the values of ω_0 and τ that maximize the long-time transition probability.

3. At time $t = 0$, the wavefunction for a one-dimensional harmonic oscillator with mass m and frequency ω is

$$\psi(0) = \frac{1}{2}\psi_{v-1} + \frac{1}{\sqrt{2}}\psi_v + \frac{1}{2}\psi_{v+1},$$

where ψ_v is the vth harmonic oscillator eigenstate. Calculate the expectation values of position and momentum, $\mathbb{x}$ and $\mathbb{p}_x$, as a function of time for $t > 0$ and interpret your result physically. *Hint*: Use the harmonic oscillator raising and lowering operators.

4. Consider the electric-dipole allowed spontaneous emission of a photon by a molecule in a crystal having a well-defined orientation in space. Choose the crystal to be oriented such that the molecule's transition dipole moment μ_{if} is directed along the laboratory z-axis. Calculate the angular distribution of propagation directions for the emitted photons, $I(\theta,\phi)$. If the goal is to detect the largest number of emitted photons and your detector can look at only a small solid angle $d\Omega$, where should the detector be placed?

5. The density of states for photons was explicitly evaluated in calculating the spontaneous emission rate. Now calculate the density of states $\rho(E)$ (states per unit energy) for a *particle* of mass m in

(a) a one-dimensional box of length L ($V = 0$ for $0 \le x \le L$, $V = \infty$ for $x < 0$ or $x > L$);

(b) a two-dimensional square box of side L;

(c) a three-dimensional cubic box of side L.

Discuss how $\rho(E)$ varies with E for each case. Note: For the two- and three-dimensional cases, you may generalize the treatment used in the text to derive the photon density of states, but be careful because the particle-in-a-box quantum numbers are defined to take only positive values.

6. The electric dipole approximation assumes that the molecule, or the part of the molecule that interacts with light, is much smaller than the wavelength of the light. For visible light (500 nm), calculate how much $\exp(\pm ikx)$ varies across the dimensions of (a) a molecule with a length of 10 Å, and (b) a gold nanoparticle with a diameter of 100 nm. How good is the assumption $\exp(\pm ikx) \approx$ constant in each case?

7. Use the definition of the z-component of the transition dipole moment, $\mu_z = \langle\phi_f|\Sigma_j q_j z_j|\phi_i\rangle$, to prove that the dipole moment of an overall neutral system is independent of the choice of origin of the coordinate system. Do this by replacing z_j by $z_j + a$, where a is an arbitrary constant, and assuming $\Sigma_j q_j = 0$.

8. In deriving Equation (3.51) from Equation (3.48), it was assumed that the absorbers are randomly oriented in three-dimensional space. Consider instead that the absorbers all lie in the yz plane, with random orientations, with the incident radiation still polarized along z. What would the analog to Equation (3.51) be in this case?

9. Calculate the three components of the transition dipole moment from the 1s ground state to each of the three 2p states of a hydrogen atom. Use the 2p wavefunctions that are the eigenstates of $\mathbb{L}_z$. Then calculate the sum of the squares of the matrix elements for each component (e.g., $|\langle 1s|\mu_x|2p_-\rangle|^2 + |\langle 1s|\mu_x|2p_0\rangle|^2 + |\langle 1s|\mu_x|2p_+\rangle|^2$ and similarly for μ_y and μ_z. Is this what you expect?

CHAPTER 4

ABSORPTION AND EMISSION OF LIGHT

4.1. EINSTEIN COEFFICIENTS FOR ABSORPTION AND EMISSION

In Chapter 3, we calculated the rates for absorption and spontaneous emission from first-order perturbation theory using the correct quantum mechanical form for the radiation field. It turns out there is a simpler, if a bit empirical, way to derive the relationship between the rates for absorption and for emission by using the properties of a system at thermal equilibrium. This result is general and often quite useful.

Consider two energy levels (eigenstates of the field-free Hamiltonian of the system), E_1 and E_2, having number densities N_1 and N_2, respectively (molecules per unit volume). Assume that this two-level system is at thermal equilibrium with a radiation field having an energy density of $\rho_{rad}(\omega)$. Let R_{12} and R_{21} be the rates of upward ($1 \rightarrow 2$) and downward ($2 \rightarrow 1$) transitions, respectively, per unit volume. The upward transition can occur only through absorption of radiation, while the downward transition can occur through either stimulated or spontaneous emission of radiation. These two rates are then given by

$$R_{12} = N_1 B_{12} \rho_{rad}(\omega) \tag{4.1a}$$

$$R_{21} = N_2 B_{21} \rho_{rad}(\omega) + N_2 A_{21}. \tag{4.1b}$$

Here, B_{12} is the rate constant for absorption, B_{21} is the rate constant for stimulated emission, and A_{21} is the rate constant for spontaneous emission. These

Condensed-Phase Molecular Spectroscopy and Photophysics, First Edition. Anne Myers Kelley.

are collectively known as the Einstein coefficients. The absorption and stimulated emission rates are proportional to the energy density at the transition frequency ω ($\hbar\omega = E_2 - E_1$), while the spontaneous emission rate is independent of the radiation. At thermal equilibrium, the ratio of the populations of the two states must obey a Boltzmann distribution,

$$\frac{N_2}{N_1} = \exp\left[\frac{-(E_2 - E_1)}{k_B T}\right] = \exp(-\hbar\omega / k_B T), \tag{4.2}$$

where k_B is Boltzmann's constant. To maintain this equilibrium, the rates of the upward and downward transitions must be balanced:

$$N_1 B_{12} \rho_{rad}(\omega) = N_2 B_{21} \rho_{rad}(\omega) + N_2 A_{21}, \tag{4.3}$$

and solving for the energy density gives

$$\begin{aligned} \rho_{rad}(\omega) &= \frac{N_2 A_{21}}{N_1 B_{12} - N_2 B_{21}} = \left(\frac{N_2}{N_1}\right) \frac{A_{21}}{B_{12} - (N_2 / N_1) B_{21}} \\ &= \frac{A_{21} \exp(-\hbar\omega / k_B T)}{B_{12} - B_{21} \exp(-\hbar\omega / k_B T)} = \frac{A_{21}}{B_{12} \exp(\hbar\omega / k_B T) - B_{21}}. \end{aligned} \tag{4.4}$$

We now recall the expression for $\rho_{rad}(\omega)$ for a blackbody source as given in Equation (2.30b),

$$\rho_{rad}(\omega) = \frac{\hbar n^3 \omega^3}{\pi^2 c^3} \frac{1}{\exp(\hbar\omega / k_B T) - 1}.$$

This expression for the energy density is in agreement with Equation (4.4) if

$$B_{12} = B_{21}, \tag{4.5a}$$

and

$$\frac{A_{21}}{B_{21}} = \frac{\hbar n^3 \omega^3}{\pi^2 c^3}. \tag{4.5b}$$

We see that the rate constants for stimulated emission and absorption are equal. This means that if the applied field is strong enough that nearly all of the emission is stimulated, the average populations in the upper and lower states should approach equality. The ratio of spontaneous to stimulated emission depends on the cube of the frequency, and as discussed before, this explains why spontaneous emission is an important process for electronic transitions that typically occur in the visible and UV region of the spectrum, and much less so for vibrational transitions that occur in the infrared.

Now we want to get expressions for the A and B coefficients in terms of molecular properties. When discussing radiation–matter interactions in

Chapter 3, we derived, from first-order time-dependent perturbation theory with the quantized radiation field, the following expression for the absorption rate valid for randomly oriented absorbers, Equation (3.51):

$$R_{12}^{(1)}(\omega) = \frac{\pi|\mu_{21}|^2}{3n^2\varepsilon_0\hbar^2}\rho_{\text{rad}}(\omega).$$

This is the rate for one molecule. The Golden Rule rate multiplied by the density of molecules should be equal to the Einstein rate of absorption defined previously:

$$N_1\frac{\pi|\mu_{21}|^2}{3n^2\varepsilon_0\hbar^2}\rho_{\text{rad}}(\omega) = N_1B_{12}\rho_{\text{rad}}(\omega), \tag{4.6}$$

and so,

$$B_{21} = \frac{\pi|\mu_{21}|^2}{3n^2\varepsilon_0\hbar^2} \tag{4.7}$$

Finally, using the relationship between the A and B coefficients in Equations (4.5) gives the spontaneous emission rate as

$$A_{21} = \frac{n\omega^3|\mu_{21}|^2}{3\pi\varepsilon_0\hbar c^3}. \tag{4.8}$$

Equation (4.8) for the spontaneous emission rate is identical to Equation (3.56), derived directly from the quantized radiation field and perturbation theory. Note that the A coefficient has the expected units for a unimolecular rate constant (s^{-1}), while the B coefficients have the somewhat nonintuitive units of $J^{-1}\cdot m^3\cdot s^{-2}$. This is because the upward rate is obtained by multiplying B by the energy density per unit frequency, whose units are $J\cdot m^{-3}/s^{-1}$.

4.2. OTHER MEASURES OF ABSORPTION STRENGTH

Physical and analytical chemists typically describe absorption strength in terms of other quantities more directly related to what is measured in the laboratory in an absorption experiment (Fig. 4.1). These are the absorption cross section, σ, and the molar extinction coefficient or molar absorptivity, ε_A. These are defined in terms of the experimentally measured quantities by

$$I(\omega,\ell) = I(\omega,0)e^{-\sigma(\omega)\ell N}, \tag{4.9}$$

and

$$I(\omega,\ell) = I(\omega,0)10^{-\varepsilon_A(\omega)\ell C_M}, \tag{4.10}$$

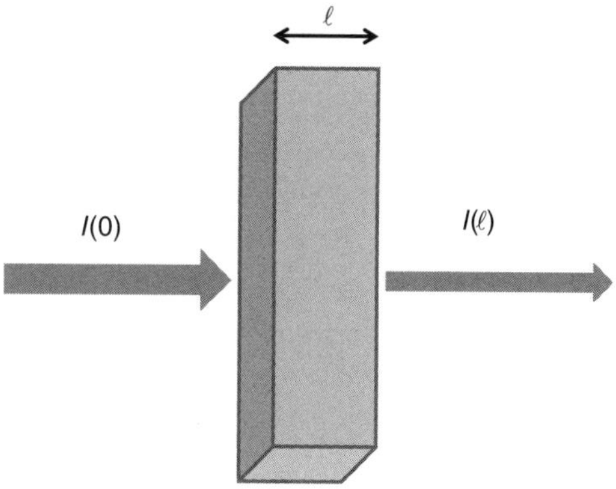

Figure 4.1. An absorption experiment. Light of intensity $I(0)$ passes through a sample of path length ℓ and emerges with a reduced intensity $I(\ell)$.

where $I(\omega,0)$ is the incident light intensity prior to entering the sample, and $I(\omega,\ell)$ is the intensity after passing through a sample of path length ℓ, with the path length given in centimeters. In Equation (4.10), C_M is the concentration of absorbers in moles per liter (standard chemical units), and ε_A has units of liters per mole per centimeter. In Equation (4.9), N is the concentration in molecules per cm^3 (units more often used by physicists), and σ has units of cm^2 per molecule. One can show by comparing these two expressions that

$$\sigma = 2.303(1000\ \mathrm{cm}^3 \cdot \mathrm{L}^{-1})\varepsilon_A / N_A, \tag{4.11}$$

where N_A is Avogadro's number (molecules per mole). Equation (4.10) can be rearranged as follows:

$$\begin{aligned} \frac{I(\omega,0)}{I(\omega,\ell)} &= 10^{\varepsilon_A(\omega)\ell c_M} \\ \log[I(\omega,0)/I(\omega,\ell)] &= \varepsilon_A(\omega)\ell c_M \\ A &= \varepsilon_A(\omega)\ell c_M, \end{aligned} \tag{4.12}$$

where the absorbance, A, is defined as $\log[I(0)/I(\ell)]$. Equation (4.12) is also known as the Beer–Lambert law or Beer's law, and is the form typically employed in analytical chemistry to determine the concentration of a substance from its optical absorbance.

Relating the absorption cross section or molar absorptivity to the Einstein coefficient for absorption is complicated by the fact that B is derived assuming a broadband radiation source and is related to the energy density per unit frequency, while σ and ε_A are defined in terms of the intensity of a pseudo-monochromatic light beam integrated over its frequency bandwidth. In order to handle this, recall $\rho_{rad}(\omega) = u(\omega)\rho(\omega)$, where $u(\omega)$ is the total energy density of the applied radiation in energy per unit volume and $\rho(\omega)$ is a normalized density of states, $\int \rho(\omega)d\omega = 1$ [see also the discussion following Eq. (3.37)]. $u(\omega)$ is the same quantity defined as u_{avg} in Equation (2.10). The intensity of the beam, $I(\omega)$ (energy per unit time per unit area), is given by $I(\omega) = u(\omega)(c/n)$, where c/n is the speed of light in the medium. dI/dx is the change in intensity per unit length, or the change in power per unit volume. This should be equal to the rate at which the molecules in that volume make transitions from Equation (4.6), multiplied by the energy of each transition:

$$\frac{-dI(\omega)}{dx} = \hbar\omega N B_{12}\rho_{rad}(\omega) = \hbar\omega N B_{12} u(\omega)\rho(\omega) = \frac{\hbar\omega n N B_{12}\rho(\omega)}{c} I(\omega).$$

This can be rearranged to give

$$-dI(\omega)/I(\omega) = [\hbar\omega n N B_{12}\rho(\omega)/c]dx.$$

and integrated from $x = 0$ to $x = \ell$, this is

$$I(\omega,\ell) = I(\omega,0)\exp\{-[\hbar\omega n B_{12}\rho(\omega)/c]N\ell\}. \tag{4.13}$$

Comparison of Equations (4.9) and (4.13) shows that B_{12} is related to the absorption cross section by

$$B_{12}\rho(\omega) = \frac{c\sigma(\omega)}{\hbar\omega n} = \frac{c}{\hbar n}\frac{\sigma(\omega)}{\omega}, \tag{4.14}$$

or to the molar absorptivity by

$$B_{12}\rho(\omega) = \frac{2.303(1000\ \mathrm{cm}^3\cdot\mathrm{L}^{-1})c}{N_A\hbar n}\frac{\varepsilon_A(\omega)}{\omega}. \tag{4.15}$$

Substituting Equation (4.14) into Equation (4.7) gives

$$B_{12}\rho(\omega) = \left(\frac{c}{\hbar n}\right)\frac{\sigma(\omega)}{\omega} = \frac{\pi|\mu_{21}|^2\rho(\omega)}{3n^2\varepsilon_0\hbar^2}.$$

We now integrate both sides over all frequencies:

$$\frac{c}{n}\int\frac{\sigma(\omega)}{\omega}d\omega = \frac{\pi|\mu_{21}|^2}{3n^2\varepsilon_0\hbar^2}\int\rho(\omega)d\omega,$$

which simplifies to

$$c\int\frac{\sigma(\omega)}{\omega}d\omega = \frac{\pi|\mu_{21}|^2}{3n\varepsilon_0\hbar}.$$

Solving for the transition dipole moment gives

$$|\mu_{21}|^2 = \frac{3\hbar cn\varepsilon_0}{\pi} \int \frac{\sigma(\omega)}{\omega} d\omega, \tag{4.16}$$

while Equations (4.15) and (4.7) similarly yield

$$|\mu_{21}|^2 = \frac{2.303(1000\, cm^3\, L^{-1})3\hbar cn\varepsilon_0}{\pi N_A} \int \frac{\varepsilon_A(\omega)}{\omega} d\omega. \tag{4.17}$$

These equations define the transition dipole moment entirely in terms of a directly measurable quantity, the integrated absorption cross section or molar absorptivity.

4.3. RADIATIVE LIFETIMES

We now consider the emission rate. In the absence of any applied radiation, state 2 decays to state 1 through first-order kinetics with a rate constant A_{21}, $-dN_2/dt = A_{21}N_2$, and so

$$N_2(\mathrm{t}) = N_2(0)\exp(-A_{21}t). \tag{4.18}$$

The natural radiative lifetime, τ_{rad}, is defined as the inverse of this rate constant: $\tau_{\mathrm{rad}} = 1/A_{21}$. For a single emitting state, the radiative lifetime can be related to the transition dipole moment using Equation (4.8):

$$\tau_{\mathrm{rad}} = \frac{1}{A_{21}} = \frac{3\pi\varepsilon_0\hbar c^3}{n\omega^3 |\mu_{21}|^2}. \tag{4.19}$$

This assumes that there is only one lower state to which the upper state can decay radiatively. More generally, if state 2 can decay to more than one state j by spontaneous emission, we have

$$-dN_2/dt = \sum_j A_{2j} N_2,$$

and

$$N_2(t) = N_2(0)\exp(-t/\tau_{\mathrm{rad}}),$$

where

$$\tau_{\mathrm{rad}} = 1\Big/\Big(\sum_j A_{2j}\Big), \tag{4.20a}$$

or equivalently,

$$\frac{1}{\tau_{\text{rad}}} = \sum_j \frac{1}{\tau_{\text{rad},j}}. \tag{4.20b}$$

Emission lifetimes will be discussed in more detail in Chapter 9.

4.4. OSCILLATOR STRENGTHS

The strength of an electronic transition is often expressed in terms of another quantity, the oscillator strength. The oscillator strength, f, of an electronic transition between states 1 and 2 is a unitless quantity defined in terms of the transition dipole moment as

$$f_{12} = \frac{2m_e\omega_{21}}{3\hbar e^2}|\mu_{12}|^2. \tag{4.21}$$

Substituting Equation (4.16) into (4.21) gives the oscillator strength in terms of a directly measurable quantity, the integrated absorption cross section:

$$f_{12} = \frac{2m_e cn\varepsilon_0}{\pi e^2}\int \sigma(\omega)\,d\omega. \tag{4.22}$$

Finally, converting from angular frequency to the more convenient wavenumber units, $\tilde{\nu} = \omega/2\pi c$, and evaluating the constants gives the numerical result

$$f_{12} = 1.13\times 10^{12}\, n\int \sigma(\tilde{\nu})\,d\tilde{\nu}, \tag{4.23}$$

where $\sigma(\tilde{\nu})$ is in units of cm^2 and $\tilde{\nu}$ is in cm^{-1}. Strong electronic transitions of molecules in solution have $\sigma \approx$ 1–5 Å^2 (1–$5 \times 10^{-16}\,\text{cm}^2$) at the maximum and span a frequency width of a few thousand cm^{-1}, making f_{12} on the order of 0.1–1. Weak transitions can have much lower oscillator strengths, in the range from 10^{-6} to 10^{-2}.

4.5. LOCAL FIELDS

In a medium, the speed of light is reduced from c to c/n, where n, a unitless quantity, is the (real) refractive index of the medium. The frequency of electromagnetic radiation does not change in a medium, but the wavelength is reduced from λ to λ/n. When spectroscopists talk about the wavelength of light used to do an experiment, they mean the wavelength *in air* (or in vacuum—these are only slightly different, but for high resolution studies, the difference matters) and not the wavelength in the medium, because the wavelength is almost always *measured* in air. As discussed briefly in Chapter 2, the refractive index is more generally a complex quantity given by $n + i\kappa$, and it depends on frequency. Far from any resonances, where the material is essentially nonabsorbing, the refractive index is essentially purely real and varies only slowly with frequency. Typical refractive indices for the visible and near-infrared region are around 1.3–1.7 for most liquids, and often larger for crystals. (Many crystals are also anisotropic, and have different refractive indices along different crystal axes.) Dilute solutions of absorbing chromophores in nonabsorbing liquids or solids are often assumed to have the same refractive index as the pure liquid or solid.

The expressions given in Chapter 3 and earlier for transition rates contain factors of the refractive index that arise from the relationship between energy density and electric field in a dielectric medium. Equations (4.8) and (4.16) show that if the transition dipole moment and transition frequency remain constant as the medium is changed, the spontaneous emission rate should scale as n while the integrated absorption cross section should scale as $1/n$. Absorption and emission strengths can also differ in different environments because of local field factors. This refers to the idea that the actual electric field felt by a chromophoric molecule at a particular site in a liquid or solid, $\mathbf{F}(t)$, is not equal to the macroscopic field $\mathbf{E}(t)$ because it has contributions from the polarizations induced by the field in all of the other molecules in the sample. Many different forms have been derived for this local field correction, which make different assumptions about both the shape of the "cavity" in the medium in which the chromophore sits and the nature of the polarization induced in the surrounding molecules. The most commonly used form, the Lorentz local field, assumes a spherical cavity and arrives at the relationship $\mathbf{F} = [(n^2 + 2)/3]\mathbf{E}$. Since $n^2 \approx 2$ for simple liquids and the absorption cross section should be proportional to the square of the local field divided by the refractive index, this implies that the apparent absorption oscillator strength in a medium should be about 6% smaller than in a vacuum. Experimentally, oscillator strengths for electronic transitions can either increase or decrease in condensed phases compared with vacuum, and it is not clear how much of this can be attributed to local field factors versus other effects. Many alternative forms for the local field correction have also been proposed, some of which take into account the anisotropic shape of most molecules. It remains unclear which of these approximate correction factors is most generally applicable.

FURTHER READING

R. C. Hilborn, Am. J. Phys. **50**, 982 (1982).

J. L. McHale, *Molecular Spectroscopy* (Prentice Hall, Upper Saddle River, NJ, 1999).

A. B. Myers and R. R. Birge, J. Chem. Phys. **73**, 5314 (1980).

PROBLEMS

1. Equation (4.21) relates the oscillator strength to the transition dipole moment, and Equation (4.17) relates the transition dipole moment to the integrated molar absorptivity.

(a) Use these two equations to derive the relationship between the oscillator strength and the molar absorptivity as a function of frequency, $\varepsilon_A(\omega)$, analogous to Equation (4.22) for the absorption cross section.

(b) Most instruments directly measure absorbance as a function of wavelength rather than frequency. Derive an expression for the oscillator strength in terms of the molar absorptivity as a function of wavelength, $\varepsilon_A(\lambda)$.

2. This problem explores the possibility of measuring the absorption spectrum of a single molecule. Equation (4.9) gives the fraction of the light transmitted in an absorption experiment as $I(\omega,\ell)/I(\omega,0) = \exp[-\sigma(\omega)\ell N]$. N is the number of molecules per cm^3 and ℓ is the path length in centimeters; thus, we can define another quantity, $\xi = \ell N$, which has units of molecules per cm^2. This can be interpreted as the number of molecules that would be intercepted by a beam with a cross-sectional area of 1 cm^2. In order to make a measurement on a single molecule, we must choose $\xi A < 1$, where A is the cross-sectional area of the light beam. Classical optics tells us that ordinary far-field optics cannot focus a light beam any more tightly than a radius of about $\lambda/2$, or about 0.25 μm for green light. Assuming the path length ℓ is small enough that the beam radius remains constant across the path length, calculate the largest value of ξ that can place only one molecule in the beam. Then make a reasonable estimate for σ for a strong transition and calculate $I(\omega,\ell)/I(\omega,0)$. Comment on the difficulty of measuring this.

3. Consider an electron bound to the origin by a three-dimensional, isotropic harmonic oscillator potential. The Hamiltonian is given by $-(\hbar^2/2m_e)\nabla^2 + (k/2)\cdot(x^2 + y^2 + z^2)$.

(a) What are the eigenvalues and eigenstates of this three-dimensional harmonic oscillator? Refer to the one-dimensional analog and notice that the Hamiltonian is separable in the three Cartesian coordinates. What are the degeneracies of the ground state and the first excited state?

(b) Assume the oscillator starts in the ground state, Ψ_0. Calculate the transition dipole moment, $\boldsymbol{\mu}_{0j} = \langle\Psi_0|e\mathbf{r}|\Psi_j\rangle$, for each upper state Ψ_j to which a transition can occur.

(c) Calculate the oscillator strength for each of these transitions as defined by Equation (4.21), $f_{0j} = (2m_e\omega_{0j}/3\hbar e^2)|\boldsymbol{\mu}_{0j}|^2$.

(d) Calculate the sum of the oscillator strengths for all of the transitions available to this electron, $\Sigma_j f_{0j}$. Use this result to comment on the meaning of oscillator strength.

4. Consider a molecule having its $v = 1$ to $v = 0$ vibrational transition at $2000\,\text{cm}^{-1}$ with a transition dipole of 0.1 D.

(a) Calculate the spontaneous radiative lifetime for the $v = 1$ state in vacuum.

(b) The absorption lineshape of a transition where the upper state decays exponentially is a Lorentzian in frequency with a half-width ($\Delta\omega$) equal to the upper state decay rate. Calculate the linewidth in wavenumbers due to this spontaneous decay.

5. For the laser dye rhodamine B dissolved in ethanol ($n = 1.36$), the integrated molar absorptivity $\int (\varepsilon_A(\omega)/\omega)d\omega$ was found to be $5937\,\text{L}\,\text{mol}^{-1}\,\text{cm}^{-1}$.

(a) Calculate the transition dipole moment of rhodamine B.

(b) Assuming the fluorescence emission appears in a narrow band centered at 580 nm, calculate the spontaneous radiative lifetime.

(c) When rhodamine B is used as a laser dye, its actual excited state lifetime under lasing conditions is much shorter than that found in (b). Give the most likely explanation for this.

6. Consider an argon ion laser emitting at 488 nm with a frequency width (full width at half maximum) of $\Delta\nu = 5\,\text{GHz}$. Approximately how many photons per unit volume are needed for the stimulated emission to be stronger than the spontaneous emission? Hint: Use the expressions for the Einstein coefficients.

7. The intensity autocorrelation function of a light source in steady-state may be defined as

$$g^{(2)}(\tau) = \frac{\langle I(t)I(t-\tau)\rangle}{\langle I(t)\rangle^2},$$

where the brackets imply averaging over the time variable t. If the radiation arises from spontaneous emission from a large number of uncorrelated emitters, as in a light bulb or the sun, $g^{(2)}(\tau) = 1$ for all values of τ, that is, there is no correlation, on average, between the intensity of the light source at one time and at another time. The intensity versus time plot shows only random noise. If, on the other hand, the emission comes from a single molecule, $g^{(2)}(\tau) = 1$ only at large values of τ. For small τ, there is a "dip" in the autocorrelation function and ideally $g^{(2)}(0) = 0$. Explain why.

8. A solution of a certain dye in methanol is prepared at a concentration of 10^{-4} mol/L. The molar absorptivity of the dye at a wavelength of 532 nm is $30{,}000\,\text{L}\,\text{mol}^{-1}\,\text{cm}^{-1}$. Light from a short-pulsed frequency-doubled Nd:YAG laser at 532 nm is incident on the sample contained in a cell of 1-mm path length. The cross-sectional area of the laser beam where it passes through the cell is $1\,\text{cm}^2$, and the energy of each pulse before it passes through the sample is 1 mJ.

(a) What is the energy of each laser pulse after passing through the sample, assuming Beer's law holds?

(b) What fraction of the dye molecules within the illuminated volume will absorb a laser photon during a single pulse? Assume that each photon is absorbed by a different molecule.

CHAPTER 5

SYSTEM–BATH INTERACTIONS

5.1. PHENOMENOLOGICAL TREATMENT OF RELAXATION AND LINESHAPES

According to the time-dependent Schrödinger equation, the time dependence of an eigenstate $|\varphi\rangle$ of the Hamiltonian $\mathbb{H}$ is given by $|\varphi(t)\rangle = \exp(-i\mathbb{H}t/\hbar)|\varphi(0)\rangle = \exp(-iEt/\hbar)|\varphi(0)\rangle$, where E is the eigenvalue. Therefore, if the system was known to be in state $|\varphi\rangle$ at time zero, the probability of finding it in this state at a later time t is $P(t) = |\langle\varphi|\varphi(t)\rangle|^2 = |\exp(-iEt/\hbar)|^2 = 1$. Eigenstates of the Hamiltonian are *stationary states*.

It is never possible to completely isolate a system from its environment, so in principle, one must propagate the system in time using the Hamiltonian for the whole universe. This is clearly not feasible, and one consequence is that the eigenstates of any *chosen* part of the universe are not stationary states. Statistical mechanics tell us that the populations in a set of eigenstates should change with time in such a way that the part of the universe we have chosen approaches a thermal Boltzmann distribution at equilibrium with the rest of the universe.

Fortunately, it is often possible to make an approximate separation into a "system" and an "environment," which interact only weakly. Then one can approximately describe the behavior of the system by treating explicitly only the part of the Hamiltonian that belongs to the system, handling the system-environment interaction terms in the Hamiltonian as a perturbation. The

Condensed-Phase Molecular Spectroscopy and Photophysics, First Edition. Anne Myers Kelley.

density matrix, discussed further later, provides a somewhat involved but formally correct way to do this. Often, one can get away with a much simpler, phenomenological treatment. Two such approaches, and their effect on spectroscopic lineshapes, will be introduced here.

One such approach is to go back to the time-dependent perturbation theory approach of Chapter 3, divide Equation (3.14) by $i\hbar$, and add a new term:

$$\frac{d}{dt}c_k(t) = c_k(t)(-iE_k/\hbar) + \sum_n c_n(t)[-iW_{kn}(t)/\hbar] - c_k(t)\gamma/2. \quad (5.1)$$

The new term involving $\gamma/2$ describes a constant rate of decay of the amplitude of state k that is independent of the applied perturbation W. In the absence of any applied perturbation, the solution to this equation is $c_k(t) = c_k(0)\exp[-(iE_k/\hbar + \gamma/2)t]$, and the probability of finding the system in state k at time t is

$$P_k(t) = |c_k(t)|^2 = |c_k(0)|^2 \exp(-\gamma t). \quad (5.2)$$

That is, the probability of finding the system still in the original state decays exponentially with time constant $\tau = 1/\gamma$. The quantity τ is known as the lifetime of the state. We encountered the concept of a lifetime in Chapter 4, but there the decay was caused by coupling of the state function for the material system to the radiation field via spontaneous emission. Here, we recognize that such decay can also occur by coupling of the initially prepared state to other states of the material system without emission of radiation.

If we now continue the derivation presented in Chapter 3, making the same substitution $c_k(t) = b_k(t)\exp(-iE_kt/\hbar)$, we end up with the analog of Equation (3.18),

$$P_{if}^{(1)}(t) = |b_f^{(1)}(t)|^2 = \left(\frac{1}{\hbar}\right)^2 e^{-\gamma t}\left|\int_0^t d\tau \exp\{(i\omega_{fi} + \gamma/2)\tau\} W_{fi}(\tau)\right|^2, \quad (5.3)$$

for a general time-dependent perturbation. Taking now the special case of a sinusoidal perturbation, $W_{fi}(t) = \mathrm{W_{fi}}(e^{i\omega t} + e^{-i\omega t})$, and keeping only the near-resonant term, we get the analog of Equation (3.21),

$$P_{\mathrm{if}}^{(1)}(t) = \left(\frac{|W_{\mathrm{fi}}|}{\hbar}\right)^2 \left\{\frac{1 - 2e^{-\gamma t/2}\cos(\omega_{\mathrm{fi}} - \omega)t + e^{-\gamma t}}{(\omega_{\mathrm{fi}} - \omega)^2 + \gamma^2/4}\right\}^2,$$

and finally, taking the long-time limit ($t >> \gamma$), we get the steady-state probability as

$$P_{\mathrm{if}}^{(1)}(t \to \infty) = \left(\frac{|W_{\mathrm{fi}}|}{\hbar}\right)^2 \left\{\frac{1}{(\omega_{\mathrm{fi}} - \omega)^2 + \gamma^2/4}\right\}. \quad (5.4)$$

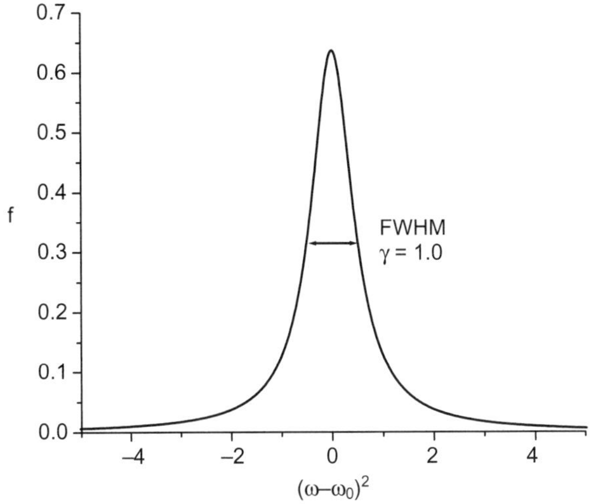

Figure 5.1. A normalized Lorentzian function of frequency, indicating the full-width at half maximum, γ.

Note that when decay of the final state is included, the probability of finding the system in the final state does not increase linearly with time as in Equation (3.23), but rather reaches a constant value. The rate at which population is put into the final state by perturbation-driven transitions from the initial state becomes balanced by the rate at which the final state population spontaneously decays. Note, too, that the frequency of the oscillatory perturbation does not have to exactly match the frequency difference between the initial and final states; the transition has now developed a linewidth because of the finite lifetime of the final state.

The full-width at half maximum, or FWHM, is the width of the transition in frequency from $-y_{max}/2$ to $+y_{max}/2$ (Fig. 5.1) and is given by γ. The half-width at half maximum, or HWHM, is the width of the transition in frequency from $-y_{max}/2$ to y_{max} or from y_{max} to $+y_{max}/2$, and is given by $\gamma/2$. The lineshape of a transition caused by exponential decay of the final state has the functional form of a Lorentzian; a normalized Lorentzian function of frequency is defined by

$$f(\omega) = \frac{1}{\pi} \frac{\gamma/2}{(\omega - \omega_0)^2 + \gamma^2/4}, \tag{5.5}$$

where γ is the FWHM and ω_0 is the center frequency. Compared with other lineshapes often encountered, such as the Gaussian or normal distribution, a Lorentzian has a comparatively sharp peak and broad "wings," that is, it does not decay to zero very quickly at frequencies far from the band center.

This is one approach to handling coupling of the quantum-mechanical system of interest to other degrees of freedom. It works best when we have a reasonably good description of the quantum-mechanical states involved in the problem, and the coupling to the environment may be considered a reasonably small perturbation. We now approach the problem in a second way that is more useful when the system is too complex to be handled by even an approximate quantum mechanical model.

We start with the expression given in Chapter 3 for the rate of making electric-dipole allowed transitions from one quantum state $|i\rangle$ to another one $|f\rangle$ under irradiation by a steady-state light source of frequency ω and polarization **e**, but here use the delta function form [Eq. (3.24)] rather than the density of states form [Eq. (3.26)]:

$$R_{if}(\omega) \propto |\langle f|\boldsymbol{\mu}\cdot\mathbf{e}|i\rangle|^2\,\delta(\omega_{fi}-\omega).$$

Here, the proportionality sign ($\propto$) is used to avoid having to carry all of the constants through each step, recognizing that here we are interested only in the lineshape of the transition, not the absolute rate. Usually what we actually measure is not the appearance of state $|f\rangle$ but the increase (for emission) or decrease (for absorption) of the intensity of the light at frequency ω. So if there's more than one transition of the material system that can contribute to the response at a given frequency, we have to sum all their contributions. If each initial state $|i\rangle$ can make a number of transitions that contribute to the response at ω, we have

$$R(\omega) \propto \sum\nolimits_f |\langle f|\boldsymbol{\mu}\cdot\mathbf{e}|i\rangle|^2\,\delta(\omega_{fi}-\omega),$$

and if the material system can exist in a number of different initial states with probabilities P_i (normally a Boltzmann distribution), we have to write this as

$$R(\omega) \propto \sum\nolimits_i P_i \sum\nolimits_f |\langle f|\boldsymbol{\mu}\cdot\mathbf{e}|i\rangle|^2\,\delta(\omega_{fi}-\omega) \tag{5.6}$$

This can be hard to evaluate directly because the number of initial states thermally occupied near room temperature can be enormous for a molecule of any significant size, and the number of final states that contribute to absorption or emission at a given frequency can be even greater. We can circumvent this through an alternative formulation of the first-order time-dependent perturbation theory result that eliminates the need to sum explicitly over eigenstates.

Start by writing the delta function as a time integral:

$$\delta(\omega_{fi}-\omega) = \frac{1}{2\pi}\int_{-\infty}^{\infty} dt\,\exp[i(\omega_{fi}-\omega)t], \tag{5.7}$$

and put it into the expression for the transition rate with the modulus squared written out explicitly:

$$R(\omega) \propto \sum_i P_i \sum_f \int_{-\infty}^{\infty} dt \exp[i(\omega_{fi} - \omega)t] \langle i|\mu \cdot e|f\rangle \langle f|\mu \cdot e|i\rangle$$
$$= \sum_i P_i \sum_f \int_{-\infty}^{\infty} dt\, e^{-i\omega t} \langle i|\mu \cdot e|f\rangle \langle f|\exp(i\omega_f t)\mu \cdot e \exp(-i\omega_i t)|i\rangle.$$

Then, noting that $\langle f|\exp(i\mathbb{H}t/\hbar) = \langle f|\exp(i\omega_f t)$ and $\exp(-i\mathbb{H}t/\hbar)|i\rangle = \exp(-i\omega_i t)|i\rangle$, where $\mathbb{H}$ is the molecular Hamiltonian, we have

$$R(\omega) \propto \sum_i P_i \sum_f \int_{-\infty}^{\infty} dt\, e^{-i\omega t} \langle i|\mu \cdot e|f\rangle \langle f|\exp(i\mathbb{H}t/\hbar)\mu \cdot e \exp(-i\mathbb{H}t/\hbar)|i\rangle.$$

Then use the closure relation, $\Sigma_f |f\rangle\langle f| = 1$, to remove the sum over final states:

$$R(\omega) \propto \sum_i P_i \int_{-\infty}^{\infty} dt\, e^{-i\omega t} \langle i|\mu \cdot e \exp(i\mathbb{H}t/\hbar)\mu \cdot e \exp(-i\mathbb{H}t/\hbar)|i\rangle.$$

An operator in the Heisenberg picture is defined as

$$\mathbb{A}(t) = \exp(i\mathbb{H}t/\hbar)\mathbb{A}\exp(-i\mathbb{H}t/\hbar). \tag{5.8}$$

Notice that expectation values of operators are the same in the ordinary (Schrödinger) and Heisenberg pictures:

$$\langle \mathbb{A}\rangle_t = \langle \psi|\exp(i\mathbb{H}t/\hbar)\mathbb{A}\exp(-i\mathbb{H}t/\hbar)|\psi\rangle = \langle \psi(t)|\mathbb{A}|\psi(t)\rangle \text{ (Schrödinger picture)}$$
$$= \langle \psi|\mathbb{A}(t)|\psi\rangle \qquad \text{(Heisenberg picture).}$$

In the Schrödinger picture, the ordinary operators (those that do not depend explicitly on time as a parameter) are assumed to be time independent, while the states evolve with time and cause the expectation values to change with time. In the Heisenberg picture, the states are considered to be constant in time while the operators evolve. The Heisenberg picture is closer to the classical picture of dynamics: a billiard ball rolling on a table remains the same billiard ball as it moves, but its position and momentum change with time.

With this definition and noting that $\boldsymbol{\mu}$ is an operator on the states of matter but $\mathbf{e}$ is not, we can write

$$\exp(i\mathbb{H}t/\hbar)\boldsymbol{\mu} \cdot \mathbf{e} \exp(-i\mathbb{H}t/\hbar) = \boldsymbol{\mu}(t) \cdot \mathbf{e},$$

where $\boldsymbol{\mu}(t)$ is the operator in the Heisenberg picture. The rate expression becomes

$$R(\omega) \propto \sum_i P_i \int_{-\infty}^{\infty} dt\, e^{-i\omega t} \langle i|\mu \cdot e\, \mu(t) \cdot e|i\rangle.$$

Finally, adopting a shorthand notation for the thermally averaged expectation value of a quantity,

$$\langle \mathbb{A} \rangle = \sum_i P_i \langle i|\mathbb{A}|i \rangle,$$

we get the following compact result,

$$R(\omega) \propto \int_{-\infty}^{\infty} dt\, e^{-i\omega t} \langle \boldsymbol{\mu} \cdot \boldsymbol{e}\, \boldsymbol{\mu}(t) \cdot \boldsymbol{e} \rangle. \tag{5.9}$$

The quantity $\langle \boldsymbol{\mu} \cdot \boldsymbol{e}\, \boldsymbol{\mu}(t) \cdot \boldsymbol{e} \rangle$ is a type of correlation function. In general, a correlation function has the form $\langle \mathbb{A}\mathbb{A}(t) \rangle$; here the operator is the component of the dipole moment operator, $\mathbb{\mu}$, along the polarization direction, **e**. A correlation function tells us what the relationship is, on average, between a dynamical variable at one time, usually chosen as $t = 0$, and at some later time. This value is usually nonzero at $t = 0$, but often becomes zero at long times, usually resulting from some random perturbation of the absorbing or emitting molecule by its environment.

Note that this result for the transition rate requires no explicit consideration of the many possible final states of the transition, $|f\rangle$. These states still matter, but all the information about them is "hidden" in the system Hamiltonian, $\mathbb{H}$, which goes into calculating $\boldsymbol{\mu}(t)$. In principle, we still need to know about all the initial states of the system and their Boltzmann populations in order to evaluate the thermally averaged correlation function. However, one often approximates this with relatively simple classical or semiclassical models.

One further useful simplification is possible for isotropic samples. Let us choose the direction of **e** to be along $\hat{z}$ (since the sample is isotropic it does not matter how we label the axes). Then,

$$\boldsymbol{\mu} \cdot \mathbf{e}\, \boldsymbol{\mu}(t) \cdot \mathbf{e} = \mu_z(0)\mu_z(t).$$

Also,

$$\begin{aligned}\langle \boldsymbol{\mu}(0) \cdot \boldsymbol{\mu}(t) \rangle &= \langle \mu_x(0)\mu_x(t) + \mu_y(0)\mu_y(t) + \mu_z(0)\mu_z(t) \rangle \\ &= 3\langle \mu_z(0)\mu_z(t) \rangle,\end{aligned}$$

for isotropic samples. Therefore,

$$\langle \boldsymbol{\mu} \cdot \mathbf{e}\, \boldsymbol{\mu}(t) \cdot \mathbf{e} \rangle = (1/3)\langle \boldsymbol{\mu}(0) \cdot \boldsymbol{\mu}(t) \rangle,$$

and

$$R(\omega) \propto \int_{-\infty}^{\infty} dt\, e^{-i\omega t} \langle \boldsymbol{\mu}(0) \cdot \boldsymbol{\mu}(t) \rangle, \tag{5.10}$$

which now contains no reference to the polarization direction of the light.

This equation says that the *lineshape* of the transition (the dependence of the transition rate on frequency) depends on a Fourier transform of the thermally averaged dipole correlation function. That is, it depends on a property

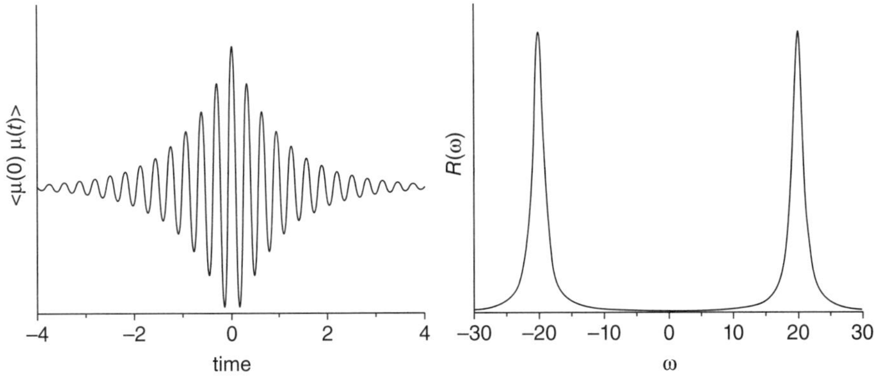

Figure 5.2. Left: dipole correlation function <**μ**(0**μ**(*t*)> = cos $\omega_0 t\exp(-\gamma|t|)$, with $\omega_0 = 20$ and $\gamma = 1$. Right: resulting spectrum, $R(\omega)$, given by Equation (5.11).

of the dynamics of the molecule *in the absence of the perturbing field*. This is a rather remarkable result.

To see how this works, look at the predicted lineshapes for a couple of very simple, classical models for the correlation function. The first is a dipole that undergoes undamped oscillation at frequency ω_0, $\boldsymbol{\mu}(t) = \boldsymbol{\mu}(0)\cos(\omega_0 t)$ for every member of the ensemble. In this case, $\langle\boldsymbol{\mu}(0)\cdot\,\boldsymbol{\mu}(t)\rangle = |\mu(0)|^2\cos(\omega_0 t)$ and the integral becomes $R(\omega) \propto |\mu(0)|^2\{\delta(\omega-\omega_0)+\delta(\omega+\omega_0)\}$. The spectrum is two sharp lines where the first represents absorption at frequency ω_0 and the second emission at frequency ω_0. Because this is a classical model, it always predicts symmetry between absorption and emission.

Now consider the case where the oscillation of the dipole is damped in time, such that $\boldsymbol{\mu}(t) = \boldsymbol{\mu}(0)\cos(\omega_0 t)\exp(-\gamma|t|)$ for every member of the ensemble (Fig. 5.2). The correlation function has its maximum value at $t = 0$, and it decays to zero at large negative or positive times. In this case, we obtain for the lineshape

$$R(\omega) \propto \frac{\gamma}{(\omega_0-\omega)^2+\gamma^2} + \frac{\gamma}{(\omega_0+\omega)^2+\gamma^2}, \qquad (5.11)$$

which is a sum of two Lorentzian functions representing absorption and emission, respectively, centered at ω_0 and with a linewidth (half-width at half maximum) of γ. Note that exponential damping of the correlation function produces a Lorentzian line in Equation (5.11) just as does exponential decay of the final state lifetime in Equation (5.4). The final state lifetime is not, however, the only way to get an exponentially decaying correlation function; interruption of the phase of the oscillation by the environment can produce the same effect, as is explored further in the Problems.

5.2. THE DENSITY MATRIX

The state vector Ψ, in principle, contains all information about a quantum mechanical system. However, since no part of the universe is ever totally isolated from the rest, that statement is rigorously true only if Ψ is the state vector for the whole universe! In order to do anything practical, we need to deal with just a part of the whole—that is, separate the universe into a "system" and a "bath." In general, we cannot ignore the bath, nor can we write the state of the universe as a simple product $\Psi_{\text{system}}\Psi_{\text{bath}}$ because there are *correlations* between system and bath. In this situation, the system degrees of freedom alone can be treated by the density matrix method, which allows the "uninteresting" bath degrees of freedom to be factored out of the problem in a correct way. First, we develop the properties of the density matrix for cases where the system *is* describable by a state vector, and show that the state vector and the density matrix methods yield the same results.

The situation where the system can be represented by a state vector Ψ is referred to as a pure case. Here, it is equally correct to use either the state vector or the density matrix. The state of the system at time t can always be expressed in some complete basis $\{u_i\}$ as

$$\Psi(t) = \sum_n c_n(t)|u_n\rangle \text{ where } c_n(t) = \langle u_n|\Psi(t)\rangle.$$

Define the pure case density operator $\rho(t)$ as

$$\rho(t) = |\Psi(t)\rangle\langle\Psi(t)|. \tag{5.12}$$

Note that $|\Psi(t)\rangle\langle\Psi(t)|$ is not the same thing as $\langle\Psi(t)|\Psi(t)\rangle$. The latter is a *number*, the inner product of a state vector with itself, equal to unity if the state vector is normalized, while the former is an *operator*. The density matrix is then just the matrix representation of this density operator:

$$\rho_{mn}(t) = \langle u_m|\rho(t)|u_n\rangle = \langle u_m|\Psi(t)\rangle\langle\Psi(t)|u_n\rangle = c_n^*(t)c_m(t). \tag{5.13}$$

Note that $e^{i\theta}\Psi(t)$, where θ is any arbitrary phase, has the same density matrix as $\Psi(t)$. The physically irrelevant phase factor does not appear in the density matrix.

For a normalized state vector, the trace of the density matrix (sum of the diagonal elements) is unity:

$$Tr\rho(t) = \sum_n \rho_{nn}(t) = \sum_n |c_n(t)|^2 = 1. \tag{5.14}$$

The density matrix can be used to calculate expectation values:

$$\langle \mathbb{A} \rangle_t = \langle \Psi(\mathrm{t})|\mathbb{A}|\Psi(\mathrm{t})\rangle = \sum_n \langle \Psi(t)|u_n\rangle\langle u_n|\mathbb{A}|\Psi(t)\rangle = \sum_n \langle u_n|\mathbb{A}|\Psi(t)\rangle\langle \Psi(t)|u_n\rangle$$
$$= \sum_n \langle u_n|\mathbb{A}\rho(t)|u_n\rangle = Tr\{\mathbb{A}\rho(t)\} = Tr\{\rho(t)\mathbb{A}\}. \tag{5.15}$$

The last step uses the result of matrix algebra that the trace of a product of matrices is independent of the order in which the matrices are multiplied.

The time evolution of the density operator can be found by using the time-dependent Schrödinger equation for the state vector and its complex conjugate separately:

$$\frac{d}{dt}\rho(t) = \left(\frac{d}{dt}|\psi(t)\rangle\right)\langle\psi(t)| + |\psi(t)\rangle\left(\frac{d}{dt}\langle\psi(t)|\right)$$
$$= \frac{1}{i\hbar}\mathbb{H}(t)|\psi(t)\rangle\langle\psi(t)| - \frac{1}{i\hbar}|\psi(t)\rangle\langle\psi(t)|\mathbb{H}(t) \tag{5.16}$$
$$= \frac{1}{i\hbar}\{\mathbb{H}(t)\rho(t) - \rho(t)\mathbb{H}(t)\} = \frac{1}{i\hbar}[\mathbb{H}(t), \rho(t)].$$

Notice that the sign is reversed from the expression for the time evolution for the expectation value of an operator,

$$\frac{d}{dt}\langle\mathbb{A}(t)\rangle = \frac{-1}{i\hbar}[\mathbb{H}(t), \mathbb{A}(t)]$$

The density operator is Hermitian, since $\rho_{\mathrm{mn}}(t) = \rho^*_{\mathrm{nm}}(t)$. Also, for a pure case, it is equal to its square: $\rho^2(t) = |\Psi(t)\rangle\langle\Psi(t)|\Psi(t)\rangle\langle\Psi(t)| = |\Psi(t)\rangle\langle\Psi(t)| = \rho(t)$. Finally, for conservative systems $\Psi(t) = \exp(-i\mathbb{H}t/\hbar)\Psi(0)$, so

$$\rho(\mathrm{t}) = |\Psi(t)\rangle\langle\Psi(t)| = \exp(-i\mathbb{H}t/\hbar)|\Psi(0)\rangle\langle\Psi(0)|\exp(i\mathbb{H}t/\hbar)$$
$$= \exp(-i\mathbb{H}t/\hbar)\rho(0)\exp(i\mathbb{H}t/\hbar) \tag{5.17}$$

The alternative to the pure case is the nonpure (statistical or mixed) case. It occurs when we cannot describe the system using a single state vector. This might arise when, for example, we would like to discuss the spectroscopy of a solvated molecule, but due to interaction between solvent and solute, we canot write a single state vector for the solute part of the problem alone. In such a case, though, we can still talk about various probabilities P_k for the system to be found in various states Ψ_k, and we can still define a density operator that will permit us to calculate the results of measurements made on the system.

If the system were known to be in state Ψ_k, then the expectation value of observable A would be $\langle\mathbb{A}\rangle = Tr\{\mathbb{A}\rho_{\mathrm{k}}\}$, where $\rho_{\mathrm{k}} = |\Psi_k\rangle\langle\Psi_k|$. If instead there are probabilities for being in a number of possible states, then the expectation value should be given by an appropriately weighted average,

$\langle \mathbb{A} \rangle = \Sigma_k P_k Tr\{\mathbb{A}\rho_k\} = Tr\{\mathbb{A}(\Sigma_k P_k\rho_k)\} = Tr\{\mathbb{A}\rho\}$, where we define the mixed-case density operator as $\rho = \Sigma_k P_k\rho_k$.

Let us explore some properties of the mixed-case density operator. It is still Hermitian and $Tr\ \rho = 1$, but unlike pure case $\rho^2 \neq \rho$ (note each probability $P_k < 1$). If each of the possible state vectors Ψ_k evolves under the usual time-dependent Schrödinger equation,

$$i\hbar \frac{d}{dt}\Psi_k(t) = \mathbb{H}(t)\Psi_k(t),$$

then the density operator at time t will be

$$\rho(\mathrm{t}) = \sum_k \mathrm{P}_k\rho_k(t) \quad \text{where} \quad \rho_k(t) = |\Psi_k(t)\rangle\langle\Psi_k(t)|. \tag{5.18}$$

The time evolution is given by

$$\frac{d}{dt}\rho(t) = \sum_k P_k \frac{d}{dt}\rho_k(t) = \frac{1}{i\hbar}\sum_k P_k[\mathbb{H}(t), \rho_k(t)] = \frac{1}{i\hbar}[\mathbb{H}(t), \rho(t)], \tag{5.19}$$

the same as for the pure case. Notice, though, that for this to hold, the Hamiltonian must be the *exact* Hamiltonian for the system *during the time interval of interest*. If the system is not in a pure state to begin with because there are interactions between the degrees of freedom of interest and something "external" to the system, then the previous expression would not exactly give the time evolution unless those interactions are negligible during the time period of interest.

We now show that a "nonpure" density matrix arises from treating only part of a composite system. Consider a system whose state vector can be written as a linear combination of products of basis vectors involving two different sets of coordinates $\{u_n(1)\}$ and $\{v_r(2)\}$, where 1 and 2 might represent space and spin, or solute and solvent. For example, assume the state vector is

$$\Psi(1,2) = \{u_1(1)v_1(2) - u_1(1)v_2(2) + u_2(1)v_1(2) + u_2(1)v_2(2)\}/2.$$

Notice that this does *not* factor into a simple product: $\Psi(1,2) \neq \psi(1)\varphi(2)$. (Try to factor it!)

Define the density operator for space 1, $\rho(1)$, as

$$\rho(1) = \sum_r \langle v_r(2)|\rho|v_r(2)\rangle = Tr_2(\rho) \tag{5.20}$$

where ρ is the total density operator and Tr_2 designates the partial trace over space 2 only. The resulting operator, $\rho(1)$, is an operator in space 1 only.

This is most easily shown by doing an example. For the total state vector $\Psi(1,2)$ given previously, the total density operator is

$$
\begin{aligned}
\rho &= |\Psi(1,2)\rangle\langle\Psi(1,2)| \\
&= \{|u_1v_1\rangle\langle u_1v_1| - |u_1v_1\rangle\langle u_1v_2| + |u_1v_1\rangle\langle u_2v_1| + |u_1v_1\rangle\langle u_2v_2| - |u_1v_2\rangle\langle u_1v_1| \\
&\quad + |u_1v_2\rangle\langle u_1v_2| - |u_1v_2\rangle\langle u_2v_1| - |u_1v_2\rangle\langle u_2v_2| + |u_2v_1\rangle\langle u_1v_1| - |u_2v_1\rangle\langle u_1v_2| \\
&\quad + |u_2v_1\rangle\langle u_2v_1| + |u_2v_1\rangle\langle u_2v_2| + |u_2v_2\rangle\langle u_1v_1| - |u_2v_2\rangle\langle u_1v_2| + |u_2v_2\rangle\langle u_2v_1| \\
&\quad + |u_2v_2\rangle\langle u_2v_2|\}/4.
\end{aligned}
$$

The density operator in space 1 is

$$
\begin{aligned}
\rho(1) &= \langle v_1|\rho|v_1\rangle + \langle v_2|\rho|v_2\rangle \text{ (this is the partial trace over space 2)} \\
&= \{|u_1\rangle\langle u_1| + |u_1\rangle\langle u_2| + |u_1\rangle\langle u_1| - |u_1\rangle\langle u_2| + |u_2\rangle\langle u_1| \\
&\quad + |u_2\rangle\langle u_2| - |u_2\rangle\langle u_1| + |u_2\rangle\langle u_2|\}/4 \\
&= \{|u_1\rangle\langle u_1| + |u_2\rangle\langle u_2|\}/2.
\end{aligned}
$$

Why is this useful? Consider an observable A that acts only in space (1), for example, a spin angular momentum in a spin-space composite system, or the dipole moment of a molecule of interest in a solvent. Calculate its expectation value:

$$
\begin{aligned}
\langle \mathbb{A}\rangle &= Tr\{\mathbb{A}\rho\} = \sum_{n,r} (\langle u_n(1)|\langle v_r(2)|)\mathbb{A}\rho(|u_n(1)\rangle|v_r(2)\rangle) \\
&= \sum_{n,r}\sum_{n',r'} (\langle u_n(1)|\langle v_r(2)|)\mathbb{A}(|u_{n'}(1)\rangle|v_{r'}(2)\rangle) \\
&\quad (\langle u_{n'}(1)|\langle v_{r'}(2)|)\rho(|u_n(1)\rangle|v_r(2)\rangle) \\
&= \sum_{n,r}\sum_{n',r'} \langle u_n(1)|\mathbb{A}|u_{n'}(1)\rangle\langle v_r(2)|v_{r'}(2)\rangle \\
&\quad (\langle u_{n'}(1)|\langle v_{r'}(2)|)\rho|u_n(1)\rangle|v_r(2)\rangle) \\
&= \sum_{n,n'} \langle u_n(1)|\mathbb{A}|u_{n'}(1)\rangle\left\{\sum_r (\langle u_{n'}(1)|\langle v_r(2)|)\rho(|u_n(1)\rangle|v_r(2)\rangle)\right\} \\
&= \sum_{n,n'} \langle u_n(1)|\mathbb{A}|u_{n'}(1)\rangle\langle u_{n'}(1)|\rho(1)|u_n(1)\rangle \\
&= \sum_n \langle u_n(1)|\mathbb{A}\rho(1)|u_n(1)\rangle = Tr\ \{\mathbb{A}\rho(1)\}.
\end{aligned}
$$

So to get the expectation value of an operator that acts only in space (1), we need only the density matrix for that space, $\rho(1)$, which generally will not be the density matrix for a pure state.

The diagonal and off-diagonal density matrix elements have particular interpretations. A diagonal element of the density matrix is $\rho_{nn} = \sum_k P_k(\rho_k)_{nn} = \sum_k P_k\langle u_n|\Psi_k\rangle\langle\Psi_k|u_n\rangle = \sum_k P_k\left|c_n^{(k)}\right|^2$. Now $\left|c_n^{(k)}\right|^2$ is a positive or zero real number giving the probability that if the system is in Ψ_k, it is also in u_n. Summing over k weighted by the probabilities P_k gives the average probability of finding the system in state u_n. ρ_{nn} is referred to as the population of state u_n. An off-diagonal element of the density matrix is $\rho_{nm} = \sum_k P_k(\rho_k)_{nm} = \sum_k P_k\langle u_n|\Psi_k\rangle\langle\Psi_k|u_m\rangle = \sum_k P_k c_n^{(k)} c_m^{(k)*} = \sum_k P_k \exp\{i\theta_{nm}^{(k)}\}\left|c_n^{(k)}\right|\left|c_m^{(k)}\right|$. These are cross terms between u_n and u_m that exist when the description of the system involves a coherent superposition of u_n and u_m. In general, ρ_{nm} is a complex number that vanishes if the phases, $\theta_{nm}^{(k)}$, are randomly distributed among the k. These off-diagonal density matrix elements are called coherences.

At thermal equilibrium, the populations are given by Boltzmann factors and the coherences are assumed to vanish. The density matrix for a thermal equilibrium system is therefore

$$\rho = \frac{1}{Z}\begin{pmatrix} \exp(-E_1 / k_B T) & 0 & \cdots \\ 0 & \exp(-E_2 / k_B T) & \\ \vdots & & \ddots \end{pmatrix}, \tag{5.21}$$

where E_n is the energy of state n and $Z = \Sigma_n \exp(-E_n/k_B T)$ is the partition function.

5.3. DENSITY MATRIX METHODS IN SPECTROSCOPY

Density matrix methods are particularly useful for describing the spectroscopy of molecules interacting with a material environment. Here one normally splits up the total Hamiltonian into three parts:

$$\mathbb{H} = \mathbb{H}_0 + \mathbb{H}_R + \mathbb{H}_{int}$$

$\mathbb{H}_0$ is the Hamiltonian for the material system undergoing the spectroscopic transition. $\mathbb{H}_{int}$ contains the matter–radiation coupling, and is the only explicitly time-dependent part of the Hamiltonian. $\mathbb{H}_R$ contains the coupling between the spectroscopic system and the environment.

The density matrix obeys the usual equation of motion,

$$\begin{aligned} d\rho(t)/dt &= (1/i\hbar)[\mathbb{H}(t), \rho(t)] = (1/i\hbar)[\mathbb{H}_0 + \mathbb{H}_{int}(t) + \mathbb{H}_R, \rho(t)] \\ &= (1/i\hbar)[\mathbb{H}_0, \rho(t)] + (1/i\hbar)[\mathbb{H}_{int}(t), \rho(t)] + (1/i\hbar)[\mathbb{H}_R, \rho(t)]. \end{aligned} \tag{5.22}$$

We assume that at some time in the distant past before the perturbation was turned on, the system was at thermal equilibrium and described by density matrix $\rho^{(0)}$ given by Equation (5.21). Turning on $\mathbb{H}_{int}(t)$ disturbs that equilibrium situation. We express $\rho(t)$ as a power series in the perturbation:

$$\rho(t) = \rho^{(0)} + \rho^{(1)}(t) + \rho^{(2)}(t) + \ldots, \tag{5.23}$$

where $\rho^{(1)}(t)$ is linear in $\mathbb{H}_{int}(t)$, $\rho^{(2)}(t)$ is quadratic in $\mathbb{H}_{int}(t)$, and so on. We then solve for each $\rho^{(i)}(t)$ in terms of itself and the perturbation acting on $\rho^{(i-1)}(t)$. That is,

$$
\begin{aligned}
&d\rho^{(0)}/dt = (1/i\hbar)[\mathbb{H}_0, \rho^{(0)}] + (1/i\hbar)[\mathbb{H}_R, \rho^{(0)}] \\
&d\rho^{(1)}(t)/dt = (1/i\hbar)[\mathbb{H}_0, \rho^{(1)}(t)] + (1/i\hbar)[\mathbb{H}_{\text{int}}(t), \rho^{(0)}] + (1/i\hbar)[\mathbb{H}_R, \rho^{(1)}(t)] \\
&d\rho^{(2)}(t)/dt = (1/i\hbar)[\mathbb{H}_0, \rho^{(2)}(t)] + (1/i\hbar)[\mathbb{H}_{\text{int}}(t), \rho^{(1)}(t)] + (1/i\hbar)[\mathbb{H}_R, \rho^{(2)}(t)] \\
&\text{etc.}
\end{aligned}
\tag{5.24}
$$

$\rho^{(0)}$ is independent of time, as shown later.

To keep things simple, we will specialize to a two-state system from now on. The two energies are E_1 and E_2. Here we can write explicitly

$$
d\rho_{11}(t)/dt = (1/i\hbar)[\mathbb{H}(t), \rho(t)]_{11} = (1/i\hbar)[\mathbb{H}_0 + \mathbb{H}_{\text{int}}(t) + \mathbb{H}_R, \rho(t)]_{11},
$$

and equivalently for ρ_{22} and $\rho_{12} = \rho_{21}^*$.

The relaxation term $\mathbb{H}_R$ is chosen to cause the system's density matrix to approach that of a thermal equilibrium system in the absence of $\mathbb{H}_{\text{int}}$. The effect of the environment is to provide a thermal bath that returns the system to equilibrium once the radiation is turned off. We define

$$
(1/i\hbar)[\mathbb{H}_R, \rho]_{11} = \gamma_{\text{down}}\rho_{22} - \gamma_{\text{up}}\rho_{11} \tag{5.25a}
$$

$$
(1/i\hbar)[\mathbb{H}_R, \rho]_{22} = -\gamma_{\text{down}}\rho_{22} + \gamma_{\text{up}}\rho_{11} \tag{5.25b}
$$

$$
(1/i\hbar)[\mathbb{H}_R, \rho]_{12} = -\Gamma\rho_{12}. \tag{5.25c}
$$

Thus, if only $\mathbb{H}_R$ were in effect, the off-diagonal terms (coherences) would decay exponentially to zero with time constant $1/\Gamma$. This is known as dephasing. The diagonal elements (populations) would decay exponentially to a constant value of $\rho_{22}/\rho_{11} = \gamma_{\text{up}}/\gamma_{\text{down}}$, and in order for this to represent thermal equilibrium, we require $\gamma_{\text{up}}/\gamma_{\text{down}} = \exp\{-(E_2 - E_1)/k_B T\}$.

The radiation–matter interaction term is the same as we have used before when discussing spectroscopy [Eq. (3.27)], which to lowest order (electric dipole) is

$$
\mathbb{H}_{\text{int}}(t) = \mathbf{E}(\mathbf{r}, t) \cdot \mu
$$

where $\mathbf{E}(\mathbf{r},t)$ is the time-dependent electric field and μ is the dipole moment operator. The required commutators for the two-state system are then

$$
[\mathbb{H}_{\text{int}}(t), \rho(t)]_{11} = \langle u_1|\mathbb{H}_{\text{int}}(t)\rho(t)|u_1\rangle - \langle u_1|\rho(t)\mathbb{H}_{\text{int}}(t)|u_1\rangle.
$$

Using closure, this becomes

$$
\begin{aligned}
[\mathbb{H}_{\text{int}}(\text{t}), \rho(t)]_{11} &= \langle u_1|\mathbb{H}_{\text{int}}(t)|u_1\rangle\langle u_1|\rho(t)|u_1\rangle - \langle u_1|\rho(t)|u_1\rangle\langle u_1|\mathbb{H}_{\text{int}}(t)|u_1\rangle \\
&\quad + \langle u_1|\mathbb{H}_{\text{int}}(t)|u_2\rangle\langle u_2|\rho(t)|u_1\rangle - \langle u_1|\rho(t)|u_2\rangle\langle u_2|\mathbb{H}_{\text{int}}(t)|u_1\rangle \\
&= \mathbf{E}(\mathbf{r}, t) \cdot \{\boldsymbol{\mu}_{11}\rho_{11}(t) - \rho_{11}(t)\boldsymbol{\mu}_{11} + \boldsymbol{\mu}_{12}\rho_{21}(t) - \rho_{12}(t)\boldsymbol{\mu}_{21}\}
\end{aligned}
$$

and the first two terms cancel, leaving

$$[\mathbb{H}_{int}(t), \rho(t)]_{11} = \mathbf{E}(\mathbf{r}, t)\cdot\{\mu_{12}\rho_{21}(t) - \rho_{12}(t)\mu_{21}\}. \tag{5.26a}$$

Similarly,

$$[\mathbb{H}_{int}(t), \rho(t)]_{22} = \mathbf{E}(\mathbf{r}, t)\cdot\{\mu_{21}\rho_{12}(t) - \rho_{21}(t)\mu_{12}\} \tag{5.26b}$$

$$[\mathbb{H}_{int}(t), \rho(t)]_{12} = \mathbf{E}(\mathbf{r}, t)\cdot\{-\rho_{11}(t)\mu_{12} + \mu_{12}\rho_{22}(t)\} \tag{5.26c}$$

$$[\mathbb{H}_{int}(t), \rho(t)]_{21} = \mathbf{E}(\mathbf{r}, t)\cdot\{-\rho_{22}(t)\mu_{21} + \mu_{21}\rho_{11}(t)\}. \tag{5.26d}$$

We now proceed to use these equations. First of all, $\rho^{(0)}$ is time-independent; recall its matrix representation is a diagonal matrix of the Boltzmann factors, and $\mathbb{H}_0$ is also a diagonal matrix of the energies, so they commute. Similarly, $\mathbb{H}_R$ commutes with $\rho^{(0)}$ by its definition. We now seek the first-order perturbation correction, $\rho^{(1)}(t)$. The diagonal elements are obtained from

$$d\rho_{11}^{(1)}(t)/dt = (1/i\hbar)\{[\mathbb{H}_0, \rho^{(1)}(t)]_{11} + [\mathbb{H}_{int}(t), \rho^{(0)}]_{11} + [\mathbb{H}_R, \rho^{(1)}(t)]_{11}\}$$

$$d\rho_{22}^{(1)}(t)/\mathrm{d}t = (1/i\hbar)\{[\mathbb{H}_0, \rho^{(1)}(t)]_{22} + [\mathbb{H}_{int}(t), \rho^{(0)}]_{22} + [\mathbb{H}_R, \rho^{(1)}(t)]_{22}\}.$$

But since $\rho^{(0)}$ has only diagonal elements, the terms involving $\mathbb{H}_{int}(t)$ are zero by reference to Equation (5.26a), so the applied field induces no change in the diagonal elements to first order.

The off-diagonal elements are found from

$$\mathrm{d}\rho_{12}^{(1)}(\mathrm{t})/\mathrm{dt} = (1/i\hbar)\{[\mathbb{H}_0, \rho^{(1)}(\mathrm{t})]_{12} + [\mathbb{H}_{int}(\mathrm{t}), \rho^{(0)}]_{12} + [\mathbb{H}_R, \rho^{(1)}(\mathrm{t})]_{12}\} \tag{5.27}$$

$\mathbb{H}_0$ has only diagonal elements and $\rho^{(1)}(t)$ has only off-diagonal elements. Working out their commutator, we get

$$[\mathbb{H}_0, \rho^{(1)}(t)]_{12} = (\mathrm{E}_1 - \mathrm{E}_2)\rho_{12}^{(1)}(\mathrm{t})$$

Plugging in for the second and third terms in Equation (5.27) from Equations (5.25) and (5.26) leaves

$$d\rho_{12}^{(1)}(t)/dt = (1/i\hbar)(\mathrm{E}_1 - \mathrm{E}_2)\rho_{12}^{(1)}(t) + (1/i\hbar)\mathbf{E}(\mathbf{r}, t)\cdot\{-\rho_{11}^{(0)}\mu_{12} + \mu_{21}\rho_{22}^{(0)}\} - \Gamma\rho_{12}^{(1)}(t).$$

If the energy separation between the two levels is sufficiently large that at thermal equilibrium only state 1 is significantly populated, then $\rho_{11}^{(0)} = 1$, $\rho_{22}^{(0)} = 0$, and

$$\begin{aligned} d\rho_{12}^{(1)}(t)/dt &= \mathrm{i}\omega_{21}\rho_{12}^{(1)}(t) + (i/\hbar)\mathbf{E}(\mathbf{r}, t)\cdot\mu_{12} - \Gamma\rho_{12}^{(1)}(t) \\ &= (\mathrm{i}\omega_{21} - \Gamma)\rho_{12}^{(1)}(t) + (i/\hbar)\mathbf{E}(\mathbf{r}, t)\cdot\mu_{12} \end{aligned} \tag{5.28}$$

where $\omega_{21} = (\mathrm{E}_2 - \mathrm{E}_1)/\hbar$.

Equation (5.28) is a first-order differential equation for the coefficient $\rho_{12}^{(1)}(t)$. The general solution to this differential equation is, setting $\rho_{12}^{(1)}(-\infty)=0$,

$$\rho_{12}^{(1)}(t)\exp\left\{\int_{-\infty}^{t} dt'(-i\omega_{21}+\Gamma)\right\}$$
$$=\int_{-\infty}^{t} dt'\left[\left(\frac{i}{\hbar}\right)\boldsymbol{E}(\boldsymbol{r},t)\cdot\boldsymbol{\mu}_{12}\right]\exp\left\{\int_{-\infty}^{t'} dt''(-i\omega_{21}+\Gamma)\right\}.$$

Plugging in the form used before for the electric field in the dipole approximation, $\mathbf{E}(\mathbf{r},t) = E_0\,\mathbf{e}(e^{i\omega t} + e^{-i\omega t})/2$, and doing the integrals leaves

$$\rho_{12}^{(1)}(t)=[(iE_0/2\hbar)\mathbf{e}\cdot\boldsymbol{\mu}_{12}]\{e^{i\omega t}/[-i(\omega_{21}-\omega)+\Gamma]+e^{-i\omega t}/[-\mathrm{i}(\omega_{21}+\omega)+\Gamma]\}.$$

Assuming ω_{21} is positive (the final state is higher in energy than the initial state), only the first term can have a small denominator. Keeping only that term yields

$$\begin{aligned}\rho_{12}^{(1)}(t)&\approx[(\mathrm{i}\mathrm{E}_0/2\hbar)\mathbf{e}\cdot\boldsymbol{\mu}_{12}]e^{i\omega t}/[-\mathrm{i}(\omega_{21}-\omega)+\Gamma]\\&=[(E_0/2\hbar)\mathbf{e}\cdot\boldsymbol{\mu}_{12}]e^{i\omega t}\{[(\omega-\omega_{21})+i\Gamma]/[(\omega-\omega_{21})^2+\Gamma^2]\},\end{aligned} \tag{5.29}$$

and finally, since the density matrix is Hermitian,

$$\rho_{21}^{(1)}(t)=\rho_{12}^{(1)*}(t). \tag{5.30}$$

We now use these density matrix elements to calculate the rate of light absorption by the system. Physically, the action of an oscillating electric field on a molecule, a collection of charged particles, is to induce an oscillating dipole in the molecule. The dipole moment is an observable, and its expectation value can be calculated from the density operator as

$$\langle\mu\rangle=Tr(\rho^{(1)}\mu)=\rho_{12}^{(1)}(t)\boldsymbol{\mu}_{21}+\rho_{21}^{(1)}(t)\boldsymbol{\mu}_{12}. \tag{5.31}$$

Assuming as is usually the case that the matrix elements of μ are real, then this is just

$$\begin{aligned}\langle\mu\rangle&=\{\rho_{12}^{(1)}(t)+\rho_{21}^{(1)}(t)\}\boldsymbol{\mu}_{12}=2\mathrm{Re}\{\rho_{12}^{(1)}(t)\}\boldsymbol{\mu}_{12}\\&=[-E_0\mathbf{e}\cdot\boldsymbol{\mu}_{12}][(\omega_{21}-\omega)\cos(\omega t)+\Gamma\sin(\omega t)]\boldsymbol{\mu}_{12}/\{\hbar[(\omega_{21}-\omega)^2+\Gamma^2]\}.\end{aligned} \tag{5.32}$$

There is a part that is in phase with the driving field [$\cos(\omega t)$] and a part that is out of phase [$\sin(\omega t)$]. The rate at which power is removed from the incident field, averaged over multiple cycles of the field, is proportional to the induced out-of-phase component. More correctly, the power per unit volume dissipated from the applied field is proportional to the time average (over an integer number of cycles of the radiation field) of $-\mathbf{E}(t)\cdot[d\mathbf{P}(t)/dt]$, where $\mathbf{P} = N\langle\mu\rangle$ is the polarization induced in the material system by the field, N being the concentration (molecules per unit volume) (Butcher and Cotter, 1990). So we have

$$\frac{d\boldsymbol{P}(t)}{dt} = N\frac{d}{dt}\langle\mu\rangle = \frac{-NE_0\boldsymbol{e}\cdot\mu_{12}\mu_{12}}{\hbar\left[(\omega_{21}-\omega)^2+\Gamma^2\right]}[-\omega(\omega_{21}-\omega)\sin(\omega t)+\omega\Gamma\cos(\omega t)], \tag{5.33a}$$

and taking the dot product with $\mathbf{E}(t) = E_0\mathbf{e}(e^{i\omega t} + e^{-i\omega t})/2 = \mathrm{E}_0\ \mathbf{e}\cos(\omega t)$ gives

$$-\boldsymbol{E}(t)\cdot\frac{d\boldsymbol{P}(t)}{dt} = \frac{\omega NE_0^2|\boldsymbol{e}\cdot\mu_{12}|^2}{\hbar\left[(\omega_{21}-\omega)^2+\Gamma^2\right]}\left[-(\omega_{21}-\omega)\sin(\omega t)\cos(\omega t)+\Gamma cos^2(\omega t)\right]. \tag{5.33b}$$

When averaged over a full cycle of the radiation field (i.e., integrated from $t = 0$ to $t = 2\pi/\omega$), the $\sin(\omega t)\cos(\omega t)$ term vanishes and the $\cos^2(\omega t)$ term averages to 1/2, leaving

$$\text{Power absorbed per unit volume} = \frac{\omega NE_0^2\Gamma|\boldsymbol{e}\cdot\mu_{12}|^2}{2\hbar\left[(\omega_{21}-\omega)^2+\Gamma^2\right]} = -\frac{dI}{dz}, \tag{5.34a}$$

where I is the light intensity (energy per unit area per unit time). As discussed in Chapter 2, the cycle-averaged light intensity is given by $I = (\varepsilon_0 ncE_0^2/2)$. So we can rewrite Equation (5.34a) as

$$-\frac{dI}{I} = \frac{\omega N\Gamma|\boldsymbol{e}\cdot\mu_{12}|^2}{\hbar cn\varepsilon_0\left[(\omega_{21}-\omega)^2+\Gamma^2\right]}dz. \tag{5.34b}$$

Averaging over a random distribution of orientations of μ_{12} relative to $\mathbf{e}$, as appropriate for an isotropic sample, gives $|\mathbf{e}\cdot\mu_{12}|^2 \to \mu_{12}^2/3$ [Eq. (3.50)], and integration of the previous expression from $z = 0$ to $z = \ell$ yields

$$I(\ell) = I(0)\exp[-\sigma_{12}N\ell], \tag{5.35}$$

where the absorption cross section, $\sigma_{12}(\omega)$, is given by

$$\sigma_{12}(\omega) = \frac{\omega\pi\mu_{12}^2}{3\hbar cn\varepsilon_0}\left\{\frac{1}{\pi}\frac{\Gamma}{(\omega_{21}-\omega)^2+\Gamma^2}\right\}. \tag{5.36a}$$

The lineshape corresponding to absorption is a Lorentzian in frequency, peaked at ω_{21} and having a HWHM of Γ. We have multiplied and divided by π in order to make the quantity in braces a normalized Lorentzian. If we divide both sides of Equation (5.36a) by ω and integrate over all frequencies, the quantity in braces integrates to unity, and we obtain

$$\int \frac{\sigma_{12}}{\omega} d\omega = \frac{\pi \mu_{12}^2}{3\hbar c n \varepsilon_0}. \tag{5.36b}$$

This is the same relationship between the transition moment and the absorption cross section as was derived in Equation (4.16) in a very different manner.

It should be noted that this procedure for calculating the rate of removal of energy from the incident field actually gives not the rate of absorption alone but the difference between the rates of absorption and stimulated emission. The latter is negligible in the limit of very weak driving fields where the perturbative approach used here is applicable.

The density matrix provides a correct but rather unwieldy way to calculate the absorption cross section. It is much more useful for describing nonlinear optical processes, as discussed in Chapter 11.

5.4. EXACT DENSITY MATRIX SOLUTION FOR A TWO-LEVEL SYSTEM

For a finite number of levels, it is possible to solve the original equations of motion for the density matrix elements exactly without recourse to the perturbation expansion. While the exact solution is very messy for more than two states, for a two-level system, the nonperturbative solution is reasonably straightforward and exposes interesting physics.

The starting point is Equation (5.22) for the time dependence of the density operator,

$$d\mathbb{p}(t)/dt = (1/i\hbar)[\mathbb{H}_0, \mathbb{p}(t)] + (1/i\hbar)[\mathbb{H}_{\text{int}}(t), \mathbb{p}(t)] + (1/i\hbar)[\mathbb{H}_R, \mathbb{p}(t)].$$

We need the three matrix elements (only three are needed since $\rho_{12} = \rho_{21}^*$) of each of the three terms. $\mathbb{H}_0$ is diagonal, so

$$[\mathbb{H}_0, \mathbb{p}(t)]_{11} = [\mathbb{H}_0, \mathbb{p}(t)]_{22} = 0$$
$$[\mathbb{H}_0, \mathbb{p}(t)]_{12} = (E_1 - E_2)\rho_{12}(t)$$

The others were worked out previously:

$$[\mathbb{H}_{\text{int}}(t), \mathbb{p}(t)]_{22} = \mathbf{E}(t)\cdot\{\boldsymbol{\mu}_{21}\rho_{12}(t) - \rho_{21}(t)\boldsymbol{\mu}_{12}\} = -[\mathbb{H}_{\text{int}}(t), \mathbb{p}(t)]_{11}$$
$$[\mathbb{H}_{\text{int}}(t), \mathbb{p}(t)]_{12} = \mathbf{E}(t)\cdot\{-\rho_{11}(t)\boldsymbol{\mu}_{12} + \boldsymbol{\mu}_{12}\rho_{22}(t)\}$$
$$(1/i\hbar)[\mathbb{H}_R, \mathbb{p}(t)]_{11} = \gamma_{\text{down}}\rho_{22} - \gamma_{\text{up}}\rho_{11}$$
$$(1/i\hbar)[\mathbb{H}_R, \mathbb{p}(t)]_{22} = -\gamma_{\text{down}}\rho_{22} + \gamma_{\text{up}}\rho_{11}$$
$$(1/i\hbar)[\mathbb{H}_R, \mathbb{p}(t)]_{12} = -\Gamma\rho_{12}.$$

Again, we assume energy level 2 is sufficiently far above level 1 that its thermal population can be neglected. This implies $\gamma_{up} = 0$, so the differential equations become (calling $\gamma_{down} = \gamma$)

$$d\rho_{11}(t)/dt = -(1/i\hbar)\mathbf{E}(t)\cdot\{\boldsymbol{\mu}_{21}\rho_{12}(t) - \rho_{21}(t)\boldsymbol{\mu}_{12}\} + \gamma\rho_{22}(t) \tag{5.37a}$$

$$d\rho_{22}(t)/dt = (1/i\hbar)\mathbf{E}(t)\cdot\{\boldsymbol{\mu}_{21}\rho_{12}(t) - \rho_{21}(t)\boldsymbol{\mu}_{12}\} - \gamma\rho_{22}(t) \tag{5.37b}$$

$$d\rho_{12}(t)/dt = (1/i\hbar)(\mathrm{E}_1 - \mathrm{E}_2)\rho_{12}(t) + (1/i\hbar)\mathbf{E}(t)\cdot\{-\rho_{11}(t)\boldsymbol{\mu}_{12} + \boldsymbol{\mu}_{12}\rho_{22}(t)\} - \Gamma\rho_{12}(t) \tag{5.37c}$$

We wish to solve these equations exactly, subject to the initial conditions $\rho_{11}(-\infty) = 1$, $\rho_{12}(-\infty) = \rho_{22}(-\infty) = 0$.

As before, we will take $\mathbf{E}(t) = \mathrm{E}_0\mathbf{e}(e^{i\omega t} + e^{-i\omega t})/2$ and assume $\boldsymbol{\mu}_{21} = \boldsymbol{\mu}_{12}$. Also, noticing $d\rho_{22}(t)/dt = -d\rho_{11}(t)/dt$ and $\rho_{11}(t) + \rho_{22}(t) = 1$ reduces the relevant equations to just two:

$$\begin{aligned} &d/dt[\rho_{11}(t) - \rho_{22}(t)] \\ &\quad = iE_0(\mathbf{e}\cdot\boldsymbol{\mu}_{12}/\hbar)(e^{i\omega t} + e^{-i\omega t})\{\rho_{12}(t) - \rho_{12}^*(t)\} - \gamma[\rho_{11}(t) - \rho_{22}(t) - 1] \end{aligned} \tag{5.38a}$$

$$d\rho_{12}(t)/dt = i\omega_{21}\rho_{12}(t) + \mathrm{i}E_0(\mathbf{e}\cdot\boldsymbol{\mu}_{12}/2\hbar)(e^{i\omega t} + e^{-i\omega t})\{\rho_{11}(t) - \rho_{22}(t)\} - \Gamma\rho_{12}(t). \tag{5.38b}$$

This is still messy to solve in general, but there is a trick we can use when we specialize to the case $\omega \approx \omega_{21}$ (near resonance excitation). Note that if the Hamiltonian consisted only of $\mathbb{H}_0$, then the equation of motion for the off-diagonal density matrix element would be [see Eq. (5.37c)] $d\rho_{12}(t)/dt = (1/i\hbar)(E_1 - E_2)\rho_{12}(t) = i\omega_{21}\rho_{12}(t)$, leading to $\rho_{12}(t) = \rho_{12}(0)\exp(i\omega_{21}t)$. We therefore define a new variable $s_{12}(t)$ such that

$$\rho_{12}(t) = s_{12}(t)e^{i\omega t}, \tag{5.39}$$

with the expectation that as long as ω is very close to ω_{21}, $s_{12}(t)$ will vary with time much more slowly than does $\rho_{12}(t)$. Then, substituting Equation (5.39) into Equation (5.38) leaves

$$\begin{aligned} &d/dt[\rho_{11}(t) - \rho_{22}(\mathrm{t})] \\ &\quad = iE_0(\mathbf{e}\cdot\boldsymbol{\mu}_{12}/\hbar)(e^{i\omega t} + e^{-i\omega t})\{s_{12}(t)e^{i\omega t} - s_{12}^*(t)e^{-i\omega t}\} - \gamma[\rho_{11}(t) - \rho_{22}(t) - 1] \end{aligned} \tag{5.40a}$$

$$ds_{12}(t)/dt = i(\omega_{21} - \omega)s_{12}(t) + iE_0(\mathbf{e}\cdot\boldsymbol{\mu}_{12}/2\hbar)(1 + e^{-2i\omega t})\{\rho_{11}(t) - \rho_{22}(t)\} - \Gamma s_{12}(t). \tag{5.40b}$$

We can then throw away the terms that oscillate at 2ω or -2ω, since these will rapidly average out to zero. This is what is known as the rotating wave approximation. This leaves

$$d/dt[\rho_{11}(t) - \rho_{22}(t)] = i\Omega\{s_{12}(t) - s_{12}^*(t)\} - \gamma[\rho_{11}(t) - \rho_{22}(t) - 1] \tag{5.41a}$$

$$ds_{12}(t)/dt = [i(\omega_{21}-\omega)-\Gamma]s_{12}(t) + i\Omega\{\rho_{11}(t)-\rho_{22}(t)\}/2 \tag{5.41b}$$

where we have defined $\Omega = E_0\,\mathbf{e}\cdot\boldsymbol{\mu}_{12}/\hbar$, the Rabi frequency. The Rabi frequency is a measure of the rate at which the system will flip back and forth between the two states when driven by an exactly resonant electromagnetic field.

By making the rotating wave approximation, the coherence $s_{12}(t)$ oscillates at the applied field frequency. We now seek the steady-state solutions: that is, set the left-hand sides of each of Equation (5.41) equal to zero. By adding and subtracting Equation (5.41b) and its complex conjugate, and then using Equation (5.41a), one finally finds

$$(\rho_{11}-\rho_{22})_{\text{avg}} = [(\omega_{21}-\omega)^2+\Gamma^2]/\{[(\omega_{21}-\omega)^2+\Gamma^2]+\Gamma\Omega^2/\gamma\} \tag{5.42a}$$

$$(\text{Im } s_{12})_{\text{avg}} = \Omega\Gamma/2\{[(\omega_{21}-\omega)^2+\Gamma^2]+\Gamma\Omega^2/\gamma\} \tag{5.42b}$$

$$(\text{Re } s_{12})_{\text{avg}} = -\Omega(\omega_{21}-\omega)/2\{[(\omega_{21}-\omega)^2+\Gamma^2]+\Gamma\Omega^2/\gamma\}. \tag{5.42c}$$

where Im and Re designate the imaginary and real parts of the matrix element, respectively.

It is interesting to look at some limits of the average difference between lower and upper state populations, $(\rho_{11} - \rho_{22})$, which is equal to 1 at thermal equilibrium (all population in the lower state). If the perturbation frequency is far from resonance, then $(\omega_{21} - \omega)$ is large compared with everything else, and the population difference approaches 1; the nonresonant field is not very effective at putting population into the upper state. For exact resonance, $\omega_{21} - \omega = 0$, the population difference becomes $\Gamma/(\Gamma + \Omega^2/\gamma)$, which again approaches 1 if $\gamma >> \Omega$ (the rate of population decay from 2 to 1 is very fast compared with the rate at which the driving field pumps population up into state 2). If on the other hand, $\Omega >> \gamma$ (the driving field is very strong), then the average population difference vanishes; actually, it oscillates rapidly between state 1 and state 2, but averages to be equal.

Finally, we can calculate the average induced polarization using Equation (5.31):

$$\begin{aligned}\langle\mu\rangle &= [\rho_{12}+\rho_{12}^*]_{\text{avg}}\boldsymbol{\mu}_{12} = [s_{12}e^{i\omega t}+s_{12}^*e^{-i\omega t}]_{\text{avg}}\boldsymbol{\mu}_{12} \\ &= \boldsymbol{\mu}_{12}\Omega\{(\omega_{21}-\omega)\cos\omega t+\Gamma\sin\omega t\}/\{[(\omega_{21}-\omega)^2+\Gamma^2]+\Gamma\Omega^2/\gamma\}.\end{aligned} \tag{5.43}$$

Notice the only difference between this result and the one we got from the perturbation approach [Eq. (5.32)] is the presence of the additional $\Gamma\Omega^2/\gamma$ term in the denominator. Therefore, following the same steps carried out previously, we end up with the absorption cross section as

$$\sigma_{12} = \frac{\omega\pi\mu_{12}^2}{3\hbar cn\varepsilon_0}\left\{\frac{1}{\pi}\frac{\Gamma}{(\omega_{21}-\omega)^2+\Gamma^2+\Gamma\Omega^2/\gamma}\right\}. \tag{5.44}$$

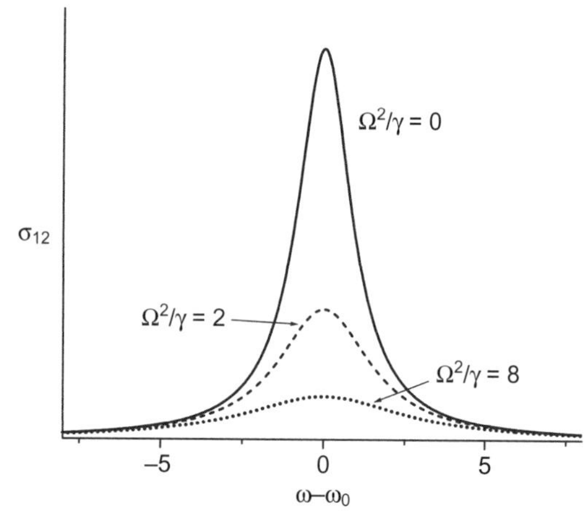

Figure 5.3. Nonperturbative solution for the absorption cross section of a two-level system from Equation (5.44), for $\Gamma = 1$ as a function of Ω^2/γ.

Because of the last term in the denominator, this is always smaller than the perturbation result in Equation (5.36). Physically, this is because the perturbation expansion does not account for the depletion of population in state 1 by the driving field, which reduces the number of molecules capable of absorbing. When the field is very strong (large Ω), absorption and stimulated emission approach equality and the absorption cross section, which describes the average rate of removal of energy from the field, approaches zero (Fig. 5.3).

REFERENCES AND FURTHER READING

P. N. Butcher and D. Cotter, *The Elements of Nonlinear Optics* (Cambridge University Press, Cambridge, 1990).

R. G. Gordon, J. Chem. Phys. **43**, 1307 (1965).

J. D. Jackson, *Classical Electrodynamics*, 2nd edition (John Wiley & Sons, New York, 1975).

PROBLEMS

1. The Kubo formulation of solvent perturbations to spectroscopic lineshapes assumes that $\langle \mu(0)\cdot\mu(t)\rangle$ decays with time due to environmentally induced fluctuations in the transition frequency. That is, $\omega(t) = \omega_0 + \delta\omega(t)$, where ω_0 is the time-averaged center frequency and $\delta\omega(t)$ is the instantaneous fluctuation away from that frequency. These fluctuations are described by a correlation function

$$\langle \delta\omega(0)\delta\omega(t)\rangle = \Delta^2 \exp(-t/\tau_c),$$

where $\Delta = \langle|\delta\omega|^2\rangle^{1/2}$ is the amplitude of the frequency perturbations and $1/\tau_c$ is the rate at which these perturbations relax. With this assumption,

$$\langle \mu(0)\cdot\mu(t)\rangle \sim \exp\{-\Delta^2\tau_c\left[t-\tau_c\left(1-e^{-t/\tau_c}\right)\right]\}.$$

The spectral lineshape is then given by the Fourier transform of this quantity,

$$I(\omega) \sim \int_{-\infty}^{\infty} dt e^{-i\omega t} \exp\{-\Delta^2\tau_c\left[t-\tau_c\left(1-e^{-t/\tau_c}\right)\right]\}.$$

(a) Use a Taylor series expansion of the final exponential to show that at very short times ($t << \tau_c$), $< \mu(0)\cdot\mu(t)>$ decays exponentially in t^2.

(b) Show that at long times ($t >> \tau_c$), $<\mu(0)\cdot\mu(t)>$ decays exponentially in t.

(c) Remember that in Fourier transform relationships, the fastest events in time determine the broadest features in frequency and vice versa. What do your results for (a) and (b) predict for the spectral lineshape in the region near the band maximum and in the wings of the spectrum far from the band maximum?

2. Assume an excited state of a molecule decays exponentially in time with a lifetime of 1 ps. This means that if the molecule is known to be in this excited state at time zero, the probability of finding it still in this state at a later time has decayed to $1/e$ at $t = 1$ ps. What is the full width at half maximum of the Lorentzian line corresponding to absorption from the ground state to this excited state? Express your result in units of frequency (ω) and wavenumber (cm^{-1}). What is the full width at half maximum of the absorption transition in nm if the transition is centered at 400 nm? What if it is centered at 800 nm?

3. Assume that the complete (space and spin) wavefunction for an electron in a molecule can be written as

$$\Psi(\mathbf{r}) = \varphi_+(\mathbf{r})\alpha + \varphi_-(\mathbf{r})\beta,$$

where α and β are the orthonormal spin 1/2 eigenstates with S_z eigenvalues $\pm\hbar/2$, respectively, and $\varphi_\pm(\mathbf{r})$ are the spatial parts of the state vector.

(a) Find the relationship between $\varphi_+(\mathbf{r})$ and $\varphi_-(\mathbf{r})$ such that the wavefunction is normalized; that is, $<\Psi|\Psi> = 1$.

(b) Write the density matrix *for the spin degrees of freedom only*, in the $\{\alpha, \beta\}$ basis. This involves extending the concept of the partial trace to a continuous (position) basis.

(c) How do $\varphi_+(\mathbf{r})$ and $\varphi_-(\mathbf{r})$ have to be related for the spin density matrix you found in (b) to represent a pure state?

(d) Explain physically what it means for the spin density matrix to represent a pure state or not.

4. Consider a molecule interacting with a bath. Let φ_a and φ_b be two nondegenerate eigenstates of the isolated-molecule Hamiltonian $\mathbb{H}$ having energy eigenvalues ε_a and ε_b, respectively. Consider three different possibilities for the density operator of this system at time $t = 0$:

$$\rho_1(0) = |\Psi\rangle\langle\Psi| \text{ where } \Psi = (\varphi_a + \varphi_b)/\sqrt{2}$$

$$\rho_2(0) = |\Psi\rangle\langle\Psi| \text{ where } \Psi = (\varphi_a - \varphi_b)/\sqrt{2}$$

$$\rho_3(0) = (|\varphi_a\rangle\langle\varphi_a| + |\varphi_b\rangle\langle\varphi_b|)/2.$$

The first two correspond to pure cases while the third is a statistical mixture.

(a) Give the 2×2 matrix representation of the density operator in the $\{\varphi_a, \varphi_b\}$ basis for each of the three situations.

(b) Calculate the population of state φ_a for each of the three situations.

(c) Assume the matrix elements of the dipole moment operator in the $\{\varphi_a, \varphi_b\}$ basis are

$$\langle\varphi_a|\mu|\varphi_a\rangle = \langle\varphi_b|\mu|\varphi_b\rangle = 0$$

$$\langle\varphi_a|\mu|\varphi_b\rangle = \langle\varphi_b|\mu|\varphi_a\rangle = \boldsymbol{\mu} \text{ (a real vector quantity).}$$

That is, there is no permanent dipole moment in either of the eigenstates of interest, but there is a transition dipole moment connecting them. Calculate the expectation value of the dipole moment, $\langle\mu\rangle$, for each of the three situations.

5. (a) Consider a system that at time $t = 0$ is in a pure state that is a linear combination of harmonic oscillator eigenstates:

$$\Psi(0) = \frac{1}{\sqrt{3}}(\psi_0 + \psi_1 + \psi_2),$$

where

$$\mathbb{H}_{HO}\psi_v = \hbar\omega(v + ½)\psi_v\,.$$

At times $t > 0$, the system continues to evolve under the influence of $\mathbb{H}_{HO}$. Calculate its density operator at time $t > 0$ and use this density operator to calculate the expectation value of the reduced position, $<\mathbb{q}(t)>$.

(b) Now consider a system that at t = 0 is in a statistical mixture of the first three harmonic oscillator eigenstates with equal weighting. That is, it has equal probabilities of being found in ψ_0,ψ_1 or ψ_2. Again, the system evolves under the influence of $\mathbb{H}_{HO}$ for times $t > 0$. Calculate its density operator at $t > 0$ and use this density operator to calculate $\langle \mathbb{q}(t) \rangle$.

6. In writing Equation (5.28), we assumed that all of the population is initially in state 1 at thermal equilibrium. If states 1 and 2 sufficiently close in energy, or the temperature is high enough, this approximation is no longer valid. Derive the analog of Equation (5.29) with the assumption that $\rho_{11}^{(0)} = F$ and $\rho_{22}^{(0)} = 1 - F$, where $0.5 < F < 1.0$. Then follow the rest of the steps carried out in the text to derive an expression for the integrated apparent absorption cross section analogous to Equation (5.36a). What is the role of stimulated emission in this situation?

7. Let $|+\rangle$, $|0\rangle$, and $|-\rangle$ represent the three eigenstates of the z-component of orbital angular momentum ($\mathbb{L}_z$) with eigenvalues $\hbar, 0,$ and $-\hbar$, respectively—for example, the $2p_+$, $2p_z$, and $2p_-$ atomic orbitals. The 2p wavefunction for an ensemble of boron atoms, which have a single 2p electron, is described by the density matrix

$$\rho = \frac{1}{4}\begin{pmatrix} 2 & 1 & 1 \\ 1 & 1 & 0 \\ 1 & 0 & 1 \end{pmatrix}.$$

(a) Does this density matrix define a pure or mixed state?

(b) What is the average (expectation) value of $\mathbb{L}_z$ for this system?

(c) What is the standard deviation in measured values of $\mathbb{L}_z$ for this system?

8. A spin 1/2 system has two eigenstates, α and β. Write the density matrix corresponding to the following states of this system:

(a) $\frac{1}{\sqrt{2}}(|\alpha\rangle + |\beta\rangle)$

(b) An equally weighted statistical mixture of $|\alpha\rangle$ and $|\beta\rangle$

(c) An equally weighted statistical mixture of the states ($|\alpha\rangle + |\beta\rangle$) and ($|\alpha\rangle - |\beta\rangle$)

Discuss physically the similarities and differences among these three situations.

9. The spin degrees of freedom of a system of two electrons can be described using the following four basis states: $\alpha(1)\alpha(2)$, $\alpha(1)\beta(2)$, $\beta(1)\alpha(2)$, and $\beta(1)\beta(2)$.

(a) The spin function corresponding to the singlet ground state of H_2 is $[\alpha(1)\beta(2) - \beta(1)\alpha(2)]/\sqrt{2}$. In the previous basis, write down the 4×4 matrix that represents the density matrix for this system. Does it represent a pure state?

(b) By carrying out the partial trace over electron 2, calculate the reduced density matrix for electron 1 in the basis $\{\alpha(1), \beta(1)\}$. Does this reduced density matrix represent a pure state?

CHAPTER 6

SYMMETRY CONSIDERATIONS

6.1. QUALITATIVE ASPECTS OF MOLECULAR SYMMETRY

Most molecules possess some type of symmetry. The simplest example is a homonuclear diatomic molecule, for example, H_2. The two hydrogen atoms are equivalent; the distribution of electron density, for example, must be the same around the two atoms. The two oxygen atoms in CO_2, the four hydrogen atoms in CH_4, and the six carbon or hydrogen atoms in benzene are other examples of symmetrically equivalent atoms in molecules. Atoms of different types are never equivalent, nor are atoms of the same type in chemically different environments, such as the ring hydrogens and the methyl hydrogens in toluene.

Symmetrically equivalent atoms must have identical measurable properties. This does not mean identical wavefunctions, as the wavefunction itself is not an observable quantity; only its absolute square is. Therefore, the square of the electronic wavefunction for H_2 must have identical values when evaluated at equivalent positions relative to each of the two atoms, but the wavefunction itself may have different signs at these two locations. That is, with respect to interchanging identical atoms "A" and "B," the wavefunction must be either symmetric (it does not change) or antisymmetric (it changes only its sign). This result is seen in the simplest quantum mechanical treatment of the H_2 molecule: the lower energy, bonding molecular orbital is an equally weighted sum

---

Condensed-Phase Molecular Spectroscopy and Photophysics, First Edition. Anne Myers Kelley.

of the 1s atomic orbitals on the two atoms, while the higher energy, antibonding molecular orbital is the equally weighted difference of the two 1s atomic orbitals.

The use of molecular symmetry can greatly simplify the task of calculating the matrix elements that are involved in determining spectroscopic transition strengths and solving many other quantum mechanical problems. Take again the specific case of H_2, and take the z-axis to lie along the internuclear axis. Consider an arbitrary matrix element $\langle\psi_A|\mathbb{O}|\psi_B\rangle$, where ψ_A and ψ_B are two arbitrary wavefunctions, and $\mathbb{O}$ is an arbitrary operator. The Dirac notation brackets imply integration over all space—for example, over x, y, and z. We know from the previous discussion that ψ_A and ψ_B are each either symmetric (even) or antisymmetric (odd) with respect to the z-coordinate. We also know that the product of two odd functions or two even functions is even, and the product of an even and an odd function is odd. Therefore, if $\mathbb{O}$ is an operator that is antisymmetric with respect to the coordinates (such as the dipole operator, $\mathbb{O} = e\mathbf{r}$), then the matrix element must be zero if ψ_A and ψ_B are either both even or both odd; it can be nonzero (but is not guaranteed to be!) only if ψ_A is odd and ψ_B is even, or vice versa. The use of symmetry often allows many matrix elements to be identified as vanishing immediately, greatly reducing the amount of work that must be done to solve the problem.

Many molecules have symmetrically equivalent atoms that come only in pairs. In these cases, all wavefunctions are either symmetric or antisymmetric with respect to interchange of the equivalent atoms, and the symmetry considerations can usually be handled quite simply. The problem becomes more complicated when there are more than two symmetrically equivalent atoms, or when there are many equivalent pairs. Spectroscopists and quantum chemists make use of the branch of mathematics known as group theory to systematize and formalize the handling of molecular symmetry.

6.2. INTRODUCTORY GROUP THEORY

Group theory classifies objects or functions into groups according to the symmetry elements they possess. For molecules, the relevant groups are point groups, in which all symmetry operations leave some point in space (the origin of the coordinate system) unchanged. Infinite crystals with translational symmetry are described in terms of space groups instead, but these will not be dealt with further here.

A symmetry element is a point, line, or plane around which one can perform a symmetry operation. The relevant symmetry operations are given in Figure 6.1 and Table 6.1.

For mathematical completeness, all groups also contain the identity element E, the operation that does nothing (analogous to 1 in multiplication). If there is more than one rotation axis, the principal axis is defined as the one with the largest n. Reflection planes are then defined as vertical (σ_v) if they include the

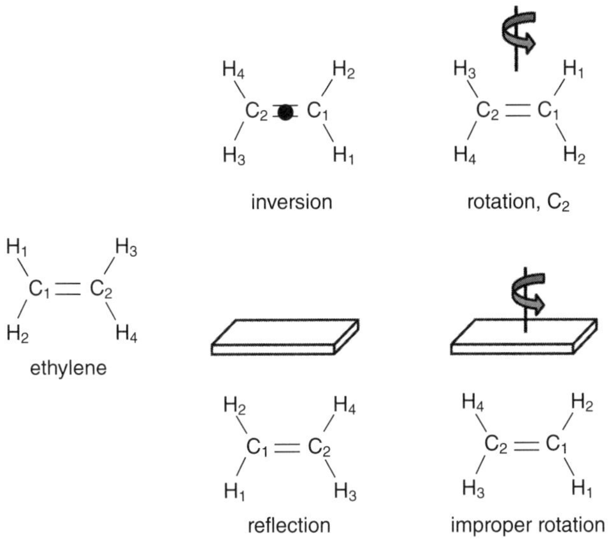

Figure 6.1. Illustration of the four types of symmetry operations for the ethylene molecule. Note that ethylene also possesses other symmetry elements not illustrated here.

TABLE 6.1. Symmetry Elements and Symmetry Operations

Element	Operation	Abbreviation
Point (always the origin)	Inversion ($x \rightarrow -x$, $y \rightarrow -y$, $z \rightarrow -z$)	i
Line	Rotation by 360/n degrees	C_n
Plane	Reflection	σ
Line and plane	Improper rotation—rotation by 360/n degrees followed by reflection in plane perpendicular to the rotation axis	S_n

principal axis and horizontal (σ_h) if they are perpendicular to the principal axis. The product of any two symmetry operations of a group is another symmetry operation that also belongs to the group.

Each group is defined by its symmetry operations and is named with a symbol. Some of the most common point groups are summarized in Table 6.2, with examples. There are many other point groups that are less common and are not listed.

Associated with any given group are one or more symmetry species or irreducible representations. Symmetry species are defined by how they change when each symmetry operation is applied. They are identified by letters and subscripts, such as A_1, B_g, and so on. Every point group has one totally

TABLE 6.2. Some Common Point Groups

Group	Symmetry Element(s)	Example
C_1	E	CFClBrI
C_2	E, C_2	H_2O_2 (gauche)
C_i	E, i	$C_2H_2Cl_2Br_2$ (all trans)
C_s	E, σ	C_2H_2ClBr
C_{2h}	E, C_2, i, σ_h	Trans-butadiene
C_{3h}	$E, C_3, \sigma_{h,} S_3$	$B(OH)_3$ (planar)
C_{2v}	$E, C_2, \sigma_v(xz), \sigma_v(yz)$	H_2O
C_{3v}	E, C_3, σ_v	NH_3
D_{2h}	$E, C_2(z), C_2(y), C_2(x), i, \sigma(xy), \sigma_v(xz), \sigma_v(yz)$	Ethylene
D_{6h}	$E, C_6, C_3, C_2, C_2', C_2'', i, S_3, S_6, \sigma_h, \sigma_d, \sigma_v$	Benzene
D_{3d}	$E, C_3, C_2, i, S_6, \sigma_d$	Ethane (staggered)
$C_{\infty v}$	$E, C(\varphi), \sigma_v$	HCN
$D_{\infty h}$	$E, C(\varphi), C_2, i, S(\varphi), \sigma_v$	H_2

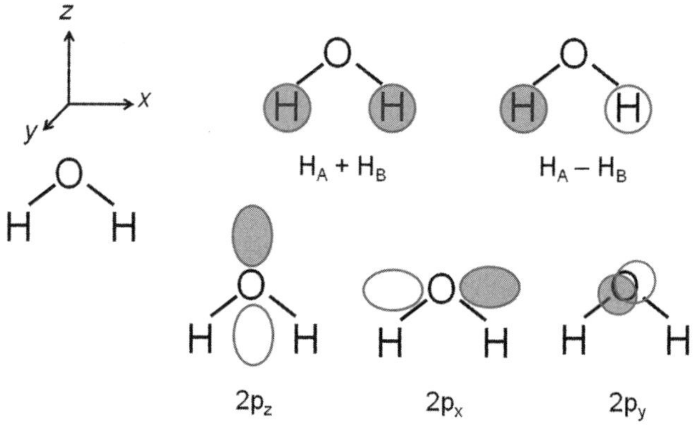

Figure 6.2. Some atomic basis orbitals of the water molecule. Gray and white indicate positive and negative signs, respectively.

symmetric species which does not change when any of the symmetry operations of the group are applied. The other species change sign—or, more generally, get multiplied by some phase factor $e^{i\varphi}$—when certain symmetry operations are applied, but remain unchanged when other operations are applied.

This is easiest to understand by reference to a particular example (Fig. 6.2). Consider the water molecule, which belongs to point group C_{2v}. According to Table 6.2, C_{2v} has four associated symmetry operations: E, C_2, $\sigma_v(xz)$, and $\sigma_v(yz)$. The C_2 axis is the axis that lies in the plane of the molecule, passing through the O atom and bisecting the H-O-H angle; we define this as the z-

axis. The x-axis is the other axis in the plane of the molecule, and the y-axis is perpendicular to the molecular plane. It is easy to see that reflection in the plane [$\sigma_v(xz)$] leaves all of the atoms in the same place, as all of the atoms lie in the plane. Reflection in the plane that bisects the H-O-H angle [$\sigma_v(yz)$] interchanges the two identical H atoms.

Now consider some molecular orbitals of H_2O approximated as linear combinations of atomic orbitals. The 1s orbitals on the two H atoms form two linear combinations, $H_A + H_B$ and $H_A - H_B$. The plus combination is positive everywhere, and rotation around the symmetry axis or reflection in either of the two symmetry planes produces a function identical to the original one. Thus, $H_A + H_B$ belongs to the totally symmetric species, which in the C_{2v} point group is called A_1. Now consider the function $H_A - H_B$. Rotation around the C_2 axis interchanges the two H atoms, so the function becomes negative where it used to be positive and vice versa. Reflection in the yz plane does the same thing. However, reflection in the xz plane does not change the 1s hydrogen orbitals. The function that is unchanged under $\sigma_v(xz)$ but changes sign under C_2 or $\sigma_v(yz)$ is called B_1.

Next, consider the p orbitals on the O atom. The p_z orbital does not change under any of the symmetry operations of the group, so it is A_1. The p_x orbital behaves like $H_A - H_B$, so it is B_1. The p_y orbital has its positive and negative lobes interchanged by C_2 rotation or by reflection in the xz plane, but reflection in the yz plane does not change it. This function is designated B_2.

The operators that correspond to rotation around each of the three axes (designated R_x, R_y, and R_z), the three Cartesian coordinates themselves (x, y, and z), and products of coordinates (x^2, yz, etc.) also all transform as one of the symmetry species of the group. For spectroscopic applications, we are usually most interested in how the coordinates transform. In the C_{2v} case, x belongs to B_2, y is B_1, and z is A_1 (just like the corresponding oxygen p orbitals).

All of this information can be conveniently put into a character table. Character tables for some common point groups are given in Appendix D, and the table for C_{2v} is reproduced as Table 6.3.

For simple point groups such as C_{2v}, the character table directly shows whether the function changes sign or not when each symmetry operation is applied.

TABLE 6.3. Character Table for the C_{2v} Point Group

C_{2v}	E	C_2	$\sigma_v(xz)$	$\sigma'_v(yz)$		
A_1	1	1	1	1	z	x^2, y^2, z^2
A_2	1	1	−1	−1	R_z	xy
B_1	1	−1	1	−1	x, R_y	xz
B_2	1	−1	−1	1	y, R_x	yz

We now address how this all relates to spectroscopy. Group theory provides a compact way to determine when the integral $<\psi_A|\mathbb{O}|\psi_B>$ must be zero for molecules that have many other types of symmetry besides just even/odd. One multiplies the characters corresponding to the symmetry species of ψ_A, ψ_B, and $\mathbb{O}$. If the resulting characters correspond to the totally symmetric species, then the integral may be nonzero (it could be zero for other reasons, but it is not zero for symmetry reasons).

Again, an example is best. Assume we want to make a transition from the ground state of H_2O, which has A_1 symmetry, to an excited state having B_2 symmetry. First, we ask whether *x*-polarized light can do this, noting that *x* belongs to B_1. Taking the product of the characters for A_1, B_1, and B_2 gives 1 1 −1 −1, which is the same as the character for A_2. This is not totally symmetric, so *x*-polarized light cannot cause this transition. If we use *y*-polarized light instead (B_2), the product of the three characters gives 1 1 1 1 which is A_1, so this transition is allowed by symmetry. Finally, if we consider z polarized light (A_1), the product of the three characters gives 1 −1 −1 1 which is B_2, so this transition cannot be produced by *z*-polarized light.

The use of character tables is a little more complicated when working with point groups that have degenerate representations (symbolized E, not to be confused with the identity operation which is also, unfortunately, labeled *E*). For example, the character table for C_{3v} is given in Table 6.4.

Assume we want to know whether a transition between a ground state of A_1 symmetry and an excited state of E symmetry is allowed. If we use *z*-polarized light (A_1), the product of the three characters is 2 −1 1, or E. The transition is not allowed. If we use *x*- or *y*-polarized light (E), the product of the characters is 4 1 0. This does not equal any single symmetry species, but we can write this character as the sum of the characters of $A_1 + A_2 + E$. Since this *contains* the totally symmetric species, we conclude that the transition is allowed with *x*- or *y*-polarized light.

In this example, it was fairly easy to see that the set of characters 4 1 0 is equal to the sum of the characters of A_1, A_2, and E. There is a more formal way to do this when it is not so easy to determine by inspection. In the previous example, the product of A_1, E, and E, having characters 4, 1, and 0, is what is known as a reducible representation. We want to find how to decompose it as a sum of irreducible representations, the symmetry species given in the character table. If $\chi_{red}(R_c)$ is the character of reducible representation Γ for

TABLE 6.4. Character Table for the C_{3v} Point Group

C_{3v}	E	$2C_3$	$3\sigma_v$		
A_1	1	1	1	z	$x^2 + y^2, z^2$
A_2	1	1	−1	R_z	
E	2	−1	0	$(x, y),(R_x, R_y)$	$(x^2 - y^2, xy)(xz, yz)$

symmetry operation R_c and $\chi_i(R_c)$ is the character of irreducible representation Γ_i for that same operation, then the number of times a_i that Γ_i contributes to Γ is given by

$$a_i = \frac{1}{g}\sum_c h_c \chi_i^*(R_c)\chi_{\text{red}}(R_c), \tag{6.1}$$

where g is the order of the group (the number of symmetry operations appearing in the character table), h_c is the number of operations of a particular type (e.g., in C_{3v}, $h_c = 1$ for E, 2 for C_3, and 3 for σ_v), and the sum is over all of the symmetry operations in the character table. Applying this formula to the previous example where $\Gamma = A_1 \otimes E \otimes E$, we have

$$a_{A_1} = \frac{1}{6}\{(1)(1)(4)+(2)(1)(1)+(3)(1)(0)\} = 1 \tag{6.2a}$$

$$a_{A_2} = \frac{1}{6}\{(1)(1)(4)+(2)(1)(1)+(3)(-1)(0)\} = 1 \tag{6.2b}$$

$$a_E = \frac{1}{6}\{(1)(2)(4)+(2)(-1)(1)+(3)(0)(0)\} = 1, \tag{6.2c}$$

and we conclude that $A_1 \otimes E \otimes E = A_1 + A_2 + E$ as was determined previously by inspection.

Note that while the electric dipole operator involves only the Cartesian coordinate along which the radiation is polarized, in most applications, the molecules are oriented randomly relative to the field, and so x, y, and z must all be considered in determining whether or not a given transition is electric dipole allowed. Similarly, although Equation (3.40b) and (3.40c) gave the matrix elements for the magnetic dipole and electric quadrupole operators as L_y and zx, respectively, this assumed a particular orientation of the molecule relative to the polarization of the incident electromagnetic field. For molecules that are randomly oriented relative to the field polarization, all three components of the angular momentum operator contribute to the magnetic dipole operator and all six products of two coordinates (x^2, y^2, z^2, xy, xz, and yz) contribute to the electric quadrupole operator.

6.3. FINDING THE SYMMETRIES OF VIBRATIONAL MODES OF A CERTAIN TYPE

We now demonstrate a more formal and systematic approach to determining the symmetries of vibrational modes of molecules. This section will address finding the symmetries of the vibrations of a particular type, while the next section will present the method for determining the symmetries of all of the vibrations of a molecule.

TABLE 6.5. Character Table for the D_{2h} Point Group

D_{2h}	E	$C_2(z)$	$C_2(y)$	$C_2(x)$	i	$\sigma(xy)$	$\sigma(xz)$	$\sigma(yz)$		
A_g	1	1	1	1	1	1	1	1		x^2, y^2, z^2
B_{1g}	1	1	−1	−1	1	1	−1	−1	R_z	xy
B_{2g}	1	−1	1	−1	1	−1	1	−1	R_y	xz
B_{3g}	1	−1	−1	1	1	−1	−1	1	R_x	yz
A_u	1	1	1	1	−1	−1	−1	−1		
B_{1u}	1	1	−1	−1	−1	−1	1	1	z	
B_{2u}	1	−1	1	−1	−1	1	−1	1	y	
B_{3u}	1	−1	−1	1	−1	1	1	−1	x	

As discussed in more detail in Chapter 7, the vibrations of a polyatomic molecule in the harmonic limit are described as normal modes. A normal mode is a motion in which all of the atoms oscillate about their equilibrium positions with generally different amplitudes but the same phase and frequency. Usually many of the atoms in the molecule are moving in any given normal mode, but often the motion of one particular type of atom dominates; for example, organic molecules have vibrations that are well described as "carbon-hydrogen stretching," even though other types of motion may also be going on in the normal mode.

As an example, we demonstrate here how to find the symmetries of the C-H out-of-plane bending modes of the ethylene molecule (Fig. 6.1). Ethylene belongs to the D_{2h} point group (Table 6.5). Ethylene has four hydrogen atoms, so we begin by creating a basis consisting of the out-of-plane bends of each of the four hydrogens. Each basis vector consists of one of the hydrogen atoms coming up out of the plane of the molecule. Call these basis vectors b_1, b_2, b_3, and b_4. The next step is to apply each of the symmetry operations of the point group to each basis vector. Each symmetry operation of D_{2h} applied to each basis vector will produce either the original basis vector or one of the other three, with or without a change in sign. Choosing two of the symmetry operations as examples,

$$C_2(z)b_1 = b_4 \quad C_2(z)b_2 = b_3 \quad C_2(z)b_3 = b_2 \quad C_2(z)b_4 = b_1$$
$$\sigma(xy)b_1 = -b_1 \quad \sigma(xy)b_2 = -b_2 \quad \sigma(xy)b_3 = -b_3 \quad \sigma(xy)b_4 = -b_4.$$

The character of the symmetry operation is defined as the sum of the diagonal coefficients, that is, the operations that returned the same basis vector multiplied by a coefficient. For $C_2(z)$, this is zero because this operation converts each basis vector to a different one. For $\sigma(xy)$ this is −4 because this operation returns the negative of each of the four basis vectors (reflection in the xy plane converts a displacement out of the plane to one going into the plane). Similarly,

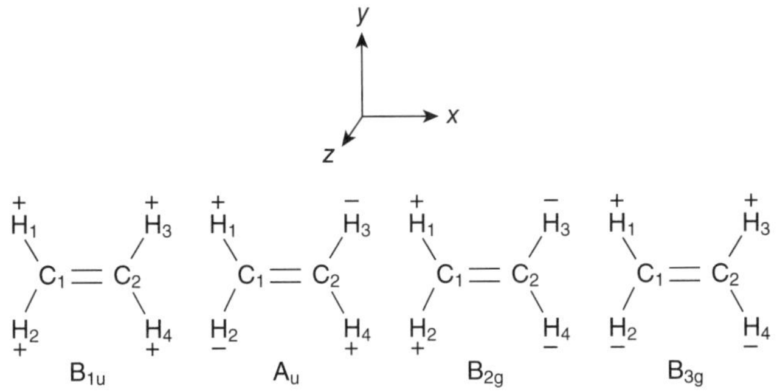

Figure 6.3. Cartoon of the four hydrogen out-of-plane motions of ethylene. The B_{3g} mode is a rotation, not a vibration.

the characters for the other operations can be shown to be +4 for E and 0 for $C_2(x)$, $C_2(y)$, i, $\sigma(xz)$, and $\sigma(yz)$. The reducible representation of the CH out of plane bends is therefore

E	$C_2(z)$	$C_2(y)$	$C_2(x)$	i	$\sigma(xy)$	$\sigma(xz)$	$\sigma(yz)$
4	0	0	0	0	−4	0	0

We then express this as a sum of irreducible representations using Equation (6.1) or by inspection, with the result $\Gamma_{red} = B_{2g} + B_{3g} + A_u + B_{1u}$. These four modes are cartooned in Figure 6.3.

This result implies that ethylene has one C-H out-of-plane vibration belonging to each of these four symmetry species. In this case that is not exactly correct, because one of the motions shown in Figure 6.3, the B_{3g}, actually corresponds to overall rotation of the whole molecule around the x-axis. Therefore it is a rotation, not a vibration. (Note that R_x transforms as B_{3g}.)

For addressing electronic states of molecules, an analogous approach can be used to find the combinations of atomic orbitals that transform as symmetry species of the molecule.

6.4. FINDING THE SYMMETRIES OF ALL VIBRATIONAL MODES

A similar approach can be used to obtain the number of vibrational modes of each symmetry species without regard for the description of the vibration. In order to do this, one starts with a basis that consists not of local stretches, bends, and so on, but of all 3N Cartesian displacements of the N atoms. This is a lengthier process because we must apply each symmetry operation (eight for ethylene) to each Cartesian displacement of each atom (18 for ethylene). The $C_2(z)$ operation moves each of the six atoms, so its character is zero when applied to the Cartesian displacements just as it was when applied to the out-of-plane bending basis vectors. The $C_2(x)$ operation, however, leaves the two carbon atoms in the same place. It does not change their x displacements, but it reverses their y and z displacements. The character of $C_2(x)$ when applied to the Cartesian displacements is therefore 2*(1 – 1 – 1) = –2. Application of all eight operations to the set of Cartesian displacements gives

E	$C_2(z)$	$C_2(y)$	$C_2(x)$	i	$\sigma(xy)$	$\sigma(xz)$	$\sigma(yz)$
18	0	0	–2	0	6	2	0

We then apply Equation (6.1) to decompose this reducible representation of Cartesian displacements into a sum of irreducible representations. Noting that $h_c = 1$ for each of the operations of D_{2h} and thus omitting it, we have

$$a_{A_g} = \frac{1}{8}\{(1)(18)+(1)(0)+(1)(0)+(1)(-2)+(1)(0)+(1)(6)+(1)(2)+(1)(0)\} = 3$$

$$a_{B_{1g}} = \frac{1}{8}\{(1)(18)+(1)(0)+(-1)(0)+(-1)(-2)$$
$$+(1)(0)+(1)(6)+(-1)(2)+(-1)(0)\} = 3$$

$$a_{B_{2g}} = \frac{1}{8}\{(1)(18)+(-1)(0)+(1)(0)+(-1)(-2)$$
$$+(1)(0)+(-1)(6)+(1)(2)+(-1)(0)\} = 2$$

$$a_{B_{3g}} = \frac{1}{8}\{(1)(18)+(-1)(0)+(-1)(0)+(1)(-2)$$
$$+(1)(0)+(-1)(6)+(-1)(2)+(1)(0)\} = 1$$

$$a_{A_u} = \frac{1}{8}\{(1)(18)+(1)(0)+(1)(0)+(1)(-2)$$
$$+(-1)(0)+(-1)(6)+(-1)(2)+(-1)(0)\} = 1$$

$$a_{B_{1u}} = \frac{1}{8}\{(1)(18)+(1)(0)+(-1)(0)+(-1)(-2)$$
$$+(-1)(0)+(-1)(6)+(1)(2)+(1)(0)\} = 2$$

$$a_{B_{2u}} = \frac{1}{8}\{(1)(18)+(-1)(0)+(1)(0)+(-1)(-2)$$
$$+(-1)(0)+(1)(6)+(-1)(2)+(1)(0)\} = 3$$

$$a_{B_{3u}} = \frac{1}{8}\{(1)(18)+(-1)(0)+(-1)(0)+(1)(-2)$$
$$+(-1)(0)+(1)(6)+(1)(2)+(-1)(0)\} = 3.$$

We thus conclude that the distribution of the 18 nuclear degrees of freedom in ethylene is $3A_g + 3B_{1g} + 2B_{2g} + B_{3g} + A_u + 2B_{1u} + 3B_{2u} + 3B_{3u}$. Three of these modes are translations and three are rotations. The character table tells us that the translations x, y, and z transform as B_{3u}, B_{2u}, and B_{1u}, respectively, while the rotations R_x, R_y, and R_z transform as B_{3g}, B_{2g}, and B_{1g}, respectively. Subtracting these six leaves the distribution of the 12 true vibrational modes as $3A_g + 2B_{1g} + B_{2g} + A_u + B_{1u} + 2B_{2u} + 2B_{3u}$. This alone does not tell us what these vibrations look like, but it can be used to tell us, for example, how many infrared-active and how many Raman-active vibrations ethylene should have.

FURTHER READING

F. A. Cotton, *Chemical Applications of Group Theory*, 2nd edition (Wiley-Interscience, New York, 1971).

R. L. Flurry, Jr., *Quantum Chemistry: An Introduction* (Prentice-Hall, Inc., Englewood Cliffs, NJ, 1983).

PROBLEMS

1. Assign point groups to the following molecules:
   (a) ethane (staggered)
   (b) ethane (eclipsed)
   (c) cyclohexane (chair form)
   (d) cyclohexane (boat form).

2. Consider the polyatomic molecules H_2O, CO_2, NH_3, CCl_4, *trans*-CH_2Cl_2, CH_3F, and CH_2O (it may help to draw Lewis dot structures to determine their shape).
   (a) Which have an inversion center?
   (b) Which have exactly one rotation axis?
   (c) Which have more than one rotation axis?
   (d) Which have at least one plane of symmetry that does not contain all of the atoms?

3. Consider the planar forms of *trans*- and *cis*-1,2-dichloroethylene.
   (a) Determine the point group of each isomer.
   (b) Each isomer has two vibrational modes that are predominantly C-H stretching, one symmetric and one antisymmetric. Assign the symmetry species of each of the C-H stretches for each isomer.

(c) Each isomer has two vibrational modes that are predominantly C-H out-of-plane bending, one where both hydrogens come above the plane together and one where one hydrogen comes above the plane at the same time the other goes below the plane. Assign the symmetry species of each of these vibrations for each isomer.

4. Formaldehyde, $H_2C{=}O$, is a planar molecule in its ground state.

(a) Determine its point group.

(b) Determine the symmetry species of the following atomic orbitals or combinations of atomic orbitals: $H_{1sa} + H_{1sb}$, $H_{1sa} - H_{1sb}$, C_{2s}, C_{2px}, O_{2py}, O_{2pz}.

5. A metal compound has the generic formula $M(CO)_4X_2$, where M is a metal and X is some ligand other than CO. Assume the metal is octahedrally coordinated.

(a) Two different isomers are possible, with the X groups being either cis or trans to one another. Determine the point group of each isomer.

(b) The C≡O stretching vibrations are relatively isolated from all other vibrations; since there are four CO groups, there are four C≡O stretching modes. For each isomer, find the four combinations of the C≡O stretches that transform as symmetry species of the molecule, give their symmetry species, and state whether or not they should be observed in the infrared spectrum.

6. A metal compound MOX_4, where M is a metal and X is some ligand other than O, may exist in either a trigonal bipyramidal geometry or a square pyramidal geometry, with the O being the atom at the apex of the pyramid. Determine the point group and the symmetry species of the M-O stretching vibration for each structure.

7. Name all of the symmetry elements possessed by:

(a) an s orbital

(b) a p orbital

(c) a d_{z^2} orbital

(d) a d_{xy} orbital.

8. Use group theory to determine which of the following electronic transitions are electric quadrupole allowed:

(a) B_2 to B_1 in C_{2v}

(b) A_{1g} to E_g in D_{3d}

(c) A_u to B_{1u} in D_{2h}.

CHAPTER 7

MOLECULAR VIBRATIONS AND INFRARED SPECTROSCOPY

7.1. VIBRATIONAL TRANSITIONS

As discussed in Chapter 1, the separation of time scales between fast electronic motion and slow nuclear motion leads to a picture in which the electron distribution defines a potential energy surface on which the nuclei move. The total molecular wavefunction can be approximately factored into a part that is a function of the electronic coordinates but depends parametrically on the nuclear coordinates, and a part that is a function of only the positions of the nuclei but is generally different for each electronic state [Eq. (1.67)]. The nuclear wavefunction can be further factored into three parts: one that describes the motion of the molecule's center of mass (translation), one that describes the orientation of the molecule in space (rotation), and one that describes the oscillations of the nuclei around their equilibrium positions relative to a translating and rotating reference frame (vibration). Translational motion is rarely interesting spectroscopically, although motion of the molecule either toward or away from the propagation direction of the light causes Doppler broadening of the spectroscopic transitions in high-resolution gas-phase experiments. In the gas phase, rotational motion is quantized and rotational spectra are rich, but in condensed phases, interactions with the environment interrupt the rotational motion on time scales short compared with rotational periods, and the rotational motion is better described classically. The spectroscopy of quantized rotations is discussed in most books on

Condensed-Phase Molecular Spectroscopy and Photophysics, First Edition. Anne Myers Kelley.

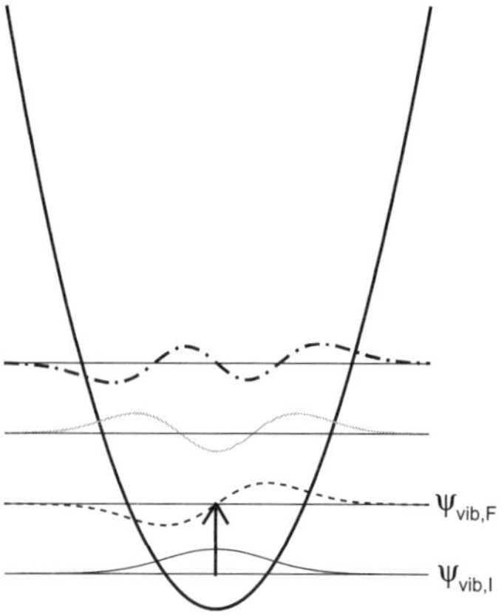

Figure 7.1. A vibrational transition from $v = 0$ to $v = 1$ of a given electronic state.

molecular spectroscopy, but not in this one, which focuses on large molecules and the condensed phase. Rotational motion in the condensed phase is further addressed in Chapter 9. For the remainder of this book, we will assume that only the vibrational part of $\psi_{nuc}(\mathbf{R})$ need be explicitly considered.

Vibrational spectroscopy refers to transitions between states that differ in their vibrational quantum numbers (initial state I, final state F), but within the same electronic state (labeled i) (Fig. 7.1):

$$\Psi_{elect,i}\psi_{vib,I} \rightarrow \Psi_{elect,i}\psi_{vib,F}. \tag{7.1}$$

Referring to Equation (3.31), the transition rate in the electric dipole approximation is given by

$$R_{IF}^{(1)}(\omega) = \frac{\pi}{2\hbar^2} E_0^2 \left|\langle \Psi_{elect,i}\psi_{vib,F} | \mu_z | \Psi_{elect,i}\psi_{vib,I} \rangle\right|^2 \rho(\omega_{FI}), \tag{7.2}$$

where we assume that the radiation is polarized along the laboratory z-direction.

In order to evaluate the matrix element, write it out explicitly as an integral over the coordinates of the electrons and nuclei:

$$\begin{aligned}
&\langle \Psi_{elect,i}\psi_{vib,F} | \mu_z | \Psi_{elect,i}\psi_{vib,I} \rangle \\
&= \int d\boldsymbol{r} \int d\boldsymbol{Q} \psi^*_{vib,F}(\boldsymbol{Q}) \Psi^*_{elect,i}(\boldsymbol{r}; Q) \mu_z \Psi_{elect,i}(\boldsymbol{r}; Q) \psi_{vib,I}(\boldsymbol{Q}) \\
&= \int d\boldsymbol{Q} \psi^*_{vib,F}(\boldsymbol{Q}) \left\{ \int d\boldsymbol{r} \Psi^*_{elect,i}(\boldsymbol{r}; Q) \mu_z \Psi_{elect,i}(\boldsymbol{r}; Q) \right\} \psi_{vib,I}(\boldsymbol{Q}),
\end{aligned}$$

where **r** and **Q** refer to the coordinates of all electrons and all nuclei, respectively. The quantity in brackets depends on the nuclear coordinates because the electronic wavefunction does, but often this dependence is fairly weak. It is therefore usual to expand it as a Taylor series (Appendix B) in the nuclear coordinates about the equilibrium nuclear geometry:

$$\left\{\int d\boldsymbol{r}\Psi^*_{\text{elect,i}}(\boldsymbol{r};\boldsymbol{Q})\mu_z\Psi_{\text{elect},i}(\boldsymbol{r};\boldsymbol{Q})\right\} = M_{zi}(Q)$$
$$\approx M_{zi}(Q_0)+\sum_j\left(\frac{\partial M_{zi}}{\partial Q_j}\right)_0(Q_j-Q_{j0})+\cdots. \quad (7.3)$$

The sum is over all the internal nuclear coordinates, and the derivatives are evaluated at the equilibrium geometry. The first term, $M_{zi}(Q_0)$, is identified as the z-component of the permanent dipole moment (evaluated at the equilibrium nuclear positions) of the molecule in electronic state i. The second term, $(\partial M_{zi}/\partial Q_j)_0$, gives the variation of that permanent dipole moment with nuclear coordinate Q_j. From now on we will drop the subscript i, with the understanding that we are referring to the dipole moment in electronic state i. The whole matrix element needed for a vibrational transition is then

$$\int d\boldsymbol{Q}\psi^*_{\text{vib,F}}(\boldsymbol{Q})\left\{M_z(Q_0)+\sum_j\left(\frac{\partial M_z}{\partial Q_j}\right)_0(Q_j-Q_{j0})+\cdots\right\}\psi_{\text{vib,I}}(\boldsymbol{Q}).$$

Now as long as we want transitions between different vibrational states (I $\neq$ F), the first term in brackets is zero since $\psi_{vib,I}$ and $\psi_{vib,F}$ are orthogonal, leaving as the leading term

$$\langle\Psi_{\text{elect,i}}\psi_{\text{vib,F}}|\mu_z|\Psi_{\text{elect,i}}\psi_{\text{vib,I}}\rangle = \sum_j\left(\frac{\partial M_z}{\partial Q_j}\right)_0\int d\boldsymbol{Q}\psi^*_{\text{vib,F}}(\boldsymbol{Q})(Q_j-Q_{j0})\psi_{\text{vib,I}}(\boldsymbol{Q}) \quad (7.4)$$

The further evaluation of this expression is slightly different for diatomic and polyatomic molecules, as developed below.

7.2. DIATOMIC VIBRATIONS

For a diatomic molecule, there is only one vibrational coordinate, the internuclear separation. The wavefunctions $\psi_{\text{vib,I}}$ and $\psi_{\text{vib,F}}$ are, to a first approximation, harmonic oscillator wavefunctions with quantum numbers v and v', respectively. The sum over j in Equation (7.4) has only one element, and the quantity $(Q_j - Q_{j0})$ is simply $(R - R_0)$, the displacement of the oscillator from its equilibrium position. For harmonic oscillators, this integral is nonzero only if v and v' differ by exactly one quantum number (Chapter 1). Thus, the selection rule for vibrational transitions is $v' = v \pm 1$. Also, the whole matrix element is nonzero only if there is a change in the permanent dipole moment of the molecule along the vibrational coordinate, that is, if $(\partial M_Z/\partial Q)_0 \neq 0$. A homonuclear diatomic molecule has no dipole moment no matter what its bond length, so homonuclear diatomics cannot have first-order dipole-allowed vibrational transitions. Since the harmonic oscillator energy in state v is $\hbar\omega_{vib}(v + (1/2))$, the transition energy for an upward transition from v to $v + 1$ (absorption), or for a downward transition from v to $v - 1$ (emission), is $\hbar\omega_{vib}$. As these energies correspond to wavelengths in the infrared region of the spectrum, direct vibrational absorption spectroscopy is synonymous with infrared spectroscopy.

7.3. ANHARMONICITY

Actual molecular vibrations do not behave exactly as harmonic oscillators. There are two kinds of anharmonicities associated with the vibration that perturb the selection rules and energy levels derived in the harmonic oscillator limit. One, sometimes called electrical anharmonicity, arises from higher terms in the nuclear coordinate dependence of the electronic transition moment. Remember, we made a Taylor series expansion of the matrix element in Equation (7.3) and dropped all terms after the linear one. If we were to keep the quadratic term, which for a diatomic molecule is simply

$$\frac{1}{2}(\partial^2 M_{zi}/\partial R^2)_0 (R - R_0)^2,$$

then transitions having $\Delta v = \pm 2$ become allowed also, since the matrix element $\langle\psi_{v'}|(R - R_0)^2|\psi_v\rangle$ is nonzero for $v' = v \pm 2$. Usually, these transitions are much weaker (roughly an order of magnitude) than the ones for which $v' = v \pm 1$, since the second derivative of the dipole moment with respect to bond length is much smaller than the first derivative in the range of bond lengths over which the wavefunctions have significant amplitude.

The other type of anharmonicity, mechanical anharmonicity, arises from the dependence of the potential energy function for nuclear motion on terms higher than quadratic in the internuclear separation. Recall the Born–Oppenheimer approximation says that the interaction of the nuclei with the averaged charge distribution of the electrons, together with the interelectronic and internuclear repulsions, gives rise to a potential energy function $V(\mathbf{Q})$ in which the nuclei move in any given electronic state. We normally expand this as a Taylor series about the equilibrium geometry to get, for a diatomic molecule,

$$V(R) = V(R_0) + (\partial V/\partial R)_0 (R - R_0) + (1/2)(\partial^2 V/\partial R^2)_0 (R - R_0)^2 + (1/6)(\partial^3 V/\partial R^3)_0 (R - R_0)^3 \ldots.$$

The first term is an arbitrary potential energy offset that can be taken to be zero. The linear term is zero by definition if R_0 is the equilibrium bond length.

The quadratic term is the first nonvanishing one, and what we did before was to truncate the series after this term, define $(\partial^2 V/\partial R^2)_0$ as the force constant k, and treat the vibrational motion as that of a harmonic oscillator.

In real molecules, the cubic and higher terms are not completely negligible, although they are usually much less important than the quadratic term and become progressively less important as one goes to higher terms. Two different approaches can be taken to include anharmonic effects. One is to treat the anharmonicity as a perturbation to the harmonic oscillator states. The other approach is to forget the Taylor series expansion altogether, get a more accurate functional form for $V(R)$ either empirically or from quantum chemical calculations, and solve for its exact eigenstates either numerically or analytically.

One particular functional form for $V(R)$, the Morse potential, has analytic (although messy) solutions for its eigenfunctions and energy eigenvalues and is also physically reasonable, at least for bond stretching vibrations. The Morse potential is defined by

$$V(R) = D_e\{1 - \exp[-\beta(R - R_0)]\}^2, \tag{7.5}$$

where D_e and β are parameters. This looks like a distorted harmonic oscillator (Fig. 7.2); it is steeper than harmonic for small internuclear separations ($R \rightarrow 0$), while at large separations ($R \rightarrow \infty$) $V(R) \rightarrow D_e$, the dissociation energy. Note that the first derivative is zero at $R = R_0$ while the second derivative is $(\partial^2 V/\partial R^2)_0 = 2\beta^2 D_e$. Equating this to the force constant k for the corresponding harmonic oscillator, and remembering that $\omega = (k/\mu)^{1/2}$, we find the "harmonic" vibrational frequency for the Morse oscillator as

$$\omega_e = \beta(2D_e/\mu)^{1/2}, \tag{7.6}$$

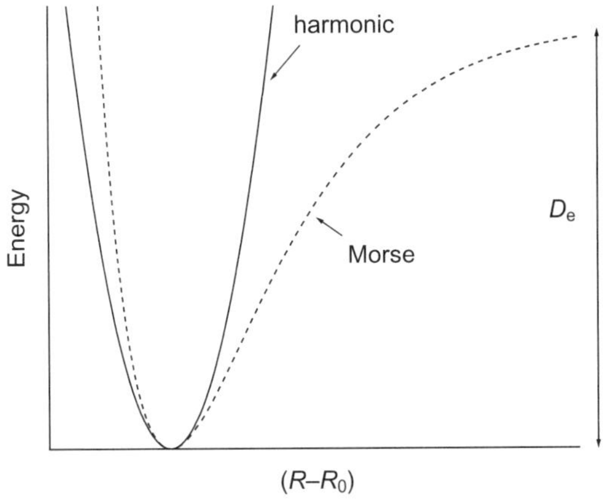

Figure 7.2. Comparison of the potential energy functions for a harmonic oscillator and a Morse oscillator having the same harmonic frequency at the potential minimum.

where the subscript "e" refers to the equilibrium geometry. The units are s^{-1} for ω_e, m^{-1} for β (if R is in m), kg for μ, and J for D_e. Vibrational spectroscopists more often express the vibrational "frequency" in units of wavenumber, $\tilde{\nu}_e = \omega_e / 2\pi c$, where c is used in units of $cm \cdot s^{-1}$ and $\tilde{\nu}_e$ is in cm^{-1}. The wavenumber is a useful unit for vibrational spectroscopy because single-quantum vibrational transitions are typically in the range of hundreds to thousands of wavenumbers.

The eigenvalues of the Morse oscillator, in wavenumbers, are given by

$$E_v = \tilde{\nu}_e \left(v + \frac{1}{2} \right) - \tilde{\nu}_e x_e \left(v + \frac{1}{2} \right)^2 \qquad v = 0, 1, 2, \ldots. \tag{7.7}$$

where $\tilde{\nu}_e x_e = \hbar \beta^2 / 4\pi c \mu$, with c in units of $cm \cdot s^{-1}$. This is just like the harmonic oscillator with one anharmonic correction term added. Note that the $v \rightarrow v + 1$ transitions do not all occur at the same energy; as v gets larger, the energy spacing between successive levels gets smaller (Fig. 7.3).

The different eigenstates of a Morse potential are still orthogonal to each other—orthogonality holds in general for nondegenerate eigenstates of any potential, and has nothing to do with the specific form of the potential. However, the selection rules for anharmonic oscillators are different because the matrix element $\langle v' | R | v \rangle$ is nonzero not only for $v' = v \pm 1$ but also for $v' = v \pm 2, \pm 3$, and so on. So even without electrical anharmonicity, real molecules have nonzero, but usually low, intensities in vibrational overtone transitions.

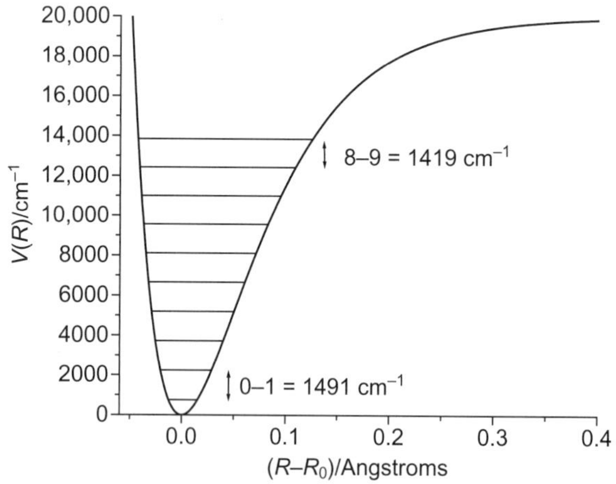

Figure 7.3. Potential energy curve and energies [Eq. (7.7)] of the $v = 0$ through $v = 9$ levels of a Morse oscillator with $\tilde{\omega}_e = 1500\, cm^{-1}$, $D_e = 20{,}000\, cm^{-1}$, and reduced mass $7.5\, g \cdot mol^{-1}$. The energy level differences for the $v = 0 \rightarrow v = 1$ and $v = 8 \rightarrow v = 9$ transitions are shown.

7.4. POLYATOMIC MOLECULAR VIBRATIONS: NORMAL MODES

Molecules with more than two atoms require more than one coordinate to define their vibrational motions. Recall that a diatomic molecule's vibration, in the harmonic limit, is described by the quantized version of the classical harmonic oscillator having the same reduced mass and force constant. Similarly, the vibrations of a polyatomic molecule are described by first finding the harmonic modes of vibration for a classically behaved body described by the same collection of masses and force constants. This is known as normal mode analysis and its result is a set of vibrational frequencies and normal coordinates (descriptions of how each atom moves in the normal mode). The essential feature of a normal mode of vibration is that all atoms oscillate about their equilibrium positions with generally different amplitudes but a common frequency and phase. If the masses are displaced from their equilibrium positions in some arbitrary way and then released, the resulting motion may be very complicated, but it can always be expressed as some linear combination of center-of-mass translation, rotation about the three inertial axes, and motion along the various normal modes of vibration. If the masses are initially displaced along just one of the normal modes, then the body will oscillate along that normal mode indefinitely (neglecting frictional damping); that is, the normal modes are the eigenfunctions of the harmonic vibrational problem.

The first step in analyzing the vibrations of a polyatomic molecule, then, is to solve for these classical normal modes of vibration. This means finding the set of coordinates in which the vibrational Hamiltonian, which is a sum of the kinetic and potential energies for all of the nuclear motions, separates into a sum of terms with no cross-terms between modes. This requires first setting up the matrix of the kinetic plus potential energy in some appropriate basis and then diagonalizing it. The eigenvalues are the vibrational frequencies, and the eigenvectors are the normal coordinates. The problem can be set up in either Cartesian or internal coordinates. First we'll consider Cartesians. Each of the N atoms in a molecule can move in three orthogonal spatial dimensions, giving $3N$ degrees of freedom for the nuclear motions. In Cartesian coordinates, the classical vibrational energy is written as

$$\begin{aligned}E_{\text{vib}} &= \text{KE}+\text{PE} = \frac{1}{2}\sum_{i=1}^{N} m_i[(dx_i/dt)^2+(dy_i/dt)^2+(dz_i/dt)^2] \\ &+\frac{1}{2}\sum_{i=1}^{N}[(\partial^2V/\partial x_i^2)_0 x_i^2+(\partial^2V/\partial y_i^2)_0 y_i^2+(\partial^2V/\partial z_i^2)_0 z_i^2] \\ &+\frac{1}{2}\sum_{i=1}^{N}\sum_{j=1}^{N}[(\partial^2V/\partial x_i\partial y_j)_0 x_i y_j+(\partial^2V/\partial y_i\partial z_j)_0 y_i z_j+(\partial^2V/\partial z_i\partial x_j)_0 z_i x_j] \\ &+\frac{1}{2}\sum_{i=1}^{N}\sum_{j<i}[(\partial^2V/\partial x_i\partial x_j)_0 x_i x_j+(\partial^2V/\partial y_i\partial y_j)_0 y_i y_j+(\partial^2V/\partial z_i\partial z_j)_0 z_i z_j].\end{aligned} \tag{7.8}$$

The first term is the kinetic energy, just $mv^2/2$ along each degree of freedom for each atom. The other two terms are the potential energy in the harmonic approximation, involving the second derivatives of the potential energy with respect to displacements of the atoms. It is the task of normal mode analysis to find the coordinate transformation—that is, a new set of coordinates that are linear combinations of these Cartesian coordinates—in which the cross terms in the third line of Equation (7.8) do not appear. While the kinetic energy is straightforward to write down in Cartesian coordinates, the potential energy is not. The force constants are not easily interpretable physically and they depend on the definition of the axis system, which is inconvenient. However, force constants in Cartesian coordinates are easily generated from electronic structure codes such as Gaussian, which is increasingly the way people are going about doing normal mode analysis. Note that while there are $3N$ degrees of freedom for an N-atom molecule, three of these correspond to the translation of the molecule's center of mass, and another three are represented by the overall rotation (for linear molecules, only two, since there is no "rotation" around the internuclear axis). Therefore, a molecule with N atoms has $3N$–6 ($3N$–5 if linear) normal modes of vibration. When the matrix of the vibrational energy is diagonalized, one will find three eigenvalues that are zero or nearly so (corresponding to center of mass translation), and three more that are very small, corresponding to the rotations.

Because of the way that the mass appears in the kinetic energy, it is most convenient to work in mass-weighted Cartesian coordinates, $\{X_i\}$, defined by

$$X_1 = \sqrt{m_1}x_1, X_2 = \sqrt{m_1}y_1, X_3 = \sqrt{m_1}z_1, X_4 = \sqrt{m_2}x_2, \ldots X_{3N} = \sqrt{m_N}z_N. \quad (7.9)$$

With this definition, we can write Equation (7.8) as

$$E_{\text{vib}} = \text{KE} + \text{PE} = \frac{1}{2}\sum_{i=1}^{3N}\dot{X}_i^2 + \frac{1}{2}\sum_{i=1}^{3N}\sum_{j\le i} b_{ij}X_iX_j, \quad (7.10)$$

where $\dot{X}_i = dX_i/dt$, and the b_{ij} now incorporate both the second derivatives of the potential energy and the atomic masses, for example,

$$b_{15} = \frac{1}{\sqrt{m_1m_2}}\left(\frac{\partial^2 V}{\partial x_1 \partial y_2}\right)_0. \quad (7.11)$$

Newton's law, $F = ma$, gives $3N$ classical equations of motion for this system, one for each atom in each Cartesian direction. For example, the force on atom i in the y direction is given by

$$F_{y,i} = m_i \frac{d}{dt}\dot{y}_i = -\frac{\partial V}{\partial y_i}, \quad (7.12)$$

which, converted to mass-weighted coordinates, gives the equation of motion of coordinate i as

$$\frac{d}{dt}\dot{X}_i + \sum_{j=1}^{3N} b_{ij} X_j = 0. \tag{7.13}$$

We now make the assumption that the time dependence of the coordinate displacement has the form $X_i = u_i e^{i\omega t}$, where u_i is a time-independent amplitude, and each atom oscillates with the same frequency ω in the normal mode. Substitution into Equation (7.13) gives

$$-\omega^2 u_i + \sum_{j=1}^{3N} b_{ij} u_j = 0. \tag{7.14}$$

There is one such equation for each of the $3N$ mass-weighted coordinates. These $3N$ equations can be expressed as a matrix equation,

$$(\mathbf{B} - \omega^2 \mathbf{I})u = 0, \tag{7.15}$$

where **B** is the matrix of the force constants defined in Equation (7.11), and **I** is the unit matrix (ones on the diagonal and zeroes everywhere else). A nontrivial solution exists only if the determinant of this matrix is zero; therefore, we find the vibrational frequencies by setting

$$|\mathbf{B} - \omega^2 \mathbf{I}| = 0, \tag{7.16}$$

and solving for the $3N$ roots, ω^2 (Appendix C). Insertion of each solution back into the original matrix equation allows determination of the eigenvector, that is, the description of the normal mode in terms of the mass-weighted Cartesian coordinates.

The other option is to work in internal coordinates. These are a set of relative motions of the atoms—bond lengths (stretches), bond angles (bends), and dihedral angles (torsions and wags)—in which it is easy to think about the force constants, and reasonable to either determine empirically or calculate force constants for one molecule and then transfer them to another molecule. In the days before high-level electronic structure calculations were feasible, vibrational analysis was done was by adjusting empirical force constants to fit experimental vibrational frequencies of small molecules. Those force constants were then transferred, perhaps with some adjustments based on chemical intuition, to other molecules that were more complex or for which experimental data were not available. This approach is still in use, but becoming less common as electronic structure computational methods become more powerful. Many different sets of internal coordinates can be used to describe the same molecule. The total number of bond lengths, bond angles, and bonded dihedral angles is usually greater than $3N$–6; such redundant coordinate systems can be used, but result in larger matrices than necessary. Procedures are available to define minimal internal coordinate systems that are complete but nonredundant. In internal coordinates, both the potential energy and the kinetic energy are nondiagonal. However, the potential energy matrix has a fairly straightforward physical interpretation, and the kinetic energy matrix, although rather complicated, is well defined; see for example, the "bible" of empirical vibrational analysis, Wilson et al. (1980) (see references). Vibrational analysis using internal coordinates most often uses the specific methods developed by Wilson, generally referred to as the Wilson FG method; F is the matrix of force constants, and G is the matrix of inverse reduced masses, upon which the kinetic energies depend, expressed in internal coordinates.

We will not delve into the details of how this is done, but will give a few examples of the normal modes and corresponding frequencies for some simple molecules. Water is a bent triatomic molecule whose geometry is defined by the two O-H bond lengths and the H-O-H bond angle. It belongs to the C_{2v} point group. It has three normal modes: the symmetric O-H stretch (A_1), the antisymmetric O-H stretch (B_2), and the bend (A_1). Cartoons showing the approximate atomic motions and the gas phase frequency of each mode are given in Figure 7.4. Note that because the bond angle in water is quite close to 90°, both the symmetric stretch and the antisymmetric stretch involve similar amounts of oxygen motion and the frequencies of both modes are similar. In contrast, in a linear triatomic molecule, such as CO_2 or CS_2, the central atom does not move at all in the symmetric stretch, and the antisymmetric stretch is nearly a factor of two higher in frequency than the symmetric stretch. Note too that in liquid water, which is strongly hydrogen bonded, both of the stretching vibrational frequencies are much lower than in the gas phase and the bands are extremely broad.

The second example, ammonia, NH_3, belongs to the C_{3v} point group. It has six normal modes. The three N-H stretches combine to form three stretching normal modes, a totally symmetric stretch belonging to symmetry species A_1 and a doubly degenerate asymmetric stretch belonging to E. There is also a doubly degenerate bend of E symmetry, and the "inversion" or "umbrella"

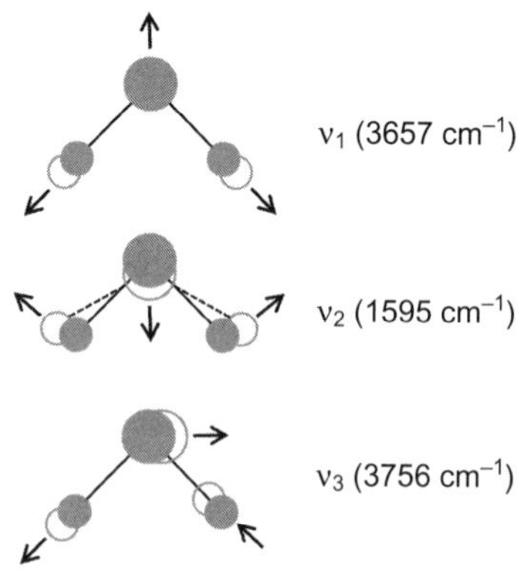

Figure 7.4. Schematic of the three vibrational normal modes of the water molecule, and the fundamental vibrational frequencies in the gas phase.

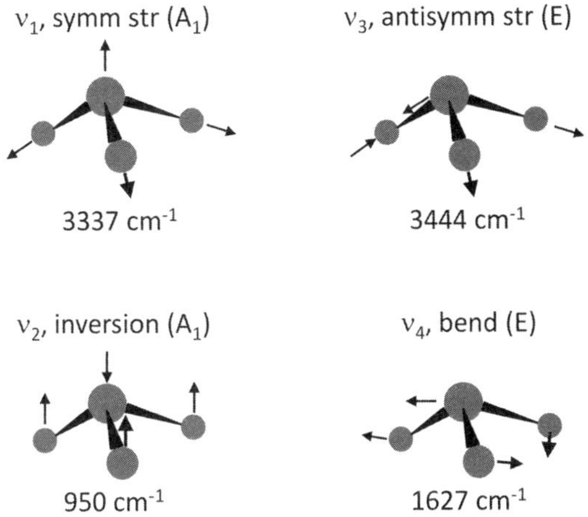

Figure 7.5. Schematic of the six vibrational normal modes (four distinct frequencies) of the ammonia molecule, and the fundamental vibrational frequencies in the gas phase.

mode of symmetry A_1. Cartoons showing the approximate atomic motions and the frequency of each mode are given in Figure 7.5.

Trans-dichloroethylene has a total of 12 normal modes. All of the atoms lie in a plane and it belongs to the C_{2h} point group. The five bonded pairs of atoms give rise to five stretching vibrations: the symmetric and antisymmetric combinations of the C-H stretches, the symmetric and antisymmetric combinations of the C-Cl stretches, and the C=C stretch. The four independent bond angles give rise to four bending vibrations, two involving mostly the C=C-H angles and two involving mostly the C=C-Cl angles. Finally, there are three vibrations that move atoms out of the plane: two of these involve mainly motions of the light H atoms out of the plane defined by the C atom to which it is bonded and the C and Cl atoms attached to the same C ("wags"), and one best described as twisting about the C=C bond ("torsion"). Cartoons of these modes and their frequencies are given in Figure 7.6.

Once the classical harmonic normal modes are known, the quantum mechanical vibrational wavefunction is obtained by imposing the same quantization conditions as were found for the one-dimensional harmonic oscillator. Since the harmonic vibrational Hamiltonian is separable when written in terms of the normal modes, the total vibrational wavefunction is just a product of the wavefunctions in each of the normal modes,

$$\psi_{vib}(Q_1, Q_2, Q_3 \ldots) = \psi_{v1}(Q_1)\psi_{v2}(Q_2)\psi_{v3}(Q_3)\ldots, \tag{7.17}$$

where $Q_i \ldots$ represents the ith normal mode and v_i is the vibrational quantum number in that mode ($v_i = 0, 1, 2, \ldots$). So a molecule with $3N$–6 normal modes

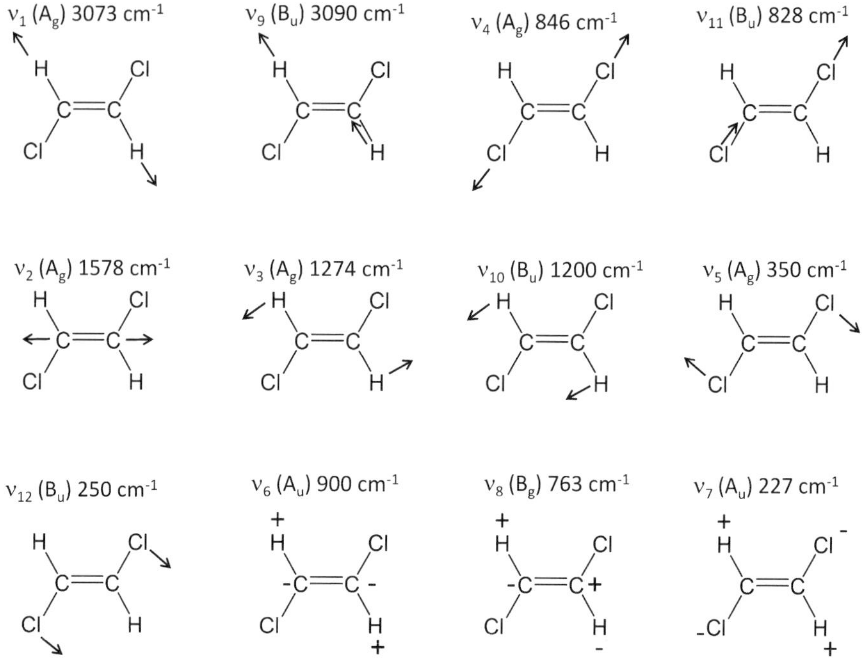

Figure 7.6. Schematic normal modes and vibrational frequencies of *trans*-1,2-dichloroethene. Only the dominant atomic motions are indicated. + and – refer to motions in and out of the plane, respectively.

needs 3*N*–6 vibrational quantum numbers to define its vibrational wavefunction. The total vibrational energy is given by

$$E_{\text{vib}}(v_1, v_2, v_3 \ldots) = \hbar\left[\omega_1\left(v_1 + \frac{1}{2}\right) + \omega_2\left(v_2 + \frac{1}{2}\right) + \omega_3\left(v_3 + \frac{1}{2}\right) + \ldots\right], \tag{7.18}$$

where ω_i is the vibrational frequency of mode *i*.

The selection rules for direct one-photon electric dipole-allowed transitions in polyatomic molecules are basically the same as in diatomics. Recall the Taylor series expansion of the electronic dipole moment from Equation (7.3):

$$M_z(Q) = M_z(0) + \sum_j \left(\frac{\partial M_z}{\partial Q_j}\right)_0 Q_j + \cdots,$$

where here we have defined $\mathbf{Q} = 0$ at the equilibrium nuclear position. The first term does not allow any changes of vibrational state. The second term allows only one normal mode at a time to change by just one vibrational quantum (called fundamentals), so the nominal vibrational selection rule is $v' = v \pm 1$ in only one mode. As for diatomic molecules, consideration of higher terms in this expansion (electrical anharmonicities) allows other transitions to occur: changes in the quantum number of more than one mode at a time (combination bands) and changes of more than one quantum in the state of a single mode (overtones).

Vibrations of polyatomic molecules also have mechanical anharmonicities. When discussing diatomic molecules, we considered the Morse oscillator as the best simple model for the vibrational motion that goes beyond the harmonic oscillator assumption. Diatomic molecules are simple because there is only one vibrational coordinate regardless of whether one considers the motion to be harmonic or anharmonic. Anharmonicity is more complicated in polyatomic molecules because the normal modes that define the vibrational coordinates do not even exist when there are cubic or higher derivatives of the potential with respect to coordinates. That is, for real anharmonic oscillators, the vibrational motion simply is not separable into contributions from $3N$–6 different modes of motion.

The usual approach taken with polyatomic molecules is to first solve for the normal modes of the system assuming harmonic potentials to get the $3N$–6 normal coordinates. Then, the vibrational energy, expressed here in wavenumber units, is written as the sum of the harmonic part and a part that contains the effect of anharmonicities:

$$\tilde{E}_{\text{vib}} = \sum_i \tilde{\nu}_i \left(v_i + \frac{1}{2} \right) + \sum_i \sum_j \tilde{x}_{ij} \left(v_i + \frac{1}{2} \right) \left(v_j + \frac{1}{2} \right) + \ldots \tag{7.19}$$

The first term is the purely harmonic energy, while the second term contains both diagonal anharmonicities (those that depend on a single vibrational coordinate, $i = j$) and off-diagonal anharmonicities (those that depend on cross terms between excitation in different vibrational modes, $i \neq j$).

In diatomic molecules, the anharmonicities are nearly always negative; the spacings between the energy levels get smaller as the vibrational quantum number increases. This is a general feature of Morse-like potentials. In polyatomic molecules, both diagonal and off-diagonal anharmonicities are usually negative, too, but there are exceptions. There is no reason to expect bending and torsional potentials to look anything like Morse potentials since the atoms do not dissociate at large values of the coordinate. Torsional potentials are better represented as sine or cosine functions since they are periodic and often do not have very high barriers to complete rotation. Torsional potentials often are quite flat near the bottom, and may, at low energies, be almost better approximated by a square-well (particle in a box) potential, which has a large positive anharmonicity.

7.5. SYMMETRY CONSIDERATIONS

Symmetry considerations based on group theory can often be used to simplify the vibrational problem considerably. Remember that all of the physically acceptable wavefunctions of a molecule must transform as one of the symmetry species of the point group to which it belongs. Furthermore, the vibrational Hamiltonian is always totally symmetric, so there can be no off-diagonal elements between coordinates that transform as different symmetry species. For this reason, it is convenient to work in a set of internal coordinates that are also symmetry coordinates—those that transform as one of the symmetry species of the point group. For example, the obvious internal coordinates for determining the normal modes of a triatomic molecule like water would be the O-H_1 bond length, the O-H_2 bond length, and the H-O-H bond angle. But it is preferable to work in coordinates that transform as the symmetry species of C_{2v} as discussed previously:

$$C1 = [r(\text{O-H}_1) + r(\text{O-H}_2)]/\sqrt{2} \qquad (\text{A}_1)$$

$$C2 = \angle(\text{H-O-H}) \qquad (\text{A}_1)$$

$$C3 = [r(\text{O-H}_1) - \text{r}(\text{O-H}_2)]/\sqrt{2}. \qquad (\text{B}_2)$$

The whole 3 × 3 matrix separates into a 2 × 2 block for the two A_1 modes and a 1 × 1 block for the B_2 mode; that is, the symmetry coordinate C3 is itself one of the normal modes, the antisymmetric stretch. The other two normal modes are linear combinations of C1 and C2, although in fact the off-diagonal elements between these two internal coordinates are small, and the true normal modes are very close to pure C1 (symmetric stretch) and C2 (bend).

To determine the symmetry species to which a given symmetry coordinate or normal mode belongs, attach arrows to each atom showing the direction in which it moves and apply the symmetry operations of the group, remembering to also reflect or rotate the arrow. Figure 7.7 depicts this for the two stretching vibrations of the water molecule.

Symmetry considerations can also be used to determine which vibrations will be allowed in the infrared spectrum, and this is one of the most powerful applications of group theory. Recall from Chapter 6 that a matrix element of the form $\langle\psi_A|\mathbb{O}|\psi_B\rangle$ may be nonzero only if the product of the characters corresponding to the symmetry species of ψ_A, ψ_B, and $\mathbb{O}$ is, or contains, the totally symmetric species. In our case, the relevant matrix element is $\langle\psi_{\text{vib,F}}|(\partial M_Z/\partial Q_k)_0 Q_k|\psi_{\text{vib,I}}\rangle$, where k labels the normal mode undergoing the transition. Assume the initial state, $\psi_{\text{vib,I}}$, is the ground vibrational state, that is, the state with zero quanta in all normal modes. Since there is no vibration along any mode, this state has the same symmetry as the molecule in its equilibrium geometry, that is, it belongs to the totally symmetric representation. If the final state, $\psi_{\text{vib,F}}$, is a fundamental (excited by one quantum in one mode), then this state has the symmetry of that normal mode (recall from Chapter 1 that the wavefunction for the v = 1 vibrational state is linear in the vibrational coordinate). The operator, $(\partial M_Z/\partial Q_K)_0 Q_k$, has the same symmetry as M_Z; the symmetry of Q_k in the numerator is canceled by ∂Q_k in the denominator. Therefore, the transition is allowed by symmetry if the product of the

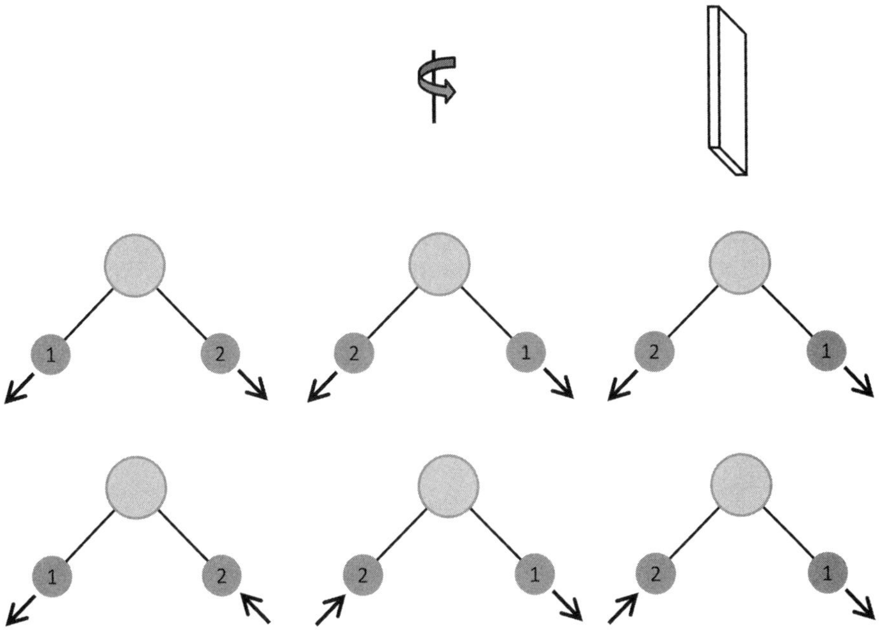

Figure 7.7. Determination of the symmetry species of the symmetric stretch (top row) and antisymmetric stretch (bottom row) of water. Rotation around the C_2 axis (middle column) or reflection in the plane that bisects the O-H bonds (right column) interchanges the identical H atoms and their directions of motion. The symmetric stretch is unchanged (A_1 symmetry) while the antisymmetric stretch changes sign (B_2).

representations of the normal mode itself and the z-component of the permanent dipole moment is or contains the totally symmetric representation.

As discussed in Chapter 6, M_Z in these equations refers to the z-component of the dipole moment in the laboratory-fixed or space-fixed coordinate system; this came from assuming that the incident radiation was polarized along the laboratory z-axis. The coordinates x, y, and z in the group theory character tables, on the other hand, refer to a molecule-fixed coordinate system. If the molecules are randomly oriented, which is usually the case in liquid-phase samples and often in solids, then the molecule-fixed coordinate system is randomly oriented relative to the space-fixed system, and z in the laboratory frame may correspond to any combination of x, y, and z in the molecular frame. Therefore, in order to determine whether a particular vibration is symmetry allowed, we need to ask whether the product of the representations of the vibration itself and either the x, y, or z coordinates transforms as the totally symmetric representation. In the case where the molecules have a definite orientation relative to the polarization of the light, as in a single crystal or other oriented media, this may not be true, and the geometry of the specific problem will have to be considered.

For a specific example, let us again consider water. We showed previously that the symmetric stretch and the bend are A_1, while the antisymmetric stretch is B_2. According to Table 6.2, the z-coordinate is A_1, x is B_1, and y is B_2. Thus, the symmetric stretch and the bend are allowed with light that is z-polarized in the molecular frame (since $A_1 \times A_1 = A_1$), and the antisymmetric stretch is allowed with light that is y-polarized in the molecular frame (since $B_2 \times B_2 = A_1$). All three vibrational fundamentals of water are predicted to show up in the infrared spectrum of a randomly oriented sample, and indeed they do. What group theory does not tell us is how strong the transitions are, as this depends on the magnitude of $(\partial M_Z/\partial Q_k)_0$. It turns out that this derivative is fairly large for all three vibrations of water since the O-H bond is quite polar, and water is a strong infrared absorber.

7.6. ISOTOPIC SHIFTS

Remember that the vibrational frequency is related to the reduced mass of the vibration by $\omega_i = (k_i/\mu_i)^{1/2}$. This relation holds for polyatomic molecules as well as diatomics, although in a polyatomic, both the effective force constant k_i and the effective reduced mass μ_i are rather complicated functions of the motions of all the atoms that participate in the normal mode. Qualitatively, however, the atoms that move the most in a given normal mode will make the largest contribution to the reduced mass of that mode. This makes isotopic substitution a powerful technique for aiding in the assignment of observed vibrational lines to particular vibrational modes and/or for checking the accuracy of calculated vibrational spectra. Isotopic substitution (most often D in place of H, or ^{13}C in place of ^{12}C) does not change the chemical properties of the atom and therefore does not change any of the force constants, but it does change the reduced masses for all vibrations that involve motion of that atom. If a sample is synthesized in which one particular carbon atom, for example, has been almost entirely replaced by the ^{13}C isotope, the vibrational transitions that shift the most between the natural abundance and isotopically substituted spectra can be assigned to normal modes that involve the greatest motion of the substituted atom.

Figure 7.8 gives an example for a simple molecule. This is not an infrared absorption spectrum but a Raman scattering spectrum, discussed further in Chapter 10. The Raman shift on the x-axis is equal to the vibrational frequency in wavenumbers.

7.7. SOLVENT EFFECTS ON VIBRATIONAL SPECTRA

Intermolecular interactions in liquids and solids cause some changes in the electron distribution and the equilibrium geometry of the molecule. These have effects on the vibrational frequencies and intensities, but usually fairly

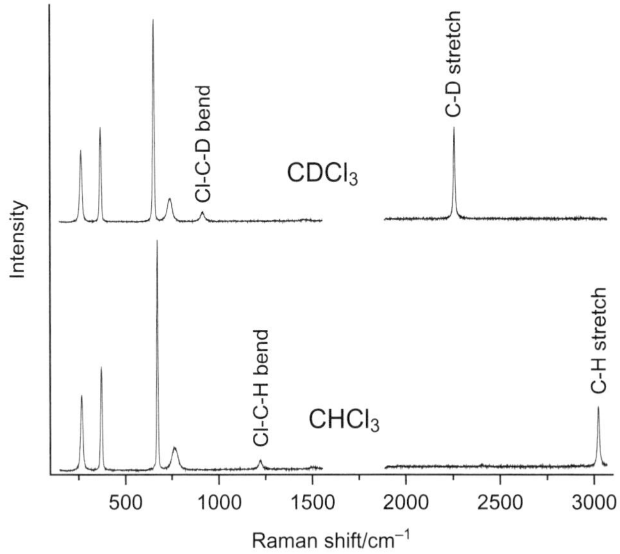

Figure 7.8. Effect of deuterium substitution on the Raman spectrum of liquid chloroform. The four lowest-frequency vibrations (between 200 and 800 cm^{-1}) arise from Cl-C-Cl bending and C-Cl stretching motions and have nearly identical frequencies in the two isotopomers. The two normal modes that involve predominantly H (D) motion shift to much lower frequencies in the heavier isotopomer.

small ones. In most cases, vibrational frequencies shift by no more than 1–2% (a few to perhaps a few tens of cm^{-1}) between the gas phase and liquid or solid phases. The atoms of the vibrating molecule of interest have interactions with the molecules in its environment that are both attractive (permanent or induced dipole–dipole) and repulsive (the hard-sphere repulsions due to overlap of electron clouds), and the net frequency shift depends on the difference between the balance of these two interactions in the vibrationally excited and ground states. It is a fairly fine balance, and either the attractive or the repulsive part of the interaction can dominate, leading to frequency shifts that may be either positive or negative and cannot always be predicted *a priori*.

Large frequency changes are usually seen only when there exist "specific" intermolecular interactions, for example, when solvent and solute can form hydrogen bonds, charge-transfer complexes, or other strong interactions. For example, O-H stretching vibrations are typically hundreds of cm^{-1} lower in solvents that can act as hydrogen bond acceptors than in non-H-bonding solvents. Frequency shifts in crystalline solids can be larger than in liquids since crystal packing forces can cause the molecule to adopt a somewhat different conformation than it does in the gas phase or in liquid solution. There are also many known cases where crystal packing forces change the point group of the molecule from what it is in the gas phase and thereby make certain vibrations allowed that were forbidden in the isolated molecule, or vice versa.

Infrared absorption intensities tend to be qualitatively similar in condensed phases and in the gas phase. Again, the largest differences are observed when strong specific interactions are present that also perturb the frequencies strongly.

The environment also has effects on the shapes and widths of the vibrational lines. Contributions to the linewidth are conveniently, if approximately, divided into two types. Homogeneous broadening affects every individual molecule in the ensemble in the same way. If you could take a spectrum of just one molecule (which can be done in some situations) the transition would still have the same width. Inhomogeneous broadening results from different molecules having slightly different resonant frequencies because they are in slightly different local environments within the liquid or solid. The observed broadening of the spectrum arises from a superposition of a large number of slightly different spectra. If you could take a spectrum of just one molecule, it would be narrower than the spectrum of the ensemble average.

The distinction between homogeneous and inhomogeneous broadening is not clear-cut, particularly in liquids where each molecule's local environment fluctuates rapidly due to the motions of the molecules making up the liquid. Whether a particular broadening mechanism acts "homogeneous" or "inhomogeneous" depends on the time scale on which the measurement is made. For linear spectroscopies, that time scale is given approximately by the inverse of the narrowest feature in the single-molecule spectrum, typically on the order of picoseconds for vibrational infrared transitions. If spectroscopically distinct environments persist for hundreds of picoseconds, then molecules in these different environments appear to have different infrared spectra, and the broadening is considered to be inhomogeneous. If the different environments interconvert on a time scale of 100 fs, then each molecule's spectrum is fluctuating during the time the spectrum is being measured, and it becomes better defined as homogeneous broadening (pure dephasing) as discussed further below.

Recall in Chapter 5 that we introduced the time-correlation function form for the absorption/emission lineshape [Eq. (5.10)]:

$$R(\omega) \propto \int_{-\infty}^{\infty} dt e^{-i\omega t} \langle \mu(0) \cdot \mu(t) \rangle.$$

For describing vibrational spectroscopy, it is appropriate to consider that the molecule's dipole moment contains (possibly) a time-independent permanent part and a time-varying part that is proportional to the vibrational coordinate Q. Thus, we need to consider the behavior of $\langle Q(0)Q(t) \rangle$. Two distinct mechanisms can cause this correlation function to decay. The first is population decay or lifetime broadening, whereby the excited vibrational level, for example, $v = 1$, decays back to $v = 0$. This may occur by spontaneous emission of an infrared photon, by intramolecular vibrational redistribution that transfers the energy from that vibration into other vibrations of the same molecule, or by a collision that transfers the vibrational energy into vibrations of the other molecule and/or rotational and translational degrees of freedom. It is easiest to interpret the correlation function $\langle Q(0)Q(t)\rangle$ in a classical sense; if one starts a molecule vibrating along coordinate Q with some amplitude at time $t = 0$, then the loss of energy from that vibration will cause the amplitude of vibration to decay with time, eventually becoming zero. The ensemble averaged quantity $\langle Q(0)Q(t)\rangle$, whether treated classically or quantum mechanically, decays exponentially in time, $\langle Q(0)Q(t)\rangle = Q^2(0)\exp(-t/T_1)$, where T_1 is the population decay time (this notation originated with magnetic resonance, but is regularly used in vibrational and electronic spectroscopies as well). The Fourier transform of this quantity gives a Lorentzian line of frequency width $1/T_1$.

Typical spontaneous emission lifetimes for vibrational infrared transitions are in the millisecond range. This is extremely long compared with other sources of population decay in liquids and most solids, so spontaneous emission usually makes a negligible contribution to vibrational population decay. When the principal contribution is from collisional energy transfer to other molecules in the environment, T_1 times can range from picoseconds to microseconds depending on the temperature and the frequencies of the vibration of interest and of the surrounding molecules. This results in a contribution to the vibrational linewidth from population decay of several cm^{-1} to much less than $0.01\,\text{cm}^{-1}$.

$\langle Q(0)Q(t)\rangle$ can also decay because of interruptions of the phase of the vibration without any net loss of energy. Since $Q(t)$ oscillates from positive to negative over the course of a vibrational period, the *average* value of $\langle Q(0)Q(t)\rangle$ will become zero if the phase of the vibration becomes scrambled, even if there has not been any overall loss of energy or amplitude from the vibration. This pure dephasing can be caused by elastic collisions between the vibrating molecule and the solvent that interrupt and randomize the phase of the vibration but leave it vibrating with the same amplitude. Quantum mechanically, this means that the collision temporarily changes the energy difference between the $v = 0$ and $v = 1$ states such that the quantity $\exp[i\omega_{10}t]$ that appears in the off-diagonal elements of the density matrix (see Chapter 5) gets changed to $\exp[i(\omega_{10} + \delta)t]$. Ensemble averaging over a random distribution of shifts δ causes this to go to zero. Remember the density matrix formulation assumes that the diagonal elements of the density matrix decay with some time constant $1/\gamma = T_1$, while the off-diagonal ones decay with a different time constant $1/\Gamma = T_2$. The total dephasing time, T_2, turns out to be related to the population decay time and the pure dephasing time, T_2^*, by

$$1/T_2 = 1/2T_1 + 1/T_2^*.$$

Pure dephasing is often much faster than population decay. In liquids T_2^* is usually somewhere between 1 and 50 ps and often makes the dominant contribution to the linewidth.

In a liquid or an amorphous solid, and to a much lesser degree in a crystal, each molecule has a slightly different arrangement of neighbors and thereby experiences different intermolecular interactions. This causes each molecule to have a slightly different set of energy levels and a slightly different spectrum. When one does an experiment on a large ensemble of molecules, as is still the norm in spectroscopy, this ensemble averaging represents a source of broadening. Inhomogeneous broadening usually results in a Gaussian distribution of transition frequencies, this being what one expects when the total perturbation is the sum of a very large number of individually small, uncorrelated perturbations. The magnitude of the inhomogeneous broadening of vibrational transitions in liquids and solids can vary from tens of cm^{-1} (even hundreds when there are strong hydrogen bonding interactions and the environment is very disordered) to hundredths of cm^{-1} or less in highly ordered crystals at low temperatures. Figure 7.9 illustrates these three different contributions to the vibrational linewidth.

Distinguishing between homogeneous and inhomogeneous broadening of vibrational transitions is not simple, particularly since the distinction depends

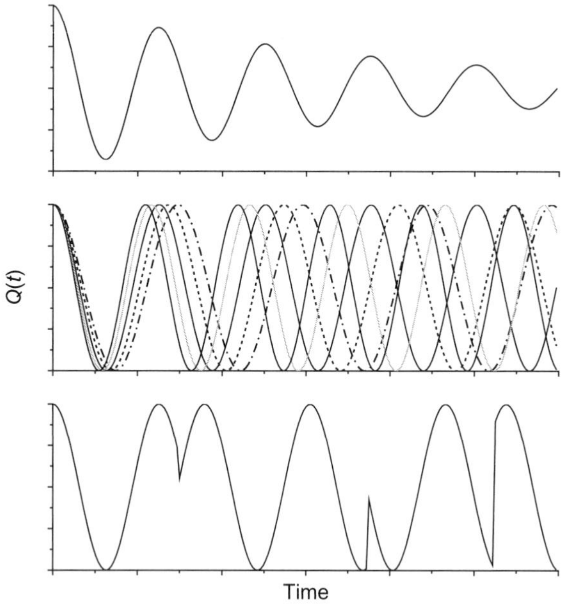

Figure 7.9. Illustration of the three principal contributions to the vibrational linewidth. Top, lifetime broadening: each oscillator has the same frequency and phase, but the vibrational amplitude decays exponentially with time. Middle, inhomogeneous broadening: each oscillator has a slightly different frequency because of its different environment. Bottom, pure dephasing: collisions cause oscillators to undergo random jumps in their phase of oscillation. All three mechanisms lead to a decay in $\langle Q(0)Q(t)\rangle$, whose Fourier transform determines the lineshape.

on the time scale of the measurement. Often, the lineshape of the transition is used as the criterion, with Gaussian lineshapes being deemed inhomogeneous and Lorentzian ones homogeneous. (When both forms of broadening are present simultaneously, which they usually are, the resulting lineshape is the convolution of the Gaussian and Lorentzian contributions, called a Voigt profile). But this is not really a valid criterion because inhomogeneous broadening does not always produce a Gaussian profile, and homogeneous dephasing produces a Lorentzian only in the limit that the inverse time scale of the fluctuations is very fast compared with the overall frequency width spanned by those fluctuations. In the opposite limit, when the fluctuations become very slow, the situation of inhomogeneous broadening is recovered. For intermediate time scales, the broadening cannot be clearly distinguished as homogeneous or inhomogeneous, and the lineshape falls between the Lorentzian and Gaussian limits. The evolution of the lineshape from Gaussian to Lorentzian as the fluctuations in the intermolecular interactions become faster is often described by stochastic theories, originally developed by Kubo in the context of magnetic resonance, that treat the modulation of the molecule's vibrational frequency by the environment as a random statistical process.

There are few experiments that can reliably distinguish homogeneous from inhomogeneous broadening except by inference from the lineshape. These include spectral hole burning, discussed in Chapter 8, and various nonlinear spectroscopies, discussed in Chapter 11.

REFERENCES AND FURTHER READING

J. L. McHale, *Molecular Spectroscopy* (Prentice Hall, Upper Saddle River, NJ, 1999).

E. B. Wilson, Jr., J. C. Decius, and P. C. Cross, *Molecular Vibrations* (Dover, New York, 1980).

PROBLEMS

1. From an analysis of the fluorescence spectrum of $^{23}Na_2$, the vibrational energy levels of the ground electronic state (in wavenumbers) were found to be given approximately by

$$E(v) = 159.12(v+1/2) - 0.725(v+1/2)^2.$$

(a) Calculate the wavenumbers of the $0 \rightarrow 1$, $0 \rightarrow 2$, and $0 \rightarrow 3$ vibrational transitions.

(b) Assuming the vibration is described by a Morse potential, calculate the dissociation energy D_e in wavenumbers.

2. The chlorine dioxide anion, $OClO^-$, is a bent molecule with three vibrational modes: the symmetric stretch at 946 cm^{-1}, the antisymmetric stretch at

1110 cm^{-1}, and the bend at 448 cm^{-1}. Within the harmonic approximation, calculate the frequencies in cm^{-1} of all possible vibrational transitions below 1600 cm^{-1} that originate from the ground state of $OClO^-$. State whether each transition is expected to have high, low, or zero intensity in the infrared absorption spectrum.

3. The benzene molecule, C_6H_6, belongs to the D_{6h} point group. Its 20 vibrational frequencies are given below in cm^{-1}. Note that all of the "e" symmetry vibrations are doubly degenerate.

a_{1g}: 993, 3074	a_{2g}: 1350	b_{2g}: 707, 990
e_{2g}: 608, 3057, 1601, 1178	e_{1g}: 847	a_{2u}: 674
b_{1u}: 1010, 3057	b_{2u}: 1309, 1150	e_{2u}: 398, 967
e_{1u}: 1038, 1484, 3064		

Calculate the zero-point vibrational energy of benzene in wavenumbers. Compare this value to the value of k_BT at room temperature (k_B = 0.695 cm$^{-1}\cdot$K^{-1}) and to the wavenumber of the lowest excited electronic transition of benzene (around 270 nm).

4. Figure 7.8 shows the Raman spectra of $CHCl_3$ and $CDCl_3$. In the $CHCl_3$ spectrum, the peaks labeled "C-H stretch" and "Cl-C-H bend" occur at 3022 and 1222 cm^{-1}, respectively, while in $CDCl_3$, the peaks labeled "C-D stretch" and "Cl-C-D bend" are at 2256 and 913 cm^{-1}, respectively. Recalling the relationship between frequency and reduced mass:

(a) Calculate the expected frequency ratio of the C-D stretch to C-H stretch, treating the vibration as a stretch of a diatomic molecule. Calculate the reduced mass of this quasi-diatomic molecule in two different ways: first as a C-H or C-D molecule ignoring the chlorine atoms, and then treating the whole CCl_3 group as a single "atom." Comment on the likely accuracy of each method.

(b) Use the observed frequency ratios of the C-H stretch to C-D stretch and Cl-C-H bend to Cl-C-D bend to calculate the apparent reduced mass of these modes. Qualitatively, what does this tell you about the amount of motion in atoms other than hydrogen in these modes?

5. In this problem, you will calculate the stretching normal modes of a linear triatomic molecule of the type A-B-A, where A and B are two different types of atom (similar to CO_2). Consider only the motions in one dimension (along the internuclear axis). Thus, **B** in Equation (7.15) will be a 3 × 3 matrix whose elements, are, for example,

$$b_{12} = \frac{1}{\sqrt{m_A m_B}} \left(\frac{\partial^2 V}{\partial x_{A1} \partial x_B} \right)_0 .$$

(a) Assume that the potential energy depends only on the lengths of the two A-B bonds independently:

$$V = \frac{1}{2}k\left(R_{\text{A1-B}}^2 + R_{\text{B-A2}}^2\right),$$

where $R_{\alpha-\beta}$ is the distance between atoms α and β. Express the bond lengths in terms of the displacements of each of the three atoms and calculate the elements of the **B** matrix (note that $b_{ij} = b_{ji}$). Solve the determinant in Equation (7.16) for the three vibrational frequencies. Then insert each of these frequencies back into Equation (7.15) and solve for the normal mode eigenvectors. Express the eigenvectors in both mass-weighted Cartesian coordinates and ordinary Cartesian coordinates. You do not need to normalize the eigenvectors.

(b) A linear triatomic molecule has two vibrations, a symmetric stretch and an antisymmetric stretch. Identify the calculated frequency and eigenvector corresponding to each of these two modes. What is the third normal mode you calculated in (a)?

6. The change in the frequency of a vibrational transition due to an applied electric field is known as the vibrational Stark effect. Recall from Chapter 3 that to lowest order in the multipole expansion, the interaction energy of a neutral molecule with an applied electric field is given by $W = \mathbf{E}\cdot\boldsymbol{\mu}$. Here $\boldsymbol{\mu} = \Sigma q_i \mathbf{r}_i$, where q_i is the charge on particle i (nucleus or electron) and $\mathbf{r}_i$ is its position.

(a) Calculate the Stark effect on the $v = 0 \rightarrow v = 1$ transition of a diatomic molecule whose atoms carry partial charges of $+q$ and $-q$ with an equilibrium bond length of r_0, treating it as a harmonic oscillator. Assume that the electric field lies along the interatomic axis. (You may be tempted to use perturbation theory, but this calculation can easily be performed exactly by writing out the harmonic oscillator Hamiltonian plus the Stark term and performing a change of variables).

(b) Do you expect to obtain the same result if the oscillator is anharmonic? Explain.

7. A vibrational overtone transition in a particular normal mode is usually considered to be the $v = 0 \rightarrow v = 2$ transition, but $v = 1 \rightarrow v = 3$, and so on can also contribute if the molecule is vibrationally hot.

(a) How should the intensity of the transition $v \rightarrow v + 2$ depend on the initial vibrational state v if the transition is allowed by electrical anharmonicity?

(b) Answer the same question if the transition is allowed by mechanical anharmonicity.

8. An O-H stretch has an infrared absorption at $3200\,\text{cm}^{-1}$. How much energy (in kJ/mol) is needed to stretch the O-H bond by an amount 0.1 Å from its equilibrium length?

9. SO_2, which has a bent equilibrium geometry, has the following vibrational frequencies in its ground electronic state:

ν_1 (symmetric S-O stretch) 1142 cm^{-1}

ν_2 (O-S-O bend) 515 cm^{-1}

ν_3 (antisymmetric S-O stretch) 1357 cm^{-1}.

(a) Which of these vibrations are expected to appear as fundamentals in the infrared absorption spectrum?

(b) SO_2 has a weak IR absorption band at 2264 cm^{-1}. What is the most probable assignment for this band?

CHAPTER 8

ELECTRONIC SPECTROSCOPY

8.1. ELECTRONIC TRANSITIONS

Before getting started, a few words are needed about nomenclature. Electronic states of molecules are designated in a variety of different ways. The convention usually followed for small molecules, although not always in larger molecules, is that the ground electronic state is called X, and excited states of the same spin multiplicity as the ground state are called A, B, C, . . . in order of increasing energy. When this ordering is not followed, it is usually for historical reasons, for example, a state was missed when the original assignments were made and the original state labels were retained after the error was found. Sometimes, but not always, a tilde (~) is placed over the letter to distinguish it from the labels for the symmetry species of some point groups. States of different spin multiplicity are usually (not always) designated with lowercase letters a, b, c, . . . Usually after the letter the term symbol for the state (spin multiplicity and symmetry species) is given, sometimes enclosed in parentheses. The ground and first excited states of Na_2, both singlet states, are named $X(^1\Sigma_g^+)$ and $A(^1\Sigma_u^+)$, respectively. Transitions between electronic states are indicated by an arrow with the higher energy state typically written first, for example, $A(^1\Sigma_u^+) \leftarrow X(^1\Sigma_g^+)$ for absorption and $A(^1\Sigma_u^+) \rightarrow X(^1\Sigma_g^+)$ for emission between these two states of Na_2.

As mentioned in Chapter 1, it is not reasonable to talk about transitions between different electronic states without also considering what happens to

Condensed-Phase Molecular Spectroscopy and Photophysics, First Edition. Anne Myers Kelley.

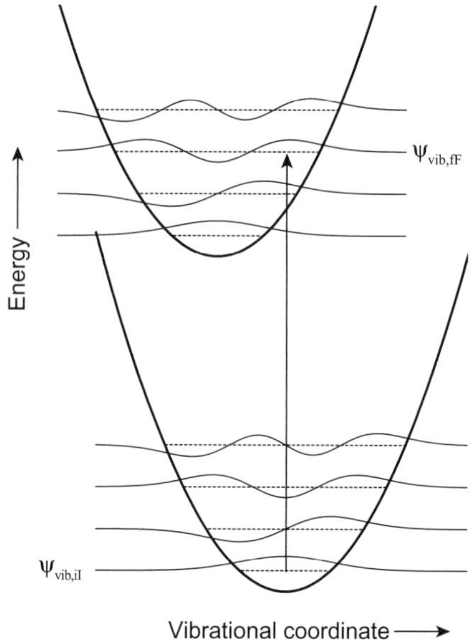

Figure 8.1. A vibronic transition from the $v = 0$ level of the ground electronic state to the $v = 2$ level of an excited electronic state.

the vibrational levels, which are different for each electronic state. So we are really concerned with vibrational–electronic ("vibronic") transitions, Figure 8.1:

$$\Psi_{\text{elect,i}}\psi_{\text{vib,iI}} \rightarrow \Psi_{\text{elect,f}}\psi_{\text{vib,fF}}. \tag{8.1}$$

Here the vibrational wavefunctions are labeled according to which electronic state they belong to, i or f, in addition to the quantum state of the vibrational oscillator, I or F, because the potential energy surfaces and thereby the vibrational wavefunctions are different for different electronic states. Separating the wavefunction into a product and writing the electronic dipole moment matrix element as a Taylor series expansion about the equilibrium nuclear geometry, as was done in Chapter 7 for vibrational transitions, yields

$$\begin{aligned}&\langle \Psi_{\text{elect,f}}\psi_{\text{vib,fF}} | \mu_z | \Psi_{\text{elect,i}}\psi_{\text{vib,iI}} \rangle \\ &= \int d\boldsymbol{Q}\psi^*_{\text{vib,fF}}(\boldsymbol{Q})\left\{M_{\text{if},z}(Q_0) + \sum_j \left(\frac{\partial M_{\text{if},z}}{\partial Q_j}\right)_0 (Q_j - Q_{j0}) + \cdots\right\}\psi_{\text{vib,iI}}(\boldsymbol{Q})\end{aligned} \tag{8.2}$$

Note that the electronic matrix element is not a permanent dipole moment (the expectation value of μ in electronic state $\Psi_{\text{elect,i}}$) but a transition dipole moment (the matrix element of μ between two different electronic states,

$\Psi_{\mathrm{elect,i}}$ and $\Psi_{\mathrm{elect,f}}$). The transition dipole moment does not have a simple classical interpretation; in particular, even in a completely nonpolar molecule, such as a homonuclear diatomic, where the initial and final states have no permanent dipole moment, the transition dipole moment can be nonzero.

8.2. SPIN AND ORBITAL SELECTION RULES

We first consider the contribution to the matrix element from the first term in curly brackets, the transition dipole moment evaluated at the equilibrium nuclear positions, $M_{\mathrm{if},z}(Q_0)$. Keeping only this term is referred to as the Condon approximation, and it usually makes the dominant contribution to the electronic transition strength unless this term is zero by symmetry. Remember that this is the first term in the Taylor series expansion of $\int d\boldsymbol{r}\Psi^*_{\mathrm{elect,f}}(\boldsymbol{r};\boldsymbol{Q})\mu_z\Psi_{\mathrm{elect,i}}(\boldsymbol{r};\boldsymbol{Q})$. The electronic wavefunction consists of both spatial and spin parts, but the dipole moment operator acts only on spatial coordinates to a first approximation. States of different spin multiplicity are orthogonal. Therefore, to a good approximation, the spin selection rule for an electronic transition is that the spin cannot change ($\Delta S = 0$); transitions between different singlet states or between different triplet states are allowed, but transitions between a singlet and a triplet, for example, are spin-forbidden.

We next consider the part of the integral that involves the spatial coordinates of the electrons. In order to evaluate it quantitatively, one needs a complete description of the electronic wavefunction as obtained, for example, from a molecular orbital calculation. However, a first step is to use symmetry to find which integrals *may* be nonzero. We will first consider diatomic molecules in some detail, and then give more general results for larger molecules. All diatomic molecules belong to the point group $D_{\infty h}$ (homonuclear) or $C_{\infty v}$ (heteronuclear). Both point groups contain symmetry species labeled Σ, Π, Δ, . . . according to the absolute magnitude of the orbital angular momentum in units of $\hbar$ (0, 1, 2, . . .) about the internuclear axis. The Σ species are split into Σ^+ and Σ^- species according to how they transform when reflected across a plane that contains the internuclear axis. In $D_{\infty h}$, each symmetry species is further split into g (even) and u (odd) pairs according to how they transform upon inversion. Each of the Σ species is nondegenerate, while each of the higher angular momentum species is twofold degenerate because of the two equivalent directions for the angular momentum.

In order to determine whether a given electronic transition is electric dipole allowed (i.e., whether $M_{\mathrm{if},z}(R_0)$ is nonzero), one imposes the usual rules of group theory: take the direct product of the symmetry species of the initial state, the final state, and the operator involved, and if that direct product is or contains the totally symmetric species (Σ^+ in $C_{\infty v}$ or Σ_g^+ in $D_{\infty h}$), the matrix element *can be* nonzero, although it may be small or zero for other reasons.

As discussed for vibrational transitions, if the molecules are randomly oriented in space, the polarization direction of the light in the laboratory frame, assumed to be z here, may correspond to any combination of x, y, and z in the molecule-fixed frame, so we need to consider all three possibilities. The character tables show that the z-coordinate transforms as Σ^+ in $C_{\infty v}$ and as Σ_u^+ in $D_{\infty h}$, while (x, y) transform as Π in $C_{\infty v}$ and as Π_u in $D_{\infty h}$. Therefore, in a heteronuclear diatomic, the matrix element of the dipole moment operator can be nonzero if the direct product of Σ^+ or Π and the symmetry species of $\Psi_{\text{elect,i}}$ and $\Psi_{\text{elect,f}}$ is or contains Σ^+. If the initial electronic state of a heteronuclear diatomic is Σ^+, as it is for all closed-shell electron configurations, then the only symmetry-allowed transitions having a transition moment along the internuclear axis (z in the molecule-fixed frame) are to other Σ^+ states. For transition dipoles perpendicular to the internuclear axis (x or y in the molecule-fixed frame), the direct product of Σ^+, Π, and Π is $\Pi + \Sigma^+ + \Sigma^-$, so the Σ^+ to Π transition is allowed with polarization perpendicular to the axis. A homonuclear diatomic molecule ($D_{\infty h}$) is just like a heteronuclear one with the addition of g or u subscripts to each symmetry species. By a similar analysis, one finds that light polarized along the internuclear axis can induce transitions from Σ_g^+ to Σ_u^+, while x- or y-polarized light causes transitions from Σ_g^+ to Π_u.

One can summarize the selection rules derived from group theory for diatomic molecules as follows, recalling that the electric dipole operator transforms as x, y, or z, the electric quadrupole operator as products of coordinates such as xy, and the magnetic dipole operator as the rotations R_x, R_y, and R_z.

8.2.1. Electric Dipole Transitions

In $D_{\infty h}$, g $\leftrightarrow$ u is allowed while g $\leftrightarrow$ g and u $\leftrightarrow$ u are forbidden. For parallel transitions (transition dipole along the internuclear axis), the angular momentum along the axis (referred to as Λ) cannot change so only $\Sigma \leftrightarrow \Sigma$, $\Pi \leftrightarrow \Pi$, $\Delta \leftrightarrow \Delta$, and so on are allowed; also, $\Sigma^+ \leftrightarrow \Sigma^+$ and $\Sigma^- \leftrightarrow \Sigma^-$ are allowed, while $\Sigma^+ \leftrightarrow \Sigma^-$ is forbidden. For perpendicular transitions (transition dipole perpendicular to the axis), the angular momentum along the axis must change by one unit so $\Sigma \leftrightarrow \Sigma$, $\Pi \leftrightarrow \Pi$, $\Delta \leftrightarrow \Delta$, and so on are forbidden, while $\Sigma \leftrightarrow \Pi$, $\Pi \leftrightarrow \Delta$, and so on are allowed.

8.2.2. Electric Quadrupole Transitions

g $\leftrightarrow$ u is forbidden, while g $\leftrightarrow$ g and u $\leftrightarrow$ u are allowed. Changes in orbital angular momentum of either 0, ±1, or ±2 are allowed, that is, transitions from Σ states to Σ, Π, or Δ are allowed. $\Sigma^+ \leftrightarrow \Sigma^+$ and $\Sigma^- \leftrightarrow \Sigma^-$ are allowed while $\Sigma^+ \leftrightarrow \Sigma^-$ is forbidden.

8.2.3. Magnetic Dipole Transitions

g $\leftrightarrow$ u is forbidden while g $\leftrightarrow$ g and u $\leftrightarrow$ u are allowed. $\Sigma^+ \leftrightarrow \Sigma^+$ and $\Sigma^- \leftrightarrow \Sigma^-$ are forbidden while $\Sigma^+ \leftrightarrow \Sigma^-$ is allowed by the $\mathbb{L}_z$ operator, and the $\mathbb{L}_x$ and $\mathbb{L}_y$

operators allow changes of one unit of angular momentum ($\Sigma \leftrightarrow \Pi$, $\Pi \leftrightarrow \Delta$, etc.).

8.2.4. Polyatomic Molecules

The orbital symmetry selection rule for polyatomic molecules is the same as for diatomics: an electronic transition is symmetry allowed in the electric dipole and Condon approximations if the direct product of the symmetry species of the initial electronic state, the final electronic state, and one of the components of the dipole operator (x, y, or z) contains the totally symmetric species. Polyatomic molecules span a wide range of point groups, and each case has to be worked out separately. One general rule is that g $\leftrightarrow$ g and u $\leftrightarrow$ u are always electric dipole forbidden for molecules having a center of symmetry.

8.3. SPIN–ORBIT COUPLING

The $\Delta S = 0$ spin selection rule is a very good approximation in light molecules. However, there are small terms in the Hamiltonian that couple spin angular momentum to orbital angular momentum, and these terms are largest for molecules containing heavy atoms (see Section 9.4). This is known as spin-orbit coupling. In a diatomic molecule, if the orbital angular momentum **L** is other than zero (a Π, Δ, etc. state) and the total spin angular momentum **S** also is not zero, then there is an additional term in the Hamiltonian proportional to $\mathbf{L}\cdot\mathbf{S}$ that must be considered. This causes the orbital and spin angular momenta to couple such that neither L nor S alone is a useful quantum number; one must instead work with the total angular momentum $\mathbf{\Omega}$, the vector sum of **L** and **S**. The effective selection rules, which are $\Delta S = 0$ and $\Delta\Lambda = 0$ or ± 1 for weak spin–orbit coupling, become $\Delta\Omega = 0$ or ± 1 when spin–orbit coupling is strong.

L and **S** independently, or their sum, also couple to the rotational angular momentum **J**. The possible coupling schemes, which depend on the relative strengths of the interactions among the different types of angular momentum, are typically divided into Hund's cases (a), (b), and (c). These are important mainly in high-resolution gas phase spectroscopy, and we will not deal with them further here.

8.4. VIBRONIC STRUCTURE

Referring back to Equation (8.2), we see that under the Condon approximation, the integral that determines the strength of the electronic transition is

$$M_{\mathrm{if},z}(Q_0)\int d\boldsymbol{Q}\psi^*_{\mathrm{vib,fF}}(\boldsymbol{Q})\psi_{\mathrm{vib,iI}}(\boldsymbol{Q}). \tag{8.3}$$

If $\psi_{\mathrm{vib,fF}}$ and $\psi_{\mathrm{vib,iI}}$ were vibrational levels of the same electronic state, as they are in a vibrational infrared transition, then the integral would be zero unless

I = F. But here $\psi_{\text{vib,iI}}$ is a vibrational eigenstate of the potential energy surface for the ground electronic state, $V(R)$, while $\psi_{\text{vib,fF}}$ is an eigenstate of a different potential energy surface characterizing the excited electronic state, $V'(R)$. In general, both the equilibrium internuclear separation and the curvature differ between the two surfaces, and so eigenstates characterized by different quantum numbers are not orthogonal; refer back to Figure 8.1. Thus, there is no reason why the term in $M_{\text{if},z}(Q_0)$ should be zero, and in fact the term in $M_{\text{if},z}(Q_0)$is usually considerably larger than the one involving the derivatives, $\partial M_{\text{if},z}/\partial Q_j$.

The integral over vibrational coordinates in Equation (8.3) is called the vibrational overlap integral. The probability of the transition depends on the square of the matrix element, $|M_{\text{if},z}(Q_0)|^2 \left|\int d\boldsymbol{Q}\psi^*_{vib,fF}(\boldsymbol{Q})\psi_{\text{vib,iI}}(\boldsymbol{Q})\right|^2$. The second factor, the square of the vibrational overlap integral, is called the Franck–Condon factor. It gives the relative strengths of the different vibrational sublevels of a given electronic transition, whose overall intensity is given by the electronic part, $|M_{\text{if},z}(Q_0)|^2$.

The Franck–Condon factors are characterized by a vibrational sum rule. The sum over all Franck–Condon factors arising from a given initial vibrational state is

$$\sum_{\text{F}}|\langle\psi_{\text{vib,fF}}|\psi_{\text{vib,iI}}\rangle|^2 = \sum_{\text{F}}\langle\psi_{\text{vib,iI}}|\psi_{\text{vib,fF}}\rangle\langle\psi_{\text{vib,fF}}|\psi_{\text{vib,iI}}\rangle = \langle\psi_{\text{vib,iI}}|\psi_{\text{vib,iI}}\rangle = 1, \tag{8.4}$$

where the quantum mechanical closure property has been used to remove the sum over F. Within the Condon approximation, the total intensity of the transition between electronic states $\Psi_{\text{elect,I}}$ and $\Psi_{\text{elect,f}}$ is a constant given by the square of the electronic transition moment. The Franck–Condon factors merely determine how that intensity is distributed among the various vibrational sublevels. This sum rule does not depend on harmonic or other approximations for the vibrational potential function.

Before continuing, we must consider more explicitly the nature of the vibrational wavefunction for a polyatomic molecule. As discussed in Chapter 7 [Eq. (7.17)], in the harmonic oscillator approximation, the vibrational wavefunctions can be factored into products of wavefunctions for each of the individual normal modes: $\psi_{\text{vib}}(\text{Q}) = \psi_{v1}(\text{Q}_1)\psi_{v2}(\text{Q}_2)\psi_{v3}(\text{Q}_3)\ldots$ This allows the $3N$–6-dimensional vibrational wavefunctions in Equation (8.3) to be factored into a product of one-dimensional wavefunctions:

$$\int d\boldsymbol{Q}\psi^*_{\text{vib,fF}}(\boldsymbol{Q})\psi_{\text{vib,iI}}(\boldsymbol{Q}) = \int dQ_1\psi^*_{v_1'}(Q_1)\psi_{v_1}(Q_1)\int dQ_2\psi^*_{v_2'}(Q_2)\psi_{v_2}(Q_2)\ldots \tag{8.5}$$

where v_i and v_i' refer to the quantum numbers for mode i in the initial and final electronic states, respectively. This assumes that the normal modes of the ground and excited electronic states are the same, which may not be the case. In general, the normal modes of the excited state are linear combinations of the normal modes of the ground state. This situation is known as the Duschinsky effect or Duschinsky rotation. Duschinsky rotation is often neglected because the normal modes of excited electronic states are not very well known and because neglecting it greatly simplifies the evaluation of vibrational overlap integrals for polyatomic molecules. In the remainder of this chapter, we will assume that the normal modes of different electronic states are the same, but the reader should be aware that in some cases this assumption may have to be relaxed.

Consider an absorption transition from an initial electronic state we will now call Ψ_g (for "ground") to a final state Ψ_e (for "excited"). The energy needed to make the transition from $|g; v_1, v_2, \ldots\rangle$ to $|e; v_1', v_2', \ldots\rangle$, in the harmonic oscillator approximation, is

$$E_{\text{trans}} = E_0 + \sum_{j=1}^{3N-6} \left\{ \hbar\omega_{ej}\left(v_j' + \frac{1}{2}\right) - \hbar\omega_{gj}\left(v_j + \frac{1}{2}\right) \right\}, \tag{8.6}$$

where E_0 is the energy separation between the minima of the potential curves for the ground and excited states (usually tens of thousands of wavenumbers for electronic transitions) and ω_{gj} and ω_{ej} are the harmonic vibrational frequencies for mode j in the ground and excited electronic states (usually hundreds to a few thousand wavenumbers) (Fig. 8.2). The transition from $(v_1, v_2, v_3, \ldots) = (0, 0, 0, \ldots)$ to $(v_1', v_2', v_3', \ldots) = (0, 0, 0, \ldots)$, called the zero–zero or electronic origin, occurs at an energy of $E_0 + \sum_{j=1}^{3N-6} \hbar(\omega_{ej} - \omega_{gj})/2$. If the ground- and excited-state vibrational frequencies are the same, the zero–zero energy

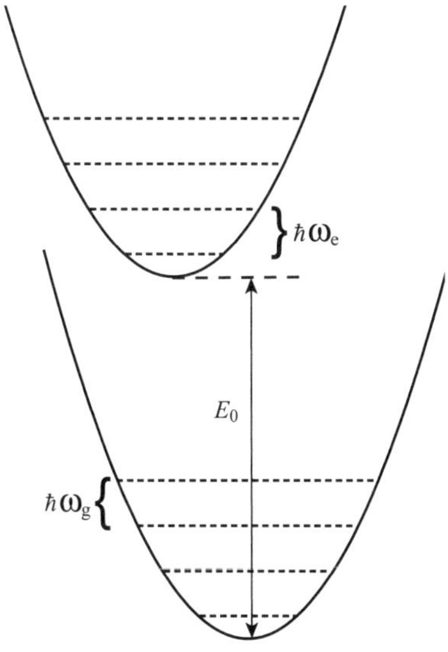

Figure 8.2. Illustration of the energies in Equation (8.6).

equals E_0. Transitions from the same initial state to states that differ in v′ for a single mode k, sometimes termed a vibrational series, appear spaced by a constant $\hbar\omega_{ek}$ in the harmonic limit; actually, as discussed before with regard to vibrational spectroscopy, vibrational anharmonicities modify this somewhat and cause transitions to successively higher v′ to become more closely spaced.

When Duschinsky rotation is neglected, the multidimensional Franck–Condon factors for a polyatomic molecule separate into products of one-dimensional factors:

$$|\langle v_1', v_2', v_3', \ldots | v_1, v_2, v_3, \ldots \rangle|^2 = |\langle v_1' | v_1 \rangle|^2 \, |\langle v_2' | v_2 \rangle|^2 \, |\langle v_3' | v_3 \rangle|^2 \ldots \tag{8.7}$$

We now discuss the properties of these one-dimensional Franck–Condon factors. If the potential curves for the two electronic states are quite similar, then the vibrational wavefunctions of the ground electronic state and of the excited state are almost the same. The Franck–Condon factors $|\langle v'|v\rangle|^2$ are large for $v = v'$ and become successively smaller as v and v' differ by successively more vibrational quanta. If the initial vibrational level is the vibrational ground state, $v = 0$, then the strongest transition in the electronic–vibrational spectrum is the zero–zero transition. The $0 \rightarrow 1$ is weaker, the $0 \rightarrow 2$ still weaker, and so on. If on the other hand, the two potential curves are very different, there may be almost zero overlap for $v = v'$, and significant overlaps only when v and v' differ by a number of quanta. The progression of vibrational transitions originating from $v = 0$ will then show a very weak $0 \rightarrow 0$ transition, successively more intense lines for $0 \rightarrow 1$, $0 \rightarrow 2$, and so on, maximizing at some $0 \rightarrow v'$, and then a further reduction in intensity for higher v' (Fig. 8.3).

The ideas above can be summarized in the following expression for the absorption spectrum of a polyatomic molecule in the Condon and non-Duschinsky approximations:

$$\sigma_{i\rightarrow f}(\omega) \sim |M_{if}(Q_0)|^2 \sum_{v_1=0}^{\infty} \cdots \sum_{v_{3N-6}=0}^{\infty} P(v_1 \cdots v_{3N-6}) \sum_{v_1'=0}^{\infty} \cdots$$
$$\sum_{v_{3N-6}'=0}^{\infty} \left(\prod_{j=1}^{3N-6} |\langle v_j' | v_j \rangle|^2 \right) \cdot \delta\left(\omega_0 - \omega + \sum_{j=1}^{3N-6} \left\{ \omega_{ej}\left(v_j' + \frac{1}{2}\right) - \omega_{gj}\left(v_j + \frac{1}{2}\right) \right\} \right) \tag{8.8}$$

Here $v_1 \ldots v_{3N-6}$ are the initial quantum numbers of each of the $3N$–6 normal modes, $P(v_1 \ldots v_{3N-6})$ is the Boltzmann probability of finding this combination of quantum states at thermal equilibrium, and $v_1' \cdots v_{3N-6}'$ are the quantum numbers of the final (excited) state. The product $\prod_{j=1}^{3N-6} |\langle v_j' | v_j \rangle|^2$ is the total vibrational Franck–Condon factor when the $3N$–6-dimensional vibrational wavefunction can be factored into $3N$–6 one-dimensional Franck–Condon factors. The delta function insures conservation of energy between the absorbed photon and the initial to final state energy difference. As discussed elsewhere, a more realistic model that accounts for the finite lifetime of the final state

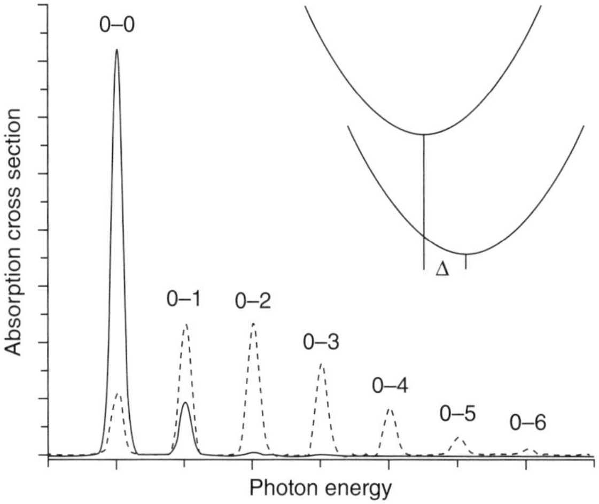

Figure 8.3. Calculated absorption spectra for a diatomic molecule having harmonic ground- and excited-state potential energy surfaces with equal vibrational frequencies. Spectra are shown for two different values of Δ, the separation between ground- and excited-state potential energy minima in units of the dimensionless vibrational coordinate q. Solid curve, $\Delta = 0.5$; dashed curve, $\Delta = 2$.

replaces the delta functions by normalized Lorentzians centered at the same frequencies.

In the special case of harmonic oscillators with equal vibrational frequencies in ground and excited states ($\omega_e = \omega_g = \omega_{vib}$) but different equilibrium bond lengths, the Franck–Condon factors for the $0 \rightarrow v'$ transitions have a particularly simple analytic form:

$$|\langle v'|0\rangle|^2 = e^{-S}\frac{S^{v'}}{(v')!}. \tag{8.9}$$

Here, $S = \Delta^2/2$, where Δ is the difference between ground and excited state potential minima in the dimensionless coordinates introduced when initially discussing the harmonic oscillator:

$$\Delta = (\mu\omega_{vib}/\hbar)^{1/2}(R_{0,e} - R_{0,g}). \tag{8.10}$$

The quantity S is also known as the Huang–Rhys factor. It equals the ratio of the $0 \rightarrow 1$ to $0 \rightarrow 0$ transition intensity. More complicated expressions exist for the Franck–Condon factors for general $v \rightarrow v'$ transitions and for harmonic potentials where $\omega_e \neq \omega_g$. Franck–Condon factors for anharmonic oscillators are generally calculated by numerical integration.

There is one more important aspect of Franck–Condon factors that depends on the symmetry of the vibrations. In the harmonic oscillator limit, all the vibrational wavefunctions have either even or odd symmetry with respect to the potential energy minimum since the potential function is even. Now, if the equilibrium geometries of the two electronic states involved in the transition belong to the same point group, then the slope of the potential energy surface along any nontotally symmetric vibration Q_j must be zero at the equilibrium geometry in both electronic states; that is, $Q_j = 0$ must be either a local minimum or a local maximum in energy. Therefore, $|v_j'\rangle$ and $|v_j\rangle$ are all either even or odd functions, and $\langle v_j'|v_j\rangle$ can be nonzero only for $v_j' = v_j, v_j \pm 2, v_j \pm 4$, and so on. That is, within the Condon approximation, only even-quantum transitions are allowed in nontotally symmetric modes (Fig. 8.4).

8.5. VIBRONIC COUPLING

So far we have considered electronic transitions in the Condon approximation, where the explicit dependence of the electronic transition moment on vibrational coordinates is suppressed. Referring to Equation (8.2), the next term in the Taylor series expansion of the electronic transition moment is

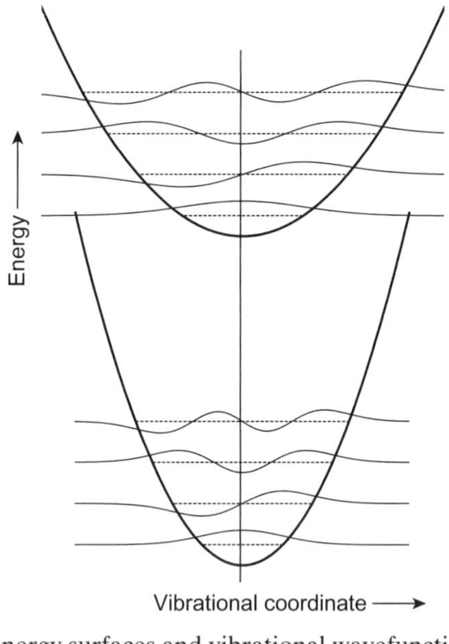

Figure 8.4. Potential energy surfaces and vibrational wavefunctions along a nontotally symmetric vibrational coordinate. Moving the atoms in the (+) or (−) directions produces indistinguishable structures that must have the same energy, so all potential surfaces must be even functions of the coordinate. Thus, the vibrational wavefunctions are all even (for v even) or odd (for v odd). The vibrational overlap integrals may be nonzero for $0 \rightarrow 0, 0 \rightarrow 2, 0 \rightarrow 4$, and so on, but must be zero for $0 \rightarrow 1, 0 \rightarrow 3$, and so on.

$\Sigma_j(\partial M_{\text{if},z}/\partial Q_j)_0 Q_j$, where the sum runs over all normal modes Q_j and we define $Q = 0$ at the equilibrium geometry. The entire matrix element is then

$$\sum_j \left(\frac{\partial M_{\text{if},z}}{\partial Q_j}\right)_0 \int d\boldsymbol{Q}\psi^*_{\text{vib,fF}}(\boldsymbol{Q})Q_j\psi_{\text{vib,iI}}(\boldsymbol{Q}),$$

which factors, in the absence of Duschinsky rotation, into

$$\sum_j \left(\frac{\partial M_{\text{if},z}}{\partial Q_j}\right)_0 |\langle v_1'|v_1\rangle|^2 |\langle v_2'|v_2\rangle|^2 \cdots |\langle v_j'|Q_j|v_j\rangle|^2 \cdots.$$

If $M_{\text{if}}(\mathbf{Q}_0)$ is nonzero, then its derivatives $(\partial M_{\text{if}}/\partial Q_j)_0$ are usually much smaller, and these "non-Condon" terms are only minor perturbations on the electronic spectrum. Things are more interesting if $M_{\text{if}}(\mathbf{Q}_0)$ is zero, for example, by the symmetry of the electronic wavefunctions.

In a linear molecule, changing the bond length does not change the point group. Therefore, $M_{\text{if}}(R_0)$ and $(\partial M_{\text{if}}/\partial R)_0$ have the same symmetry, and the inclusion of higher terms in the series expansion of the transition moment cannot change the symmetry selection rules. All they can do is change the magnitude of the electronic transition moment due to the dependence of the electronic wavefunction on internuclear separation.

Linear molecules having three or more atoms also belong to either $D_{\infty h}$ or $C_{\infty v}$, but the non-Condon terms $(\partial M_{\text{if}}/\partial Q_j)_0$ can now be nonzero even when $M_{\text{if}}(Q_0)$ is zero if Q_j is a nontotally symmetric vibrational coordinate. The way to look at this is that moving the nuclei along a nontotally symmetric coordinate changes the point group of the molecule and makes transitions allowed that otherwise would not be. The term $(\partial M_{\text{if}}/\partial Q_j)_0$ can give rise to an electric dipole-allowed transition if the product of the symmetry species of the initial state $\Psi_{\text{elect,i}}$, the final state $\Psi_{\text{elect,f}}$, the dipole operator μ, and the vibrational coordinate Q_j is or contains the totally symmetric representation. As a concrete example, consider a $\Sigma_g^+ \leftrightarrow \Pi_g$ singlet electronic transition of CO_2. This transition is not electric dipole allowed under x, y, or z within the Condon approximation. However, the antisymmetric stretching vibration ν_3 belongs to Σ_u^+. The direct product of Σ_g^+, Π_g, Σ_u^+, and x or y (Π_u) gives $\Delta_g + \Sigma_g^+ + \Sigma_g^-$, so the transition is allowed through the non-Condon term involving the antisymmetric stretch. This is easy to understand through a simple argument: $\Sigma^+ \leftrightarrow \Pi$ would be allowed if it were not g $\leftrightarrow$ g, but distortion along the antisymmetric stretch removes the center of symmetry, so g and u are no longer good symmetry labels, and that selection rule goes away.

As discussed above and depicted in Figure 8.4, if Q_j is a nontotally symmetric coordinate, $|v_j'\rangle$ and $|v_j\rangle$ must each have either even or odd symmetry with respect to the vibrational coordinate. Then, since Q_j is odd, $|\langle v_j'|Q_j|v_j\rangle|^2$ can be nonzero only when v_j' and v_j differ by an odd number of quanta. The zero–zero transition is therefore forbidden, and the electronic spectrum induced by non-Condon effects will consist of Franck–Condon progressions in various modes, all built on one quantum (i.e., the $0 \to 1$ transition) of this nontotally symmetric promoting mode. There may be more than one such promoting mode, although usually one mode is dominant.

Let us now look more closely at the origin of the nuclear coordinate dependence of the transition dipole moment. The quantity $M_{if}(Q)$ is the matrix element $\langle \Psi_f(\mathbf{r};\mathbf{Q})|\mu|\Psi_i(\mathbf{r};\mathbf{Q})\rangle$, where the electronic wavefunctions Ψ_i and Ψ_f are functions of the electronic coordinates **r** and depend parametrically on the vibrational normal coordinates **Q**. Let us consider expanding the electronic Hamiltonian as a Taylor series around the equilibrium geometry, $\mathbf{Q}_0$:

$$\mathbb{H}_{el}(\mathbf{r};\mathbf{Q}) = \mathbb{H}_{el}(\mathbf{r};\mathbf{Q}_0) + \sum_k \left(\frac{\partial \mathbb{H}_{el}}{\partial Q_k}\right)_0 Q_k + \cdots. \tag{8.11}$$

We first solve for the electronic wavefunctions at the equilibrium geometry, $\mathbf{Q}_0$, and then use perturbation theory (Chapter 1) to find the electronic wavefunctions at any other geometry as a linear combination of these basis functions:

$$\Psi_f(\mathbf{r};\mathbf{Q}) = \Psi_f(\mathbf{r};\mathbf{Q}_0) + \sum_{j\neq f}\left\{\frac{H'_{jf}}{\varepsilon_f - \varepsilon_j}\right\}\Psi_j(\mathbf{r};\mathbf{Q}_0) + \cdots, \tag{8.12}$$

and analogously for Ψ_i. Here ε_f and ε_j are the energies of states Ψ_f and Ψ_j at the equilibrium nuclear geometry, and H'_{jf} is the matrix element of the perturbation between these states,

$$H'_{jf} = \langle \Psi_j(\mathbf{r};\mathbf{Q}_0)|\sum_k \left(\frac{\partial \mathbb{H}_{el}}{\partial Q_k}\right)_0 Q_k|\Psi_f(\mathbf{r};\mathbf{Q}_0)\rangle \tag{8.13}$$

Abbreviating ($\mathbf{r}$;$\mathbf{Q}_0$) as the superscript 0, the dependence of the final state electronic wavefunction on nuclear coordinates is seen to be

$$\Psi_f(Q) = \Psi_f^0 + \sum_{j\neq f}\sum_k \langle \Psi_j^0|\frac{\partial \mathbb{H}_{el}}{\partial Q_k}|\Psi_f^0\rangle \frac{Q_k}{\varepsilon_f - \varepsilon_j}\Psi_j^0. \tag{8.14}$$

This is known as Herzberg–Teller coupling. It says that if the electronic Hamiltonian varies with nuclear coordinate Q_k, then motion along that coordinate has the effect of mixing other electronic states into the state of interest. Since $(\partial \mathbb{H}_{el}/\partial Q_k)$ has the same symmetry as the vibrational mode Q_k, the states that can be mixed into Ψ_f are those for which the direct products of the symmetry species of Ψ_j^0, Q_k, and Ψ_f^0 contains the totally symmetric representation.

As always with perturbation theory, the extent to which two states can mix depends on the difference between their unperturbed energies, $\varepsilon_f - \varepsilon_j$. Therefore, Herzberg–Teller coupling is usually much more important for excited electronic states than for ground electronic states because there is usually a large energy gap between the ground state and the next higher state, whereas the different excited states tend to lie much closer in energy.

We now use Equation (8.14) to calculate the transition dipole moment from Ψ_i to Ψ_f, assuming that Ψ_i is the ground electronic state and is unaffected by Herzberg–Teller coupling:

$$\begin{aligned}\langle \Psi_f(\mathbf{Q})|\mu|\Psi_i(\mathbf{Q})\rangle &= \langle \Psi_f^0|\mu|\Psi_i^0\rangle + \sum_{j\neq f}\sum_k \langle \Psi_j^0|\frac{\partial \mathbb{H}_{el}}{\partial Q_k}|\Psi_f^0\rangle \frac{Q_k}{\varepsilon_f-\varepsilon_j}\langle \Psi_j^0|\mu|\Psi_i^0\rangle \\ &= M_{fi}^0 + \sum_{j\neq f}\sum_k \left(\frac{\partial H_{jf}}{\partial Q_k}\right)\frac{Q_k}{\varepsilon_f-\varepsilon_j}M_{ji}^0. \end{aligned} \tag{8.15}$$

The transition dipole moment contains a nuclear coordinate-independent part, M_{fi}^0, plus terms linear in the various vibrational normal modes Q_k just as we assumed when expanding $\mu(\mathbf{Q})$ as a Taylor series around Q_0 in Equation (8.2). But now we can see where it comes from: motion along Q_k mixes into the electronic state Ψ_f small pieces of other states Ψ_j, which have different transition moments from the ground state. Thus the vibrations that are likely to produce the largest non-Condon effects are those that have the right symmetry to couple the excited electronic state of interest to other nearby electronic states that themselves have large transition dipole moments. When the excited state of interest has a small or zero transition moment from the ground state, most or all of the transition intensity may come from this mechanism, known as intensity borrowing.

8.6. THE JAHN–TELLER EFFECT

Molecules belonging to point groups with fairly high symmetry often have degenerate electronic states. Consider, for example, octahedral transition-metal complexes of the general formula ML_6. The five metal-centered d orbitals, degenerate in the isolated atom, are split by ligand-field interactions into three degenerate orbitals (derived from the d_{xy}, d_{xz}, and d_{yz} atomic orbitals) having t_{2g} symmetry at lower energy, and two degenerate orbitals (derived from the $d_{x2\text{-}y2}$ and d_{z2} AOs) having e_g symmetry at higher energy. If, however, the two ligands along the z-axis are moved out slightly to make those two metal-ligand distances slightly longer than the other four metal-ligand distances, reducing the symmetry of the complex to D_{4h}, then the degeneracy is partially lifted; the d_{xz} and d_{yz} orbitals go down in energy slightly relative to the d_{xy}, and the d_{z2} comes down in energy relative to the d_{x2-y2}. If, in the octahedral complex, either the t_{2g} orbitals or the e_g orbitals are only partly filled, then distortion to a D_{4h} geometry allows the electrons to be placed in lower-energy orbitals and thus stabilizes the complex (Fig. 8.5). This is a classic manifestation of the theorem of Jahn and Teller that a nonlinear symmetric

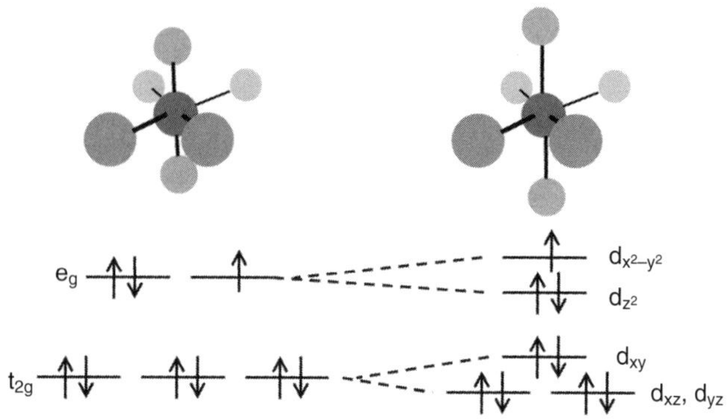

Figure 8.5. The Jahn–Teller effect for an octahedral d_9 complex, for example, $Cu(H_2O)_6$. Lengthening of the two metal-ligand bonds along the z-axis breaks the degeneracy of the d_{z^2} and $d_{x^2-y^2}$ orbitals and lowers the total energy.

molecule in an orbitally degenerate state will be unstable with respect to distortion to a lower symmetry state. The lifting of degeneracy due to distortion along a nontotally symmetric normal mode is known as the Jahn–Teller effect. Its most obvious spectroscopic consequence is the splitting of otherwise degenerate electronic transitions.

8.7. CONSIDERATIONS IN LARGE MOLECULES

The total intensity of an electronic transition, given in the Condon approximation by $|M_{if}(\mathbf{Q}_0)|^2$, is distributed among all possible vibrational sublevels according to their Franck–Condon factors. The number of possible vibrational transitions increases very rapidly as the number of atoms increases.

Consider the situation where all of the initial population is in the ground vibrational level of the ground state. In a diatomic molecule, the electronic transition will be spread out over all the $0 \rightarrow v'$ levels that have Franck–Condon factors significantly different from zero—in a "typical" case, perhaps 10 transitions. In a triatomic molecule having three normal modes, the transition will be spread over all $(0_1, 0_2, 0_3) \rightarrow (v_1', v_2', v_3')$ levels having significant Franck–Condon factors; if 10 transitions in each normal mode must be considered, this makes $10^3 = 1000$ vibrational sublevels of the electronic transition that carry significant intensity. Clearly, the spectrum can get crowded very quickly as the molecule gets large (Fig. 8.6). This is mitigated somewhat by the fact that there are usually only a few normal modes along which there is a large potential surface change between the two electronic states, so only a few of the $3N$–6 modes have large Franck–Condon factors for many different v_j'. Also, larger molecules usually tend to have smaller changes in vibrational

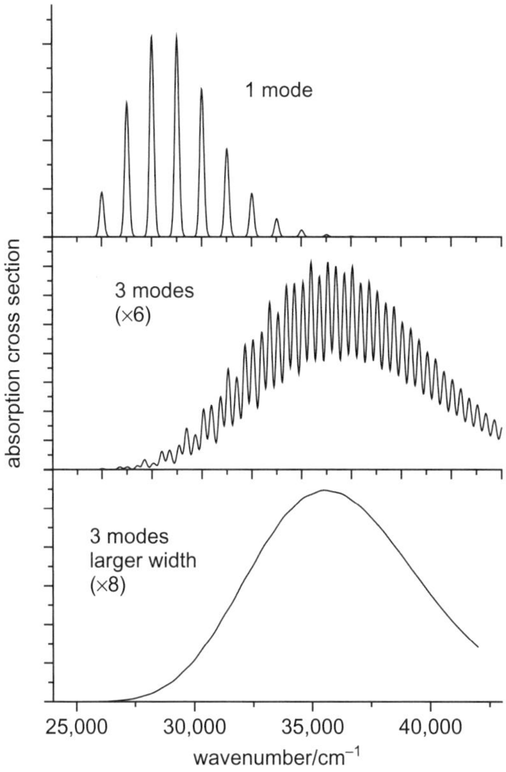

Figure 8.6. Illustration of spectral congestion caused by multiple vibronically active modes. Top: absorption spectrum of a molecule with one vibronically active mode having $\Delta = 2.4$ [see Eqs. (8.9) and (8.10)] and a narrow linewidth. Individual vibronic transitions are clearly resolved. Middle: three modes with different frequencies, each with $\Delta = 2.4$, at the same linewidth. Individual vibronic transitions are only partly resolved. Bottom: same three modes, but a larger linewidth. Now none of the underlying vibronic structure is apparent; the spectrum appears to be a single broad band.

potential function when the electronic state changes, as the change in electron density is spread out over more atoms. On the other hand, larger molecules tend to have at least some vibrations that are quite low in frequency, and if the sample is not very cold, there may be many different *initial* vibrational levels (v_1, v_2, v_3, . . .) having Boltzmann population at the start of the experiment, allowing even more transitions to contribute to the spectrum. These are the reasons why large molecules usually do not exhibit sharp-line electronic spectra unless they are extensively cooled, and sometimes not even then.

To calculate the entire absorption spectrum for a large molecule, one must sum over all $\{v_j\}$ for all normal modes j that are thermally populated in the initial state, weighted by their Boltzmann factors, and over all important $\{v_j'\}$, weighted by the Franck–Condon factor for each transition. The Franck–Condon factors for polyatomic molecules are easy to calculate if the harmonic oscillator approximation is made *and* if the normal modes of the ground and excited electronic states are the same (no Duschinsky rotation). Even so, there are often thousands of transitions with significant Franck–Condon intensity arising from each initial state in a large molecule.

Many diatomic and small polyatomic molecules have all of their vibrations at frequencies higher than 600 cm^{-1} or so. Since $k_B T$ is about 200 cm^{-1} at room temperature, the Boltzmann factors for all vibrational states other than the ground state are fairly small; for a frequency of 600 cm^{-1}, the ratio of $v = 1$ to $v = 0$ population at room temperature is $e^{-3} = 0.05$. Most of the population at thermal equilibrium is therefore in the vibrational ground state. In contrast, large molecules usually have at least some bending or torsional vibrations at rather low frequencies. For example, phenanthrene, a relatively rigid planar aromatic molecule with formula $C_{14}H_{10}$, has 66 normal modes of which 13 are below 600 cm^{-1}: 585, 546, 537, 500, 495, 438, 428, 401, 392, 244, 239, 226, 99, and 97 cm^{-1}. There are six different vibrational states (combinations of quantum numbers in different vibrational modes), including the ground state, that have energies below 200 cm^{-1} (relative Boltzmann population >36%), and 25 states below 400 cm^{-1} (relative population >13%). So at room temperature, many dozens of different vibrational states have significant thermal populations, and any vibrational or vibronic transition involving a given nominal change in vibrational quantum numbers is actually a superposition of dozens of different transitions involving the same quantum number changes but starting from different initial states, and therefore slightly shifted in frequency due to the anharmonicities.

A related problem with molecules that have many "floppy" degrees of freedom, such as torsions, is the presence of multiple conformations that interconvert too rapidly for them to be chemically isolated, such as *trans* versus *gauche* 2-butene. This is really just a special case of having many thermally populated initial states, except that different conformers really do act in a sense like different molecules from a spectroscopic point of view (different vibrational frequencies and, usually, slightly different electronic state energies).

8.8. SOLVENT EFFECTS ON ELECTRONIC SPECTRA

Electronic spectra tend to shift in frequency much more between vapor phase and condensed phases than do vibrational spectra. Shifts of several thousand cm^{-1} are not unusual even when there are no specific solvent-solute interactions, such as complex formation or hydrogen bonding. For this reason, it is often assumed that solvation changes the energy difference between the potential minima of the ground and excited electronic states without changing the shapes of those potential surfaces, that is, the vibrational frequencies. Solution phase spectra are often quite broad as discussed below, and individual vibronic transitions usually cannot be resolved (refer to Fig. 8.6). Thus, one typically talks about absorption or fluorescence frequencies in solution as the frequency or wavelength of the maximum of the absorption or fluorescence spectrum (λ_{max}). This can be a little ambiguous to compare with a gas-phase spectrum that consists of sharp lines.

The shift of an absorption or emission spectrum in a condensed phase environment depends on the difference between the energetic perturbation of the upper and lower states by the solvent. Usually, both the ground and the excited states are stabilized (lowered in energy) by solvation. In many cases, the excited state is lowered more than the ground state because excited states are usually more polarizable than ground states (the electrons are not bound as tightly), and the attractive, induced dipole-induced dipole (dispersive) interaction is more favorable in the excited state. Then the absorption and fluorescence spectra shift to lower energies or longer wavelengths (red-shift) upon increasing the polarizability of the environment. This is usually the case for pi-electron transitions of conjugated organic molecules. For polar molecules in polar solvents, it is the interaction energy among the permanent dipoles of the chromophore and the molecules in its environment that dominates the solvation energy. As long as the solvent molecules are free to rotate as in a liquid, the solvent dipoles will orient themselves to have the most favorable interaction energy with the solute's ground-state dipole moment. If the excited state has a dipole moment of greater magnitude than the ground state but about the same direction, then the solvent dipoles will lower the energy of the excited state more than that of the ground state, leading to a red shift of the optical absorption. This is the most common case. If, however, the excited state is less polar than the ground state or has a different direction for its dipole moment, the optical transition will blue-shift upon increasing the polarity of the environment. In some cases, particularly with small molecules, the solvent shift is dominated by the change in size of the molecule upon excitation. This occurs in Rydberg transitions where the excited state has an electron in a very diffuse orbital far from the nucleus. These transitions are often very strongly perturbed by solvation, generally blue shifted, because of overlap between the electrons of the solvent molecules and the excited electron in the Rydberg orbital.

The Born–Oppenheimer and Condon approximations say that electronic excitations are very fast compared with nuclear motions—that is, the electrons change their state while the nuclei are basically frozen. This applies to the nuclei of not only the chromophore, but also the molecules in its environment. Their electron distributions can respond essentially instantaneously to the change in electronic state of the chromophore, but their nuclei cannot. While the induced dipole-induced dipole interactions that dominate the solvent–solute interaction in nonpolar molecules involve mainly the rearrangement of

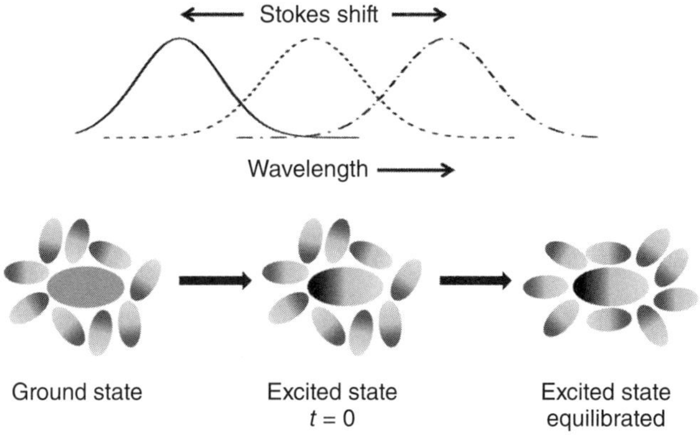

Figure 8.7. Solvent reorganization. A ground state having little charge separation (solid gray oval) will be solvated by polar solvent molecules (smaller two-toned ovals) in fairly random orientations. If optical excitation creates a more polar excited state, the solvent molecules are initially in an energetically unfavorable configuration, and they will rotate and translate to a more favorable configuration. The fluorescence at very short times (dashed curve) is shifted to slightly longer wavelengths than the absorption (solid curve), but the equilibrated fluorescence (dash-dot curve) is shifted much more. The difference between the absorption and equilibrated emission maxima is the Stokes shift.

the electron distributions, interactions involving permanent dipole moments require reorientation of the whole solvent molecule. Thus, when electronic excitation changes the magnitude and/or direction of a chromophore's dipole moment, the surrounding solvent molecules are no longer in the right arrangement to minimize the energy (Fig. 8.7). They have to translate and/or rotate to new positions, and this takes time. This process is known as solvent reorganization and is further discussed in Chapter 12 in the context of electron transfer reactions, where it is particularly important. The energy change between the solvent in its initial arrangement (equilibrated with the solute's ground-state charge distribution) and its final arrangement (equilibrated with the excited state charge distribution) is called the reorganization energy. The same process occurs with the nuclear motions (vibrations) of the chromophore itself; since the excited state generally has a different equilibrium geometry than the ground state, the molecular geometry has to change to reach the potential energy minimum in the excited state, and as the nuclei move toward this minimum, the vibrational energy is transferred to the solvent, lowering the energy of the chromophore itself. This process is often known as internal or vibrational reorganization, or (using terminology from transition metal complexes) as inner-sphere reorganization, while the corresponding process involving the solvent is outer-sphere reorganization. They are really the same thing except that the vibrations of the chromophore are typically high in frequency (high compared with k_BT), which means that in quantitative treatments, the vibrational motion has to be treated quantum mechanically. The solvent motions (translations and hindered rotations) are usually very low in frequency and can be treated using classical mechanics.

The reduction in energy of the photoexcited chromophore due to solvent and internal reorganization causes the maximum of the fluorescence spectrum to occur at lower energy than the maximum of the absorption spectrum. This shift between λ_{max} for absorption and for emission is called the Stokes shift (Fig. 8.7). The internal contribution may range from nearly zero to thousands of cm^{-1}, or sometimes greater than $10^4 cm^{-1}$ when the change in molecular geometry is really large. The contribution from the solvent is usually $1000 cm^{-1}$ or less for nonpolar solvents, and several thousand cm^{-1} for polar solvents when the solute undergoes a fairly large change in dipole moment. A variety of theories with different levels of sophistication can be used to calculate the solvent reorganization energy corresponding to a particular change in solute charge distribution.

Solvent reorganization requires some time to occur. These times are usually in the picosecond range while most fluorescence lifetimes are nanoseconds, so it is usually a pretty good approximation to assume that solvent reorganization is complete prior to emission. However, this may not be true if the emission is quite short-lived. It is particularly not true when the fluorescence is time-resolved by performing the experiment with short-pulsed excitation and fast detection, usually with fluorescence upconversion or time-correlated single photon counting. In this way, it is possible to measure the Stokes shift as a function of time, the dynamic Stokes shift.

Very weak electronic transitions often undergo significant intensity changes in condensed phases. Symmetry-forbidden transitions that are made allowed by non-Condon effects may be much stronger in solution or in a crystal if the environment tends to distort the molecular geometry away from its gas-phase symmetric structure. Spin-forbidden electronic transitions can often be made more intense by putting the molecule into a solvent that contains heavy atoms (e.g., methyl iodide) with strong spin-orbit coupling (the external heavy-atom effect). Fully allowed electronic transitions may become either stronger or weaker in condensed phases than in the gas phase, by factors of up to ±50%. The origin of such intensity changes is not generally clear. A number of theories based on local field effects (Chapter 4) have been proposed but rarely have been shown to work well for a large number of systems even where specific solvent-solute interactions appear unimportant.

Electronic absorption and emission spectra are subject to the same broadening mechanisms described in Chapter 7 for vibrational spectra: inhomogeneous broadening, pure dephasing, and lifetime broadening. However, the magnitude of the total broadening is almost always much greater for electronic than for vibrational transitions. In liquid or amorphous solid phases, electronic transitions are usually broadened by several hundred cm^{-1} relative to the gas phase, meaning that individual vibronic transitions cannot be resolved

(Fig. 8.6). Often, particularly for polar molecules in polar solvents, the broadening is so severe that the whole absorption spectrum degenerates into one broad blob with a single maximum. In most cases, lifetime broadening makes only a small contribution to the total width; even a very short lifetime, say around 10 ps, contributes only about 1 cm^{-1} to the width. Inhomogeneous broadening and pure dephasing, with the usual caveat about there not being a clear distinction between the two, make the major contributions to the broadening of electronic transitions in disordered environments. In well-ordered crystals, the electronic linewidth may be much smaller (only a few cm^{-1}) but is still usually dominated by the inhomogeneous and/or pure dephasing mechanisms. As the temperature is lowered, all molecular motions slow down, and regardless of whether the medium is an amorphous solid or a crystal, "pure dephasing" tends to turn into "inhomogeneous broadening."

The technique of spectral hole burning provides one way to distinguish between homogeneous and inhomogeneous broadening and to partially eliminate the effects of inhomogeneous broadening from an experiment. It is based on the idea that if the sample is composed of collections of molecules having slightly different spectra, then excitation with a very narrow-bandwidth laser, as narrow as a spectral line of a single molecule, is resonant with only a small subset of all the molecules in the sample and will excite only those molecules that absorb that particular frequency. If the excitation frequency is chosen to fall near the low-energy edge of the absorption to a particular electronic state, then essentially only those molecules that have their electronic zero–zero transitions at that precise laser frequency will be excited.

This selective absorption by a limited subset of the inhomogeneous distribution may be detected in several ways. One is by fluorescence. Since all of the molecules that were excited have their electronic zero–zero transitions at almost exactly the same frequency, the fluorescence will not be subject to the electronic inhomogeneous broadening. The only source of inhomogeneous broadening will be the distribution of vibrational frequencies among molecules that have the same electronic transition frequency, and this is usually quite small, a few cm^{-1} or less as discussed above. This increase in resolution of a narrowband-excited fluorescence spectrum compared with a broadband-excited spectrum is known as fluorescence line-narrowing (Fig. 8.8). The other way to detect absorption by a subset of the inhomogeneous distribution is to make use of situations where the photoexcited molecules do something that removes them from the absorption spectrum. They may undergo a photochemical reaction to produce a species that absorbs in a completely different frequency region (photochemical hole-burning), or photoexcitation of the chromophore may result in a rearrangement of the environment such that, upon returning to its electronic ground state, the molecule no longer absorbs at the same frequency (photophysical or nonphotochemical hole burning). When using pulsed lasers, one can even make use of a radiationless transition to some long-lived excited state, such as a triplet state, which may last for milliseconds before returning to the ground state—this is often known as

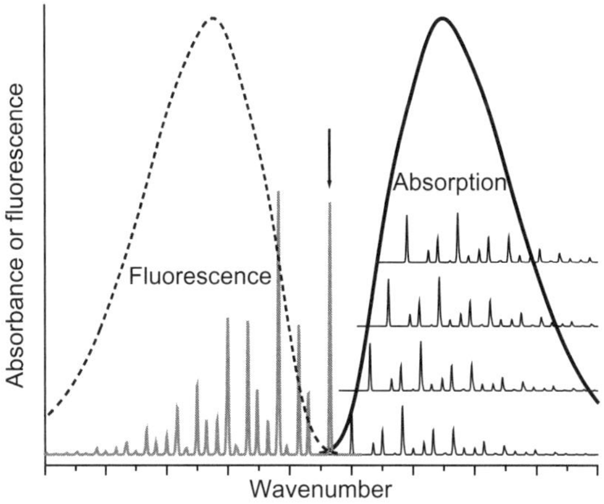

Figure 8.8. Fluorescence line narrowing. An inhomogeneously broadened absorption spectrum (thick solid curve) is composed of individually sharp spectra. The fluorescence spectrum of the ensemble is also broad (dashed). However, excitation near the low-energy edge of the absorption (arrow) excites only a small subset of molecules that emit with a sharp spectrum (gray).

transient hole burning. In any case, if one measures the absorption spectrum of the sample at a later time, there will be a "hole" or "dip" in the spectrum at the laser frequency because some of the molecules that used to absorb that frequency no longer do. From the shape, width, and depth of the hole, one can infer the true homogeneous and inhomogeneous broadening parameters of the system.

Both of these techniques work well only at low temperatures—at least below 100 K, and preferably below 10 K. This is because the inhomogeneous distribution must remain fixed on the time scale of the experiment, which is the fluorescence lifetime for fluorescence line-narrowing (typically nanoseconds) or the time delay between burning and reading the hole for hole burning (anywhere from microseconds to minutes depending on how the experiment is done). At higher temperatures, there is enough molecular motion that a given molecule's transition frequency jumps around some, and by the time the fluorescence is emitted or the hole is probed, the well-defined energy distribution defined by the exciting laser has become scrambled. Much can be learned about the dynamics of the matrix by looking at the time dependence of hole "refilling" in hole burning experiments.

FURTHER READING

J. L. McHale, *Molecular Spectroscopy* (Prentice Hall, Upper Saddle River, NJ, 1999).

PROBLEMS

1. The ground state of the C_2 molecule has symmetry ${}^1\Sigma_g^+$. The first three excited states are ${}^1\Pi_u$, ${}^1\Delta_g$, and ${}^1\Sigma_g^+$, respectively. Which of these transitions are allowed through the electric dipole, magnetic dipole, and electric quadrupole terms, respectively? Which do you expect to show the strongest absorption transition?

2. Consider two electrons confined to the same infinite one-dimensional particle-in-a-box potential: $V(x) = 0$ for $0 \le x \le a$, $V(x) = \infty$ for $x < 0$ or $x > a$.

(a) Write the properly antisymmetrized and normalized total wavefunction, including both space and spin degrees of freedom, for the ground state of this system.

(b) Write the properly antisymmetrized and normalized total wavefunctions that correspond to both the singlet and the triplet terms of the lowest excited state of this system.

(c) Consider the probability of finding both electrons at the same location in the box, for both the singlet state and the triplet state found in part (b). Recognizing that the Coulombic repulsion between electrons is $e^2/(4\pi\varepsilon_0 x_{12})$, should the singlet state or the triplet state be lower in energy?

3. The SO_2 molecule has a bent ground state geometry (C_{2v} point group), and its ground electronic state is ${}^1A_1$. Its first three singlet excited states have symmetries of ${}^1B_1$, ${}^1A_2$, and ${}^1B_2$.

(a) Which of these states are electric dipole allowed from the ground state, and for those that are, what is the direction of the transition dipole moment in the molecular frame? You will need to use a character table for the C_{2v} point group.

(b) A weak absorption between 280 and 340 nm is attributed to the transition ${}^1A_1 \rightarrow {}^1A_2$. Comment on the probable mechanism whereby this transition is allowed.

4. Consider vibronic transitions of O_2 in the Condon approximation. The ground state vibrational frequency is $1555\,\text{cm}^{-1}$, and the equilibrium bond length in the ground electronic state ($X^3\Sigma_g^-$) is 1.22 Å. The three lowest excited triplet states of O_2 are $A^3\Sigma_u^+$, $B^3\Sigma_u^-$, and $C^3\Delta_u$. The equilibrium bond lengths in these states are, respectively, 1.53, 1.60, and 1.48 Å.

(a) Which of these excited electronic states have allowed transitions from the ground state in the electric dipole approximation?

(b) Assume that the potential surfaces for nuclear motion in the ground and excited electronic states have the same force constant (same vibrational frequency) but different potential minima as given above. If Δr is the separation between the two potential minima along the internuclear coordinate, the Franck–Condon factor between the ground vibrational level of the ground electronic state and vibrational level $|v'\rangle$ of the excited electronic state is

$$|\langle v'|0\rangle|^2 = \exp(-\Delta^2/2)\{(\Delta^2/2)^{v'}/v'!\} \quad \text{where } \Delta = (\mu\omega/\hbar)^{1/2}(\Delta r).$$

Calculate the Franck–Condon factors for $v' = 0, 5,$ and 10 for the transition(s) you concluded were allowed in part a. For each allowed transition, find the v′ state that has the largest Franck-Condon factor.

5. The perylene molecule ($C_{20}H_{12}$) belongs to the D_{2h} point group. The $S_1 \rightarrow S_0$ transition is symmetry allowed. Both the infrared absorption spectrum and the $S_1 \rightarrow S_0$ fluorescence spectrum (in a low-temperature matrix) show many vibrational frequencies. Some of the infrared frequencies are very close to the fluorescence frequencies (within a few cm^{-1}). Is it likely that they correspond to the same normal modes? Explain.

6. It is often assumed that the absorption spectrum corresponding to the lowest-energy electronic transition of a molecule and the fluorescence spectrum arising from emission from that same state should be mirror images of one another as shown, for example, in the plot. Discuss the conditions required for this mirror image symmetry to be observed.

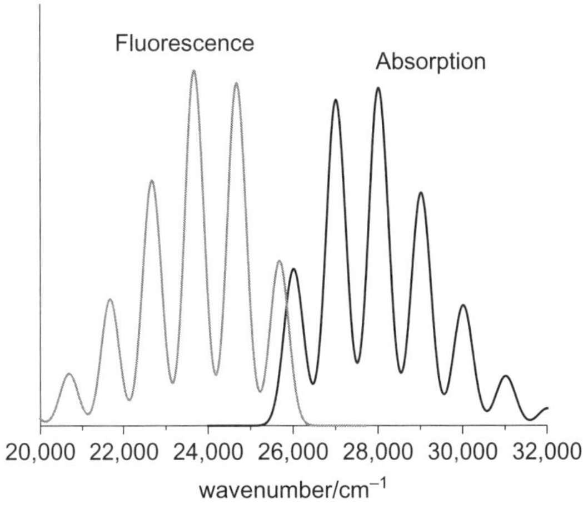

7. The organic molecule pyrene is famous for forming excimers. An excimer is a molecular dimer that is stable only in an excited electronic state. It is formed when an excited pyrene monomer encounters a ground-state pyrene monomer, forming a weakly bound dimer in which the excitation is shared between the two monomers:

$$P + P \xrightarrow{h\nu} P + P^* \rightarrow [PP]^*.$$

(a) Based on the fact that the dimer is stable only in the excited state, draw a rough plot of the ground state and first excited state potential energy surfaces for two pyrene molecules along the intermolecular distance axis.

(b) How do you expect the fluorescence spectrum of the excimer, $[PP]^*$, to differ from that of the monomer?

8. With reference to Figure 8.5, show the expected energies of the five d orbitals for a 5-coordinate square pyramidal complex of the type ML_5, where M is a metal and L is some ligand. Explain how this can be considered as an extreme case of the Jahn–Teller effect.

9. Consider the three-dimensional particle in a box problem that was mentioned in Chapter 1. Take this as a model for the molecular orbitals containing the valence electrons in a three-dimensional molecule. Assume that the molecule has a total of five valence electrons. Because they are fermions, a maximum of two electrons, with opposite spins, can go into a single orbital.

(a) First assume that the box is cubic. Each side has a length of L and the volume of the cube is $V = L^3$. In terms of L, calculate the first few energy levels and find the lowest-energy configuration of the five electrons. Find the total electronic energy as a sum of the energies of the five electrons.

(b) Now let one of the sides of the box be a different length: two of the sides have length A and one has length B. Now the volume is $V = A^2B$, and we require the volume to be the same as it was in part (a), that is, $A^2B = L^3$. Find the lowest energy levels, put the electrons into them in the lowest-energy configuration, and calculate the total electronic energy. What values of A and B minimize the energy? Is this energy lower than the energy found in part (a)?

(c) Discuss the relationship between this simple model problem and Jahn–Teller distortions in octahedral molecules.

CHAPTER 9

PHOTOPHYSICAL PROCESSES

Absorption of electromagnetic radiation leaves the molecule in a higher-energy state. If the radiation is in the visible or ultraviolet region of the spectrum, light absorption places a great deal of energy into its electronic degrees of freedom: the energy of a visible photon is about 100 times greater than k_BT near room temperature. Thus, following light absorption, the absorber must redistribute its energy, both within its own internal degrees of freedom and to the environment, in order to reestablish thermal equilibrium.

Here we begin with a simple, largely kinetics-based description of photophysical processes. We then introduce a formal quantum mechanical description of some of these processes. Finally, we discuss some aspects of photophysics that are specifically connected to the reorientation of molecules in liquid solution.

9.1. JABLONSKI DIAGRAMS

The processes that a molecule can undergo following light absorption are conveniently discussed with reference to a generic energy level scheme known as a Jablonski diagram (Fig. 9.1). The one shown is most appropriate to large organic molecules but may be applied with appropriate modifications to other systems.

Condensed-Phase Molecular Spectroscopy and Photophysics, First Edition. Anne Myers Kelley.

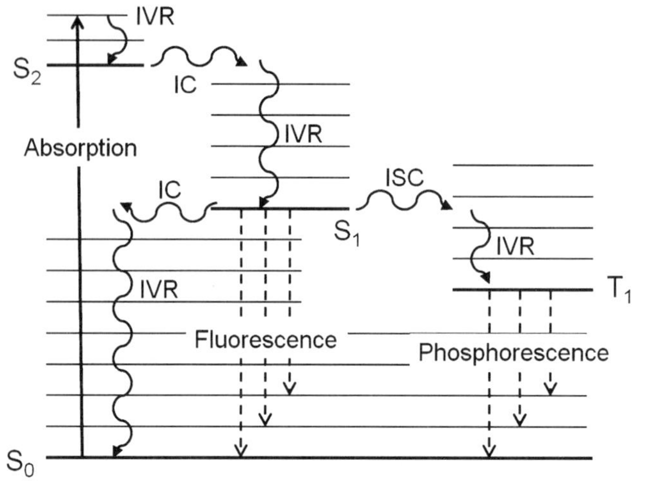

Figure 9.1. A generic Jablonski diagram for an organic molecule excited to its second excited singlet state, S_2. Shown are absorption, fluorescence, phosphorescence, intramolecular vibrational redistribution (IVR), internal conversion (IC), and intersystem crossing (ISC).

A generic organic molecule at room temperature is usually found in its electronic ground state, which is usually a closed-shell singlet, S_0. Most of its vibrational modes are also in their ground states, although large molecules typically have some low-frequency vibrations that have reasonable populations in higher levels at thermal equilibrium. Light absorption can promote the molecule to a higher electronic state; assume for concreteness that the photon energy is chosen such that absorption reaches an excited vibrational level of the second excited singlet state, S_2.

Once it has absorbed the photon, there are a number of things the molecule can do with the excess energy. We will focus on the most probable ones. One process that is usually fast in large molecules is intramolecular vibrational redistribution or intramolecular vibrational relaxation (IVR). This is a process whereby the molecule remains in the same electronic state (here, S_2) but *redistributes* its vibrational energy from the initially excited mode to other vibrational modes and/or *relaxes* by dropping to a lower vibrational energy level, dumping the extra energy into the environment. In large molecules in condensed phases near room temperature, vibrational redistribution normally occurs in less than a picosecond to a few picoseconds, and relaxation back to a thermal distribution of vibrational levels of the electronic state occurs in a few to tens of picoseconds.

Another usually fast process that immediately follows or competes with IVR is internal conversion. This is a transition between two different electronic states with the same spin multiplicity. Since energy must be conserved, the transition is from low vibrational levels of S_2 to higher vibrational levels of S_1 that have the same total energy. Generally, when two electronic states of the same spin multiplicity are fairly close in energy, internal conversion is fast (picoseconds). The reason will be discussed later. Once in the S_1 state, IVR can occur in this state to take the molecule to a thermal distribution of vibrational levels of S_1. At this point, it is possible to again undergo internal conversion to highly excited vibrational levels of S_0, but such processes are usually much slower than S_2–S_1 internal conversion because the energy separation between S_0 and S_1 is usually much greater than between S_1 and S_2. Thus S_1–S_0 internal conversion requires making a transition between vibrational levels with very different vibrational quantum numbers, and such processes are typically slow (see next section).

Thus, other processes can compete effectively with the relatively slow S_1–S_0 internal conversion. One of these is fluorescence from S_1 to S_0. Often, the S_1–S_0 radiative rate is much faster than the S_1–S_0 internal conversion rate, and fluorescence is the dominant mechanism for getting rid of the excess energy. In general, the S_2 state can also fluoresce, but because S_2 does not last very long before undergoing internal conversion to S_1, only a small fraction of excited molecules lose their energy in this way. This leads to Kasha's rule: no matter which electronic state is initially excited, essentially all (99+%) of the fluorescence originates from S_1. There are exceptions to Kasha's rule, molecules that show fairly strong S_2–S_1 or S_2–S_0 fluorescence; the best-known example is azulene, which happens to have an unusually large S_2–S_1 energy gap. If the final state in the fluorescence process is an excited vibrational level of S_0, this state will relax to a thermal distribution via IVR as discussed previously.

Organic molecules also have triplet excited states that are usually lower in energy than the corresponding singlets. As discussed previously, optical transitions between singlets and triplets are usually very weak because of the spin selection rules. Nonradiative transitions are also usually slow, but they can occur. A nonradiative transition between electronic states of different spin multiplicity is called intersystem crossing, and intersystem crossing from S_1 to excited vibrational levels of T_1 is another mechanism that can depopulate S_1. The time scales for such intersystem crossing processes vary greatly, from picoseconds in some cases to microseconds or longer. Once in T_1, IVR occurs rapidly just as it does in S_1. Finally, once the system has reached a thermal vibrational distribution in T_1, further relaxation back to S_0 may occur nonradiatively, through another intersystem crossing process. Alternatively, it may occur radiatively through phosphorescence, the term given to radiative transitions between electronic states of different spin multiplicity. Since both intersystem crossing and phosphorescence are slow processes, triplet states tend to have long lifetimes—typically microseconds near room temperature, and even longer at low temperatures.

Finally, there are a number of other processes that occur only in some molecules. Nonradiative processes that convert the original molecule to a completely different chemical species are known as photochemistry. We will also consider the special case of electron transfer processes in Chapter 12. Energy transfer, a process whereby the originally excited molecule transfers its excitation to another molecule, is discussed below.

9.2. QUANTUM YIELDS AND LIFETIMES

In most situations, the photophysical processes discussed previously can be modeled with a simple reaction kinetics scheme. All of the previous processes—IVR, internal conversion, intersystem crossing, fluorescence, phosphorescence—are unimolecular processes. Recall from chemical kinetics that an elementary unimolecular reaction exhibits exponential kinetics: the concentration of reactants decays exponentially with time, $N_r(t) = N_0e^{-kt}$, and the concentration of products rises exponentially, $N_p(t) = N_0(1-e^{-kt})$. In the context of photophysics, if we use a short light pulse or some other very fast perturbation to create N molecules in the S_1 state at time zero and the only thing they can do is fluoresce to return to S_0, then the numbers of S_1 and S_0 molecules at a later time will be $N_1(t) = Ne^{-k_f t}$ and $N_0(t) = N\left(1-e^{-k_f t}\right)$ where k_f is the radiative rate constant for the $S_1 \rightarrow S_0$ fluorescence. The lifetime, τ, of the S_1 state is defined by analogy to reaction kinetics as the time at which the concentration has decayed to $1/e$ of its original value: $\tau = 1/k_f$. When the only available decay routes involve light emission, the lifetime is referred to as the natural radiative lifetime as discussed in Chapter 4.

Now consider the case where we use a short light pulse to create N molecules in S_1 and they can do three different things: $S_1 \rightarrow S_0$ fluorescence with rate constant k_f, S_1–S_0 internal conversion with rate constant k_{ic}, and S_1–T_1 intersystem crossing with rate constant k_{isc}. In this case, the concentration of S_1 decays exponentially in time with the *sum* of the three rate constants: $N_1(t) = N\exp[-(k_f + k_{ic} + k_{isc})t]$. The lifetime of S_1 is now $\tau = 1/(k_f + k_{ic} + k_{isc})$. Even if a state can decay through many different processes, as long as each process alone gives exponential decay, the overall decay will still be single exponential.

The quantum yield, φ_X, for any photophysical process "X" is defined as the number of "X" events divided by the number of photons initially absorbed. For example, the quantum yield for fluorescence is the number of fluorescent photons emitted divided by the number of incident photons absorbed. Let us calculate the fluorescence quantum yield for the example of the previous paragraph. The rate at which photons are emitted through $S_1 \rightarrow S_0$ fluorescence is the same as the rate of forming S_0 molecules through $S_1 \rightarrow S_0$ fluorescence: $dN_0/dt = k_f N_1 = k_f N\exp[-(k_f + k_{ic} + k_{isc})t]$, and assuming $N_0 = 0$ at $t = 0$, this integrates to $N_0(t) = \{k_f N/(k_f + k_{ic} + k_{isc})\}\{1 - \exp[-(k_f + k_{ic} + k_{isc})t]\}$. At long times, this becomes $N_0(\infty) = N\{k_f/(k_f + k_{ic} + k_{isc})\}$, and the fluorescence quantum yield is $\varphi_f = N_0(\infty)/N = k_f/(k_f + k_{ic} + k_{isc})$. That is, the quantum yield for the process of interest is the rate constant for that process divided by the sum of the rate constants of that process and all others that compete with it.

If, for example, $k_f = 10^{-8}$ s^{-1}, $k_{ic} = 10^{-7}$ s^{-1}, and $k_{isc} = 10^{-6}$ s^{-1}, then the quantum yields for the three processes are $\varphi_f = 0.90$, $\varphi_{ic} = 0.09$, and $\varphi_{isc} = 0.01$. More generally, we can write the yield for process "X" as

$$\varphi_X = \frac{k_X}{\sum_i k_i} = \frac{1/\tau_X}{\sum_i 1/\tau_i}, \tag{9.1}$$

where the denominator contains a sum over the rate constants for all processes that deplete the excited state, assuming they are unimolecular in the excited molecule.

9.3. FERMI'S GOLDEN RULE FOR RADIATIONLESS TRANSITIONS

In Chapter 3, we derived the Fermi golden rule for the rate of transitions between two different eigenstates of matter, driven by a time-dependent perturbation of frequency ω [Eq. (3.26)]:

$$R(\omega) = \frac{2\pi}{\hbar^2}|W_{avg}(\omega)|^2 \rho(\omega).$$

There is no restriction on the frequency of the perturbation, and we may set it to zero to derive the rate at which a *constant* perturbation W induces transitions between an initial state $|i\rangle$ and a group of final states $\{|f\rangle\}$:

$$R_{i\to f} = \frac{2\pi}{\hbar^2}|W_{if,avg}|^2 \rho(\omega_f = \omega_i) = \frac{2\pi}{\hbar}|W_{if,avg}|^2 \rho(E_f = E_i). \tag{9.2}$$

Here $W_{if,avg}$ is the average matrix element between $|i\rangle$ and the isoenergetic final states $|f\rangle$, and ρ is the density of energy-conserving final states; the units of ρ(ω) and ρ(*E*) are inverse frequency and inverse energy, respectively, and the two are related by ħ. In the next few sections, we further develop some specific applications of Fermi's golden rule.

9.4. INTERNAL CONVERSION AND INTERSYSTEM CROSSING

We now discuss in more detail the mechanisms of internal conversion and intersystem crossing. Physical chemists tend to think about spectroscopy in terms of nuclei moving on Born–Oppenheimer potential energy surfaces defined by the electronic wavefunctions. But the molecule just has one wavefunction that does not separate exactly into products of nuclear, electronic, and spin coordinates. Therefore, the nominal electronic states are not actually stationary states of the system, and transitions between them can occur due to small but non-negligible terms in the complete Hamiltonian.

Let us go back and look at the Born–Oppenheimer approximation (Chapter 1) in more detail. The total Hamiltonian for a molecule can be written as

$$\mathbb{H} = \mathbb{T}_N(\mathbf{Q}) + \mathbb{T}_e(\mathbf{r}) + \mathbb{U}_{eN}(\mathbf{r};\mathbf{Q}) + \mathbb{V}_N(\mathbf{Q}) + \mathbb{H}_{so}, \tag{9.3}$$

where the first two terms are the kinetic energies for nuclear and electronic motion, the third term contains the electron–electron repulsions and electron–nuclear attractions, the fourth term is the internuclear repulsions, and the last term is the spin–orbit operator. The first four terms correspond to the first four terms in Equation (1.55) for the H_2^+ molecule; spin–orbit coupling was neglected for H_2^+. In the usual Born–Oppenheimer approximation, we find electronic wavefunctions by solving

$$[\mathbb{T}_e(\mathbf{Q}) + \mathbb{U}_{eN}(\mathbf{r};\mathbf{Q})]|\Psi_j(\mathbf{r};\mathbf{Q})\rangle = \varepsilon_j(\mathbf{Q})|\Psi_j(\mathbf{r};\mathbf{Q})\rangle, \tag{9.4}$$

where the $|\Psi_j(\mathbf{r};\mathbf{Q})\rangle$ are electronic wavefunctions calculated at a particular nuclear coordinate and the $\varepsilon_j(\mathbf{Q})$ are the electronic energies as a function of nuclear coordinates, which also constitute (together with $\mathbb{V}_N$) the potential energy surfaces for nuclear motion in the electronic states $|\Psi_j\rangle$. The Born–Oppenheimer approximation says that we can define electronic wavefunctions that are functions only of electronic coordinates, but it does not imply that these solutions will be the same at all nuclear positions.

The actual total wavefunction for the molecule, which is a function of both electronic and nuclear coordinates, can then be written as a linear combination of these Born–Oppenheimer electronic states weighted by nuclear coordinate-dependent factors:

$$|\Psi(\mathbf{r},\mathbf{Q})\rangle = \sum\nolimits_j |\Psi_j(\mathbf{r};\mathbf{Q})\rangle|\chi_j(\mathbf{Q})\rangle. \tag{9.5}$$

The total wavefunction must satisfy the total Schrödinger equation:

$$\mathbb{H}|\Psi(\mathbf{r},\mathbf{Q})\rangle = E|\Psi(\mathbf{r},\mathbf{Q})\rangle \tag{9.6}$$

where E is the total energy of the molecule. Substituting Equations (9.3) and (9.5) into Equation (9.6) gives

$$[\mathbb{T}_N(\mathbf{Q}) + \mathbb{T}_e(\mathbf{r}) + \mathbb{U}_{eN}(\mathbf{r};\mathbf{Q}) + \mathbb{V}_N(\mathbf{Q}) + \mathbb{H}_{so} - E]\sum\nolimits_j |\Psi_j(\mathbf{r};\mathbf{Q})\rangle|\chi_j(\mathbf{Q})\rangle = 0. \tag{9.7}$$

Now take the inner product with a specific one of the Born–Oppenheimer states, $\langle\Psi_m(\mathbf{r};\mathbf{Q})|$, and use the fact that

$$\langle\Psi_m(\mathbf{r};\mathbf{Q})|[\mathrm{T}_e(\mathbf{r}) + \mathrm{U}_{eN}]|\Psi_j(\mathbf{r};\mathbf{Q})\rangle = \varepsilon_m(\mathbf{Q})\delta_{mj} \text{ (definition of electronic B-O states)} \tag{9.8a}$$

$$\langle\Psi_m(\mathbf{r};\mathbf{Q})|\mathrm{V}_N(\mathbf{Q})|\Psi_j(\mathbf{r};\mathbf{Q})\rangle = V_N(\mathbf{Q})\delta_{mj}, \tag{9.8b}$$

(since $V(\mathbf{Q})$ does not operate on electronic wavefunctions)

$$\langle\Psi_m(\mathbf{r};\mathbf{Q})|E|\Psi_j(\mathbf{r};\mathbf{Q})\rangle = E\delta_{mj}, \tag{9.8c}$$

to obtain

$$[\varepsilon_m(\mathbf{Q})+\mathbb{V}_N(\mathbf{Q})-E]|\chi_m(\mathbf{Q})\rangle + \sum_j \langle\Psi_m(\mathbf{r};\mathbf{Q})|\mathbb{T}_N(\mathbf{Q})+\mathbb{H}_{so}|\Psi_j(\mathbf{r};\mathbf{Q})\rangle|\chi_j(\mathbf{Q})\rangle = 0,$$

or, separating the sum over j into its diagonal ($j = m$) and off-diagonal ($j \neq m$) parts,

$$\begin{aligned}&[\varepsilon_m(\mathbf{Q})+\mathbb{V}_N(\mathbf{Q})-E+\langle\Psi_m|\mathbb{T}_N(\mathbf{Q})|\Psi_m\rangle+\langle\Psi_m|\mathbb{H}_{so}|\Psi_m\rangle]|\chi_m(\mathbf{Q})\rangle\\ &= -\sum_{j\neq m}\langle\Psi_m(\mathbf{r};\mathbf{Q})|\mathbb{T}_N(\mathbf{Q})+\mathbb{H}_{so}|\Psi_j(\mathbf{r};\mathbf{Q})\rangle|\chi_j(\mathbf{Q})\rangle.\end{aligned} \tag{9.9}$$

The nuclear kinetic energy operator for a polyatomic molecule is given by

$$\mathbb{T}_N(\mathbf{Q}) = -\Sigma_k(\hbar^2/2\mu_k)\nabla_k^2 = -\Sigma_k(\hbar^2/2\mu_k)(\partial/\partial Q_k)(\partial/\partial Q_k),$$

where the sum is over all $3N$–6 normal modes and μ_k is the reduced mass of mode k. Therefore,

$$\begin{aligned}\mathbb{T}_N(\mathbf{Q})|\Psi_j(\mathbf{r};\mathbf{Q})\rangle|\chi_j(\mathbf{Q})\rangle &= -\sum_k(\hbar^2/2\mu_k)(\partial/\partial Q_k)\{(\partial|\Psi_j\rangle/\partial Q_k)|\chi_j(\mathbf{Q})\rangle\\ &\quad+|\Psi_j\rangle(\partial|\chi_j(\mathbf{Q})\rangle/\partial Q_k)\}\\ &= -\sum_k(\hbar^2/2\mu_k)\{(\partial^2|\Psi_j\rangle/\partial Q_k^2)|\chi_j(\mathbf{Q})\rangle\\ &\quad+2(\partial|\Psi_j\rangle/\partial Q_k)(\partial|\chi_j(\mathbf{Q})\rangle/\partial Q_k)\\ &\quad+|\Psi_j\rangle(\partial^2|\chi_j(\mathbf{Q})\rangle/\partial Q_k^2)\},\end{aligned}$$

and so

$$\begin{aligned}&\sum_{j\neq m}\langle\Psi_m|\mathrm{T}_N(\mathbf{Q})|\Psi_j\rangle|\chi_j(\mathbf{Q})\rangle\\ &= -\sum_k(\hbar^2/2\mu_k)\sum_{j\neq m}\{(\langle\Psi_m|(\partial^2/\partial Q_k^2)|\Psi_j\rangle)|\chi_j(\mathbf{Q})\rangle\\ &\quad+2(\langle\Psi_m|\partial/\partial Q_k|\Psi_j\rangle)(\partial|\chi_j(\mathbf{Q})\rangle/\partial Q_k)\}\\ &= \sum_{j\neq m}\Big\{(\langle\Psi_m|\mathbb{T}_N(\mathbf{Q})|\Psi_j\rangle)|\chi_j(\mathbf{Q})\rangle\\ &\quad-\sum_k(\hbar^2/\mu_k)(\langle\Psi_m|\partial/\partial Q_k|\Psi_j\rangle)(\partial|\chi_j(\mathbf{Q})\rangle/\partial Q_k)\Big\},\end{aligned}$$

and Equation (9.9) becomes

$$[\varepsilon_m(\mathbf{Q})+\mathbb{V}_N(\mathbf{Q})-E+\langle\Psi_m|\mathbb{T}_N(\mathbf{Q})|\Psi_m\rangle+\langle\Psi_m|\mathbb{H}_{so}|\Psi_m\rangle]|\chi_m(\mathbf{Q})\rangle$$
$$=-\sum_{j\neq m}\{\langle\Psi_m|\mathbb{T}_N(\mathbf{Q})|\Psi_j\rangle|\chi_j(\mathbf{Q})\rangle-\sum_k(\hbar^2/\mu_k)\langle\Psi_m|\partial/\partial Q_k|\Psi_j\rangle(\partial|\chi_j(\mathbf{Q})\rangle/\partial Q_k)$$
$$+\langle\Psi_m|\mathbb{H}_{so}|\Psi_j\rangle|\chi_j(\mathbf{Q})\rangle\},$$

which simplifies to

$$\begin{aligned}&[\varepsilon_m(\mathbf{Q})+\mathbb{V}_N(\mathbf{Q})-E]|\chi_m(\mathbf{Q})\rangle\\&=-\sum_j\{\langle\Psi_m|\mathbb{T}_N(\mathbf{Q})|\Psi_j\rangle|\chi_j(\mathbf{Q})\rangle-\sum_j\langle\Psi_m|\mathbb{H}_{so}|\Psi_j\rangle|\chi_j(\mathbf{Q})\rangle\\&+\sum_{j\neq m}\sum_k(\hbar^2/\mu_k)\langle\Psi_m|\partial/\partial Q_k|\Psi_j\rangle(\partial|\chi_j(\mathbf{Q})\rangle/\partial Q_k).\end{aligned} \tag{9.10}$$

Now if the right-hand side of Equation (9.10) is negligible, we have just

$$[\varepsilon_m(\mathbf{Q})+\mathbb{V}_N(\mathbf{Q})]|\chi_m(\mathbf{Q})\rangle=E|\chi_m(\mathbf{Q})\rangle, \tag{9.11}$$

which is the usual eigenvalue equation for the Born–Oppenheimer vibrational wavefunctions $|\chi_m(\mathbf{Q})\rangle$. It describes motion of the nuclei on a potential energy surface defined by $\mathbb{V}_m(\mathbf{Q}) = \varepsilon_m(\mathbf{Q}) + \mathbb{V}_N(\mathbf{Q})$, which is different for each electronic state Ψ_m. Remember $\varepsilon_m(\mathbf{Q})$ is the energy of the Born–Oppenheimer electronic state as a function of nuclear coordinates, and $\mathbb{V}_N(\mathbf{Q})$ is the internuclear repulsion. But if the terms on the right-hand side of that equation are not negligible, then this simple picture of nuclear motion on fixed Born–Oppenheimer surfaces breaks down because the three terms on the right-hand side can cause transitions between these electronic basis states.

The one involving $\mathbb{H}_{so}$ is the spin–orbit coupling that induces transitions between electronic states that have different spin multiplicities. This is intersystem crossing. The other two terms involve the first and second derivatives of the electronic states with respect to nuclear coordinates. These are the terms responsible for internal conversion—transitions between electronic states of the same spin multiplicity.

The rate at which some initially prepared Born–Oppenheimer electronic–vibrational state $|\Psi_j(\mathbf{r};\mathbf{Q})\rangle|\chi_j(\mathbf{Q})\rangle$ will make transitions to some other state $|\Psi_m(\mathbf{r};\mathbf{Q})\rangle|\chi_m(\mathbf{Q})\rangle$ through internal conversion is given, if the coupling is weak, by the Fermi golden rule, Eq. (9.2):

$$\text{Rate}=(2\pi/\hbar)|\langle\Psi_m(\mathbf{r};\mathbf{Q})|\langle\chi_m(\mathbf{Q})|\mathbb{H}'|\Psi_j(\mathbf{r};\mathbf{Q})\rangle|\chi_j(\mathbf{Q})\rangle|^2\rho(\varepsilon_j=\varepsilon_m), \tag{9.12}$$

where $\mathbb{H}'$ is the part of the total Hamiltonian that couples different Born-Oppenheimer states. $\mathbb{H}' = \mathbb{T}_N(\mathbf{Q}) + \mathbb{H}_{so}$ according to the development previously, but $\mathbb{H}_{so}$ couples only states of different spin so we neglect it for treating internal conversion. $\mathbb{T}_N(\mathbf{Q})$ is an operator on both the electronic and the nuclear wavefunctions, and as worked out previously,

$$\langle\chi_m|\langle\Psi_m|\mathbb{T}_N|\Psi_j\rangle|\chi_j\rangle$$
$$=\langle\Psi_m|\mathbb{T}_N|\Psi_j\rangle\langle\chi_m|\chi_j\rangle-\sum_k(\hbar^2/\mu_k)\langle\Psi_m|\partial/\partial Q_k|\Psi_j\rangle\langle\chi_m|(\partial|\chi_j\rangle/\partial Q_k). \tag{9.13}$$

So to a good first approximation, the electronic Born–Oppenheimer states that will be most strongly coupled by internal conversion will be those for which the matrix element $\langle\Psi_m|\partial/\partial Q_k|\Psi_j\rangle$ or $\langle\Psi_m|\partial^2/\partial Q_k^2|\Psi_j\rangle$ is largest. [Remember $\mathbb{T}_N(\mathbf{Q})=-\Sigma_k(\hbar^2/2\mu_k)(\partial^2/\partial Q_k^2)$]. Normally the first term on the right-hand side of Equation (9.13) is neglected relative to the second because the second derivative of the electronic state with respect to nuclear coordinates is small compared with the first derivative. So we are left with

$$\langle\chi_m|\langle\Psi_m|\mathbb{T}_N|\Psi_j\rangle|\chi_j\rangle=-\sum_k(\hbar^2/\mu_k)\langle\Psi_m|(\partial/\partial Q_k)|\Psi_j\rangle\langle\chi_m|(\partial/\partial Q_k)|\chi_j\rangle. \tag{9.14}$$

For a polyatomic molecule, the vibrational states $|\chi_j(\mathbf{Q})\rangle$ are functions of all $3N$–6 normal modes, while the derivative acts on only one mode at a time. It is simplest to make the same approximation we did when discussing electronic spectroscopy of polyatomic molecules, which is to assume that the normal modes of the ground state are the same as those of the excited state (no Duschinsky rotation). Then we can write

$$\langle\chi_m|(\partial/\partial Q_k)|\chi_j\rangle=\langle\chi_m|\chi_j\rangle_1\langle\chi_m|\chi_j\rangle_2\cdots\langle\chi_m|(\partial/\partial Q_k)|\chi_j\rangle_k\cdots\langle\chi_m|\chi_j\rangle_{3N\text{-}6},$$

where the subscripts 1, 2,…($3N$–6) designate the normal mode. We can therefore write the matrix element more explicitly as

$$\langle\chi_m|\langle\Psi_m|\mathbb{T}_N|\Psi_j\rangle|\chi_j\rangle$$
$$=-\sum_k\left\{(\hbar^2/\mu_k)\langle\Psi_m|(\partial/\partial Q_k)|\Psi_j\rangle\langle\chi_m|(\partial/\partial Q_k)|\chi_j\rangle_k\prod_{k'}\langle\chi_m|\chi_j\rangle_{k'}\right\}, \tag{9.15}$$

where the product over k' designates all normal modes *except* the one labeled k. Finally, let us look at the specific case where $|\chi_j\rangle$ is the zero-point (ground) vibrational level, $|0_j\rangle$:

$$(\partial/\partial Q_k)|0_j\rangle_k=(\partial/\partial Q_k)\{(\alpha/\pi)^{1/4}\exp(-\alpha Q_k^2/2\mu_k)\}\sim|1_j\rangle_k.$$

So, dropping some constants for simplicity,

$$\langle\chi_m|\langle\Psi_m|\mathbb{T}_N|\Psi_j\rangle|\chi_j\rangle\approx\sum_k\left\{\langle\Psi_m|\partial/\partial Q_k|\Psi_j\rangle\langle\chi_m|1_j\rangle_k\prod_{k'}\langle\chi_m|0_j\rangle_{k'}\right\},$$

and the rate of transitions is proportional to the modulus squared of this quantity, summed over all the states $|\Psi_m\rangle|\chi_m\rangle$ that are isoenergetic with the initial state $|\Psi_j\rangle|0_j\rangle$.

The electronic matrix element clearly has symmetry restrictions. For example, if the two states belong to different symmetry species, then the vibrational normal mode Q_k that couples them must be nontotally symmetric. Often there is one vibrational mode that is much more effective than all others in coupling the two electronic states, and it may reasonably be considered as the only mode k that matters. This is known as a promoting mode. The vibrational part of the matrix element then depends on products of vibrational overlaps, which usually involve large changes in quantum number, since usually internal conversion couples electronic states that have very different zero-point energy levels (e.g., S_1 and S_0 are tens of thousands of cm^{-1} apart). Then there have to be at least some normal mode(s) k' for which the overlap $\langle \chi_m | 0_j \rangle_{k'}$ is significantly different from zero even though $\langle \chi_m |$ is a high v state ($v = 5$ to 10 or more), which can only happen if the two coupled electronic states have very different potential energy surfaces along $Q_{k'}$. Otherwise, the dominant "accepting" modes $Q_{k'}$ will tend to be the ones with the highest frequencies since then fewer vibrational quanta are required to achieve energy matching with the initial level $|0_j\rangle$ (Fig. 9.2). The fact that it is hard to get good vibrational overlaps between states of very different v is what gives rise to the "energy gap law," which says, roughly (there are many versions of different sophistication), that the internal conversion rate goes down as the energy gap between the initial and final electronic states increases—counterintuitively, the rate slows down as the "reaction" becomes more energetically favorable. This also explains why deuteration of a molecule with hydrogen atoms tends to slow down the internal conversion rate. Deuterium substitution lowers the frequencies of all hydrogen-containing modes, and so more quanta are required to make an energy-conserving transition, and these larger-quantum transitions have poorer vibrational overlaps.

The term in the previous development that depends on $\langle \Psi_m | \mathbb{H}_{so} | \Psi_j \rangle$ gives rise to radiationless transitions between states of different spin multiplicity (intersystem crossing). The spin–orbit operator for a polyatomic molecule is (Klessinger)

$$\mathbb{H}_{so} = \frac{e^2}{8\pi\varepsilon_0 c^2 m_e^2} \sum_j \sum_\mu \frac{Z_\mu}{r_{j\mu}^3} \mathbb{L}_j^\mu \cdot \mathbb{S}_j \tag{9.16}$$

where $\mathbb{L}_j^\mu$ is the orbital angular momentum of electron j with respect to nucleus μ, $\mathbb{S}_j$ is the spin angular momentum of electron j, $r_{j\mu}$ is the distance from electron j to nucleus μ, and Z_μ is the atomic number of nucleus μ. The dependence on Z explains why spin–orbit coupling is generally most important for molecules containing heavy atoms (e.g., iodine). The rate of intersystem crossing is described by a Golden Rule expression just like that written for internal conversion previously.

The spin–orbit operator couples electronic states that are related by a simultaneous change of spin angular momentum and orbital angular momentum. This is why, in conjugated organic molecules, spin–orbit coupling is often

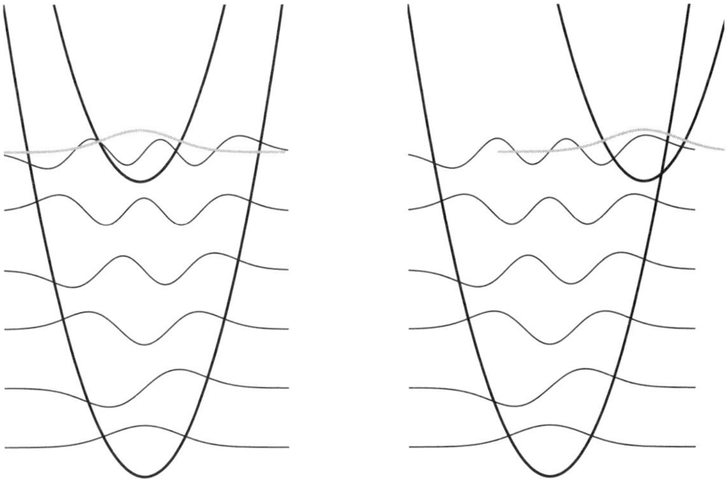

Figure 9.2. Internal conversion from the $v = 0$ level of an excited electronic state (grey curve) to the isoenergetic $v = 5$ level of the ground state. If the two surfaces have little displacement along the vibrational coordinate (left), the two vibrational wavefunctions are nearly orthogonal. Good overlap is obtained only if the two surfaces have a large displacement along the vibrational mode (right).

significant between a $^1(\pi, \pi^*)$ state and a $^3(n, \pi^*)$ state or from $^1(n, \pi^*)$ to $^3(\pi, \pi^*)$, but not from $^1(\pi, \pi^*)$ to $^3(\pi, \pi^*)$ or $^1(n, \pi^*)$ to $^3(n, \pi^*)$. Consider a carbonyl compound like formaldehyde, $H_2C{=}O$. If we define the z-axis to be along the C=O bond and the y-axis to be out of the plane of the four atoms, then the π and π^* orbitals are plus and minus combinations of the p_y atomic orbitals on C and O while the nonbonding orbital is the oxygen p_x orbital. The matrix element $\langle p_y(O)|\Sigma_j \mathbb{L}_j^O|p_x(O)\rangle$ has a nonzero z-component, which can be quite large due to the $|r|^{-3}$ factor since both orbitals are on the same atom. Since the z-component of the matrix element of $\mathbb{S}_j$ is also nonzero when the states differ in their spin multiplicity, the whole matrix element of $\mathbb{L}_j^\mu \cdot \mathbb{S}_j$ has a nonzero z component. On the other hand, since the orbital angular momentum does not change in a $^1(\pi, \pi^*)$ to $^3(\pi, \pi^*)$ or $^1(n, \pi^*)$ to $^3(n, \pi^*)$ transition, the matrix element of $\mathbb{H}_{so}$ should vanish for these transitions.

9.5. INTRAMOLECULAR VIBRATIONAL REDISTRIBUTION

If the Born–Oppenheimer and harmonic approximations were exact, then the vibronic states, labeled by the electronic potential surface to which they belong and by the number of quanta excited in each vibrational normal mode, would be the true eigenstates of the molecular Hamiltonian. If we were to excite a particular vibronic state using light, the only way for population to leave that state would be by emission of radiation to go to some lower state. But the true molecular Hamiltonian contains non-Born–Oppenheimer and anharmonic terms, so these simple vibronic states are not the true molecular eigenstates but merely a useful basis. Therefore, when one such basis state (a "bright state") is excited with light, after a time, there will be some population in other basis states that are approximately isoenergetic with it but are not coupled to the initial state through the electromagnetic field ("dark states"). This process is known as intramolecular vibrational redistribution (IVR).

We will first examine in some detail how this works for just two coupled states (a two-level system), and then extend these ideas to the more common case where there are many coupled levels.

Consider two basis states that are the eigenstates of some zeroth-order Hamiltonian $\mathbb{H}_0$, for example, the Born–Oppenheimer and harmonic molecular Hamiltonian:

$$\mathbb{H}_0|\varphi_1\rangle = E_1|\varphi_1\rangle, \qquad \mathbb{H}_0|\varphi_2\rangle = E_2|\varphi_2\rangle, \qquad \text{and} \qquad \langle\varphi_i|\varphi_j\rangle = \delta_{ij}.$$

Now add a time-independent small perturbation $\mathbb{W}$, such that $\mathbb{H} = \mathbb{H}_0 + \mathbb{W}$. The matrix of $\mathbb{W}$ is

$$\begin{pmatrix} 0 & W \\ W^* & 0 \end{pmatrix};$$

it couples the eigenstates of $\mathbb{H}_0$, but has no diagonal part. This might be a non-Born–Oppenheimer term that couples different electronic states, or an anharmonic term that couples different vibrational states of the same electronic surface. The total Hamiltonian in the $\{|\varphi_1\rangle, |\varphi_2\rangle\}$ basis is therefore

$$\mathbb{H} = \begin{pmatrix} E_1 & W \\ W^* & E_2 \end{pmatrix}, \tag{9.17}$$

and we wish to solve for its eigenvalues and eigenvectors. Diagonalizing this 2×2 Hermitian matrix, assuming $E_1 > E_2$, gives

$$E_\pm = E_\mathrm{m} \pm (\Delta^2 + |W|^2)^{1/2} \text{ for the two eigenvalues} \tag{9.18a}$$

$$|\Psi_+\rangle = \cos(\alpha/2)e^{-i\beta/2}|\varphi_1\rangle + \sin(\alpha/2)e^{i\beta/2}|\varphi_2\rangle \text{ (higher eigenstate)} \tag{9.18b}$$

$$|\Psi_-\rangle = -\sin(\alpha/2)e^{-i\beta/2}|\varphi_1\rangle + \cos(\alpha/2)e^{i\beta/2}|\varphi_2\rangle \text{ (lower eigenstate),} \tag{9.18c}$$

where

$$E_m = (E_1 + E_2)/2 \tag{9.19a}$$

$$\Delta = (E_1 - E_2)/2 \tag{9.19b}$$

$$W = |W|e^{i\beta} \tag{9.19c}$$

$$\tan(\alpha) = |W|/\Delta. \tag{9.19d}$$

First, let us explore the dependence of the energies on Δ (the difference between the energies of the uncoupled basis states) and W (the coupling strength) for a fixed E_m (the average energy of the states). For $|\Delta| >> |W|$, $E_\pm \approx E_m \pm \Delta = E_1$ or E_2. That is, the perturbation has only a small effect when it is small compared with the energy separation. For $|\Delta| << |W|$, $E_\pm \approx E_m \pm |W|$. As Δ approaches zero, E_+ and E_- are degenerate in the absence of the perturbation, but they couple and split apart when W is turned on. The larger Δ is (the farther from degenerate the unperturbed levels are), the less the perturbation matters.

Next, explore how the eigenstates depend on these parameters. For $|\Delta| >> |W|$, $\alpha \approx 0$, $\cos(\alpha/2) \approx 1$, and $\sin(\alpha/2) \approx 0$. Thus,

$$|\Psi_+\rangle \approx e^{-i\beta/2}|\varphi_1\rangle \tag{9.20a}$$

$$|\Psi_-\rangle \approx e^{i\beta/2}|\varphi_2\rangle. \tag{9.20b}$$

The eigenstates are nearly identical to the unperturbed basis states to within an unimportant pure phase factor. Therefore, if only one of the basis states (say φ_1) has a nonzero transition dipole moment from the ground state (φ_1 is bright and φ_2 is dark), then only one allowed transition will be observed, to Ψ_+. In the other extreme, where $|\Delta| << |W|$, $\alpha \approx \pi/2$ and $\cos(\alpha/2) \approx \sin(\alpha/2) \approx 1/\sqrt{2}$. Thus,

$$|\Psi_+\rangle \approx \{e^{-i\beta/2}|\varphi_1\rangle + e^{i\beta/2}|\varphi_2\rangle\}/\sqrt{2} \tag{9.21a}$$

$$|\Psi_-\rangle \approx \{-e^{-i\beta/2}|\varphi_1\rangle + e^{i\beta/2}|\varphi_2\rangle\}/\sqrt{2}. \tag{9.21b}$$

The eigenstates become about equal mixtures of the two basis states. Now, even if only the φ_1 basis state has an allowed transition from the ground state, two transitions of about equal intensity will be observed in the spectrum because both Ψ_+ and Ψ_- contain about equal amounts of the φ_1 basis state. In intermediate cases, there are two transitions, but of unequal intensity.

Finally, look at the dynamics of a coupled two-level system. Specifically, we ask the following question: if the system is prepared in one of the $\mathbb{H}_0$ eigenstates at time $t = 0$, what will be the probabilities of finding the system in each of those two states at a later time? The eigenstates of $\mathbb{H}_0$ are not stationary states since $\mathbb{W}$ couples them.

In the $\{|\varphi_1\rangle, |\varphi_2\rangle\}$ basis, the state vector at time t is, in general,

$$|\Psi(t)\rangle = c_1(t)|\varphi_1\rangle + c_2(t)|\varphi_2\rangle = \langle\varphi_1|\Psi(t)\rangle|\varphi_1\rangle + \langle\varphi_2|\Psi(t)\rangle|\varphi_2\rangle.$$

To find $c_1(t)$ and $c_2(t)$, first write $|\Psi(0)\rangle$ in terms of eigenstates of the total Hamiltonian, $\mathbb{H} = \mathbb{H}_0 + \mathbb{W}$:

$$|\Psi(0)\rangle = \lambda|\Psi_+\rangle + \mu|\Psi_-\rangle.$$

Then the time dependence is straightforward:

$$|\Psi(t)\rangle = \lambda \exp(-iE_+t/\hbar)|\Psi_+\rangle + \mu \exp(-iE_-t/\hbar)|\Psi_-\rangle,$$

and we can get $c_1(t)$ and $c_2(t)$ by projecting $|\Psi(t)\rangle$ onto $\langle\varphi_1|$ and $\langle\varphi_2|$.

Assume we start in $|\Psi(0)\rangle = |\varphi_1\rangle$ (e.g., we prepare an excited electronic state with a short light pulse and φ_1 is a bright state, very close in energy to a dark state φ_2). Solving Equation (9.18) for φ_1 in terms of Ψ_+ and Ψ_- gives

$$|\varphi_1\rangle = e^{i\beta/2}\{\cos(\alpha/2)|\Psi_+\rangle - \sin(\alpha/2)|\Psi_-\rangle\},$$

so

$$|\Psi(t)\rangle = e^{i\beta/2}\{\cos(\alpha/2)\exp(-iE_+t/\hbar)|\Psi_+\rangle - \sin(\alpha/2)\exp(-iE_-t/\hbar)|\Psi_-\rangle\}.$$

The amplitude for finding the system in the other basis state at a later time t is then

$$\begin{aligned}\langle\varphi_2|\Psi(t)\rangle &= c_2(t)\\ &= e^{i\beta/2}\{\cos(\alpha/2)\exp(-iE_+t/\hbar)\langle\varphi_2|\Psi_+\rangle - \sin(\alpha/2)\exp(-iE_-t/\hbar)\langle\varphi_2|\Psi_-\rangle\}\\ &= e^{i\beta}\cos(\alpha/2)\sin(\alpha/2)\{\exp(-iE_+t/\hbar) - \exp(-iE_-t/\hbar)\}.\end{aligned}$$

So the probability of finding the system in $|\varphi_2\rangle$ at time t is

$$\begin{aligned}|\langle\varphi_2|\Psi(t)\rangle|^2 = |c_2(t)|^2 &= \frac{1}{2}\sin^2\alpha\{1-\cos[(\omega_+ - \omega_-)t]\}\\ &= \frac{1}{2}\sin^2\alpha\left\{1-\cos\left(2\sqrt{\Delta^2+|W|^2}\,t/\hbar\right)\right\}.\end{aligned} \tag{9.22}$$

The probability of finding the system in the initially unoccupied state oscillates between zero and a maximum value that depends on the value of α. The oscillation frequency is given by the energy difference between the true molecular eigenstates (the eigenstates of the total Hamiltonian), which depends on both the magnitude of the perturbation and the energy separation between the unperturbed basis states as already discussed. The depth of the modulation also depends on these two factors through α. As the perturbation strength $|W|$ gets larger compared with the energy separation between the states, the amplitude of the oscillation gets larger; for either a very strong perturbation or initially degenerate states, there will be some later times at which *all* the amplitude has been transferred to the other state.

One experiment to which these results are relevant is the observation of "quantum beats" in molecular fluorescence. Consider three states of a molecule, $|g\rangle$, $|b\rangle$, and $|d\rangle$ (standing for "ground," "bright," and "dark," respectively). Assume there is an electric dipole allowed transition between $|g\rangle$ and $|b\rangle$ but not from $|g\rangle$ to $|d\rangle$. However, $|b\rangle$ and $|d\rangle$ are coupled by some other nonradiative term in the total Hamiltonian. For example, $|d\rangle$ might be a triplet state coupled weakly to $|b\rangle$ via spin–orbit coupling.

If a system in its ground state $|g\rangle$ is excited by a very short light pulse arriving at $t = 0$, the amplitudes in states $|b\rangle$ and $|d\rangle$ will be proportional to the matrix elements $\langle b|\mu|g\rangle$ and $\langle d|\mu|g\rangle$ respectively; assuming the latter is zero, all the amplitude at time $t = 0$ is in $|b\rangle$ (actually most of it is still in $|g\rangle$ but we do not have to worry about that part here). But since $|b\rangle$ and $|d\rangle$ are coupled nonradiatively, at a later time the excited part of the system will be in some linear combination of the two states,

$$|\Psi(t)\rangle = c_b(t)|b\rangle + c_d(t)|d\rangle.$$

This excited state can emit light, and the emitted light intensity $I(t)$ will be proportional to the square of the matrix element,

$$|\langle g|\mu|\Psi(t)\rangle|^2 = |c_b(t)\langle g|\mu|b\rangle + c_d(t)\langle g|\mu|d\rangle|^2 = |c_b(t)|^2 |\langle g|\mu|b\rangle|^2.$$

Using Equation (9.22),

$$P_b(t) = 1 - P_d(t) = 1 - \frac{1}{2}\sin^2\alpha\left\{1 - \cos\left(2\sqrt{\Delta^2 + |W|^2}\,t/\hbar\right)\right\}. \tag{9.23}$$

So the emitted light intensity should oscillate with time (Fig. 9.3). From the magnitude and frequency of these oscillations, one can gain information about the strengths of the perturbations and the energies of dark states that are not spectroscopically observable themselves but couple to other states that are.

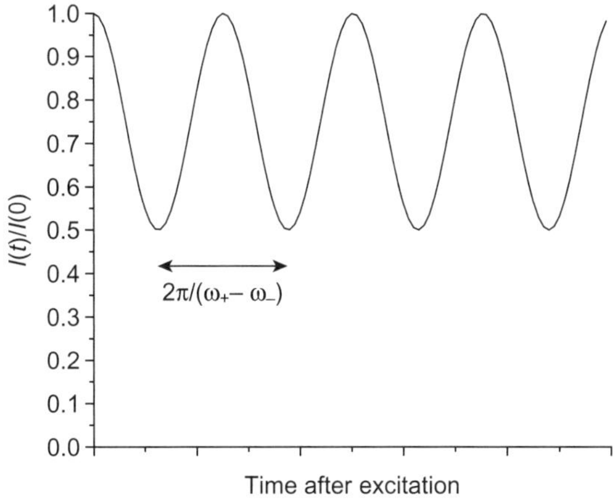

Figure 9.3. Emission intensity as a function of time for a system with two coupled basis states, one bright and one dark [Eq. (9.23)]. Here, the strength of the perturbation coupling the two basis states, W, is equal to half the energy separation between the uncoupled states, Δ.

In small molecules, the densities of states are small enough that often a given bright state is significantly coupled to only one dark state. The steady-state spectra then exhibit well-defined splittings of expected transitions into pairs of lines, and the time-resolved emission shows clear quantum beats. As the molecule gets larger and more dark states are coupled to a given bright state, the spectra become more complicated "forests" of lines and the quantum beats wash out due to the superposition of many different oscillation frequencies. In the large-molecule limit, where every bright state is coupled to essentially a dense continuum of dark states, the spectral "lines" become Lorentzian bands and the emission exhibits an irreversible decay. The coupling to this continuum of dark states has just become a pathway for exponential decay of the initial state, given by Fermi's Golden Rule, Equation (9.2).

Radiative transitions between different vibrational levels of the same electronic state correspond to frequencies in the infrared and are quite slow in spontaneous emission due to the ω^3 dependence of the rate. Vibrational relaxation within a single electronic state is usually dominated by nonradiative processes. In an isolated molecule in the gas phase, the only type of vibrational relaxation that can occur is intramolecular (IVR). The total vibrational energy from the initially prepared zeroth-order harmonic oscillator state is redistributed into other normal modes of the molecule through anharmonic terms in the Hamiltonian. In the gas phase when collisions are possible, vibrational energy can also be lost from the molecule into the other molecules with which it collides. Vibrational relaxation in the gas phase is usually discussed in terms of isolated binary collision models, which assume that the total vibrational relaxation rate is a product of the collision frequency and the probability of transferring energy per collision. The latter depends on the amount of energy to be transferred, the nature of the vibrational mode, the relative kinetic energy of the colliding partners, the relative orientations of the colliding partners, and the details of the intermolecular interactions. Vibrational energy may be transferred into vibrations, rotations, and/or translations of the collision partner, the only absolute restrictions being that total energy and total angular momentum must be conserved.

In condensed phases, intermolecular interactions are sufficiently strong and frequent that the concept of "isolated" binary collisions begins to break down; that is, the "duration" of a single collision becomes comparable with the time between collisions. It becomes more useful to consider a combined molecule-solvent system that can transfer energy from the molecular vibrations to the solvent degrees of freedom through off-diagonal terms in the total system Hamiltonian. To satisfy conservation of energy, if the molecule loses vibrational energy, then the solvent must gain it. The largest matrix elements are generally found for processes that involve the smallest changes in quantum number. Therefore, vibrational relaxation is usually fastest when the solvent has a vibrational mode at the same frequency as the molecule, so the molecule can vibrationally relax by transferring its full one quantum of vibrational energy directly to one vibrational mode of the solvent. The next best situation is a 2:1 resonance, where one quantum of molecular vibration is isoenergetic with two quanta of a solvent mode or vice versa. If there is no exact energy match between the internal vibrations of molecule and solvent, then some of the excess energy has to be taken up by translations and rotations of the solvent, and the matrix elements for these processes tend to be much smaller. This explains why vibrational relaxation tends to be quite slow for molecules in monatomic liquids, such as supercritical Xe; the solvent has no vibrations or rotations, so for the molecule to lose one quantum of vibrational energy, it has to transfer hundreds or thousands of cm^{-1} into translational modes of the solvent, a very unlikely process. Also, vibrational relaxation tends to be fastest for vibrations that have frequencies similar to those of the solvent; for example, C-H stretching or bending vibrations tend to relax rapidly (a few to tens of ps) in solvents containing C-H groups, while C=O stretches, which fall in a rather sparse frequency range (1800–2200 cm^{-1}), may relax quite slowly in hydrocarbon solvents that have no vibrational fundamentals in this frequency region.

9.6. ENERGY TRANSFER

Another way in which molecules can transfer energy is through a process usually known as resonant energy transfer. The most common mechanism, involving nonradiative coupling between the transition dipole moments on two different chromophores, was first worked out by Förster. It is often abbreviated as FRET, where the "F" may stand for either Förster or fluorescence; the actual mechanism of the energy transfer does not involve fluorescence, but the result is often detected as a transfer of fluorescence from one chromophore to another. There is a related mechanism, Dexter energy transfer, which involves concerted exchange of electrons between the two chromophores and requires overlap between their wavefunctions. The rate of the Dexter mechanism falls off much more quickly with interchromophore distance than does the Förster process.

Förster energy transfer requires two molecules, a donor D and an acceptor A, which may or may not be the same chemical species. If D is electronically excited to form an excited state D*, it may transfer its energy to A through the process

$$D^* + A \rightarrow A^* + D.$$

Unlike the situation with vibrational energy relaxation, here D* and A do not have to "collide" or be extremely close to each other. Resonant energy transfer can occur over distances of tens of Å. The net result is the same as if D had emitted a fluorescence photon and A had absorbed it, but no actual emission is involved; it is coupling between the transition dipole moments on A and D.

The formal quantum mechanical description of this process assumes that the rate obeys the Golden Rule formula, Equation (9.2):

$$k_{\text{FRET}} = \frac{2\pi}{\hbar}\left|\langle \text{D}^*\text{A}|\mathbb{H}_{\text{int}}|\text{DA}^*\rangle\right|^2 \rho(E_{\text{D}^*\text{A}} = E_{\text{DA}^*}).$$

The interaction Hamiltonian $\mathbb{H}_{\text{int}}$ is given by the interaction energy between the transition dipoles of the two molecules (Stone, 1996),

$$\mathbb{H}_{\text{int}} = \frac{\boldsymbol{\mu}_{\text{A}} \cdot \boldsymbol{\mu}_{\text{D}}}{4\pi\varepsilon_0 n^2 R_{\text{AD}}^3} - \frac{3(\boldsymbol{\mu}_{\text{A}} \cdot \boldsymbol{R}_{\text{AD}})(\boldsymbol{\mu}_{\text{D}} \cdot \boldsymbol{R}_{\text{AD}})}{4\pi\varepsilon_0 n^2 R_{\text{AD}}^5}. \tag{9.24}$$

Here $\boldsymbol{\mu}_{\text{A}}$ and $\boldsymbol{\mu}_{\text{D}}$ are the transition dipole moment vectors of molecules A and D, and $\mathbf{R}_{\text{AD}}$ is the vector that runs from A to D. In this equation, the solvent is treated as a simple dielectric continuum with refractive index n, n^2 being the relative dielectric constant at optical frequencies; this may not be a good assumption when the distance between donor and acceptor is not large relative to molecular dimensions, or when the intervening medium is a highly heterogeneous material, such as a protein. The dipole-dipole energy depends on both the distance between donor and acceptor and the relative orientations of their dipole moments. If we lump all the orientational dependence into a dimensionless parameter κ, this can be rewritten as

$$\mathbb{H}_{\text{int}} = \kappa\mu_{\text{A}}\mu_{\text{D}}/4\pi\varepsilon_0 n^2 R_{\text{AD}}^3, \tag{9.25}$$

where μ_{A} and μ_{D} now refer only to the magnitudes of the transition dipole moments. We can then write the Golden Rule rate as

$$k_{\text{FRET}} = (\kappa^2/8\pi\varepsilon_0^2\hbar n^4 R_{\text{AD}}^6)\mu_{\text{A}}^2\mu_{\text{D}}^2\rho(E_{\text{D}^*\text{A}} = E_{\text{DA}^*}). \tag{9.26}$$

The most important part of this expression is the dependence on internuclear separation to the inverse sixth power. This means that the rate of resonant energy transfer is a very sensitive function of donor-acceptor distance.

The previous expression assumes that the states D, D*, A, and A* are single eigenstates with well-defined energies. For large molecules at finite temperatures, excited donor molecules and acceptor molecules generally exist as a Boltzmann distribution of many vibrational states, each of which can make transitions to a very large number of other vibrational states that have nonzero Franck–Condon factors. Furthermore, different molecules have different energies due to inhomogeneous broadening. Each particular D*→D transition can only transfer energy to an A → A* transition with which it is isoenergetic, but there are many different energies for which such a match can occur. For useful applications, Equation (9.26) is usually written in the following simplified form,

$$k_{\text{FRET}} = (1/\tau_{\text{obs}})(R_0/R_{\text{DA}})^6, \tag{9.27}$$

where τ_{obs} is the observed lifetime of D* in the absence of A. The quantity R_0 is a distance that depends on the extent of overlap between the absorption spectrum of A and the emission spectrum of D*. The significance of writing the rate constant in this way is that when $R_{DA} = R_0$, then $k_{FRET} = 1/\tau_{obs}$, meaning that the rate constants for energy transfer and for decay of D* by all other pathways are equal. Thus, when $R_{DA} = R_0$, exactly half of the excited donor molecules will transfer their energy to A during their excited-state lifetime.

Resonant energy transfer is widely used in biophysical chemistry for measuring the distance between two groups on, for example, a protein which can be labeled with appropriate dyes to play the roles of donor and acceptor (Stryer, 1978). The rate of energy transfer is usually inferred by measuring the quenching of fluorescence originating from the initially excited D* and/or the appearance of fluorescence from the product A* (Fig. 9.4). For "typical" pairs of D and A, R_0 values range from 10 to 50 Å or so. The parameter κ^2 can be

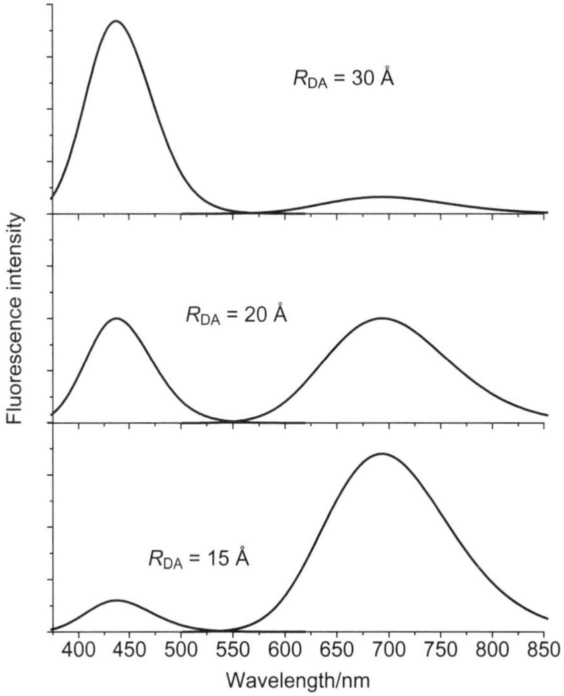

Figure 9.4. Fluorescence spectra of an energy transfer pair as a function of donor-acceptor distance, assuming $R_0 = 20$ Å [see Eq. (9.27)]. As the donor-acceptor distance is reduced, the fluorescence from the originally excited donor (centered near 440 nm) is quenched and the fluorescence from the acceptor (centered near 700 nm) increases. Because of the dependence of the rate on the sixth power of the distance, the amounts of donor and acceptor fluorescence are very sensitive to the distance.

calculated if the orientations of donor and acceptor transition moments are known (its value can range from 0 to 4). When the orientations are not known, or when they are expected to be fluctuating, the orientationally averaged value of 2/3 is often used (Stryer, 1978).

9.7. POLARIZATION AND MOLECULAR REORIENTATION IN SOLUTION

The probability that a molecule absorbs or emits light depends on, among other factors, the dot product of the polarization vector of the light with the transition dipole vector of the molecule. This may not matter when working with isotropic materials, such as ordinary liquids or amorphous or polycrystalline solids, where it is often adequate to replace these dot products by averages over a random distribution as done, for example, to go from Equations (3.48) to (3.51). However, one cannot necessarily neglect the vector properties of the transition dipole moment when describing processes that involve more than one radiation–matter interaction, such as fluorescence induced by optical excitation. If polarized light is used to perform the excitation, then the initially created distribution of excited states is no longer randomly oriented in space—those molecules that have their absorbing dipoles oriented parallel to the polarization of the incident light are more likely to be excited than those that are oriented perpendicular. This is generally known as photoselection (Fig. 9.5), and it can have a number of consequences for any subsequent optical process. Emission experiments often use a polarized laser and detect emission into a particular direction (Fig. 9.6). Even if the polarization of the emission is not deliberately analyzed, because propagation and polarization directions must be orthogonal, there will usually be rotational effects on the emission intensity too.

For an isolated molecule in the gas phase, angular momentum is conserved and rotational motion is quantized. In the solid state, usually the arrangement of neighboring molecules is sufficiently tightly packed and stable that rotation

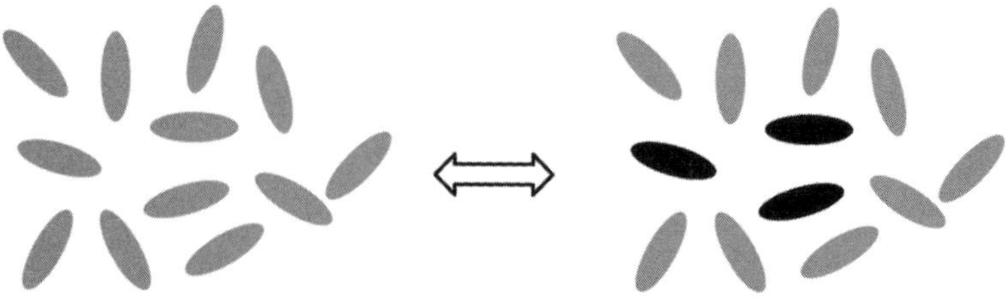

Figure 9.5. Photoselection. Excitation of a randomly oriented ensemble of anisotropic molecules with a horizontally polarized laser preferentially excites those molecules that have their transition dipole moments parallel to the laser polarization.

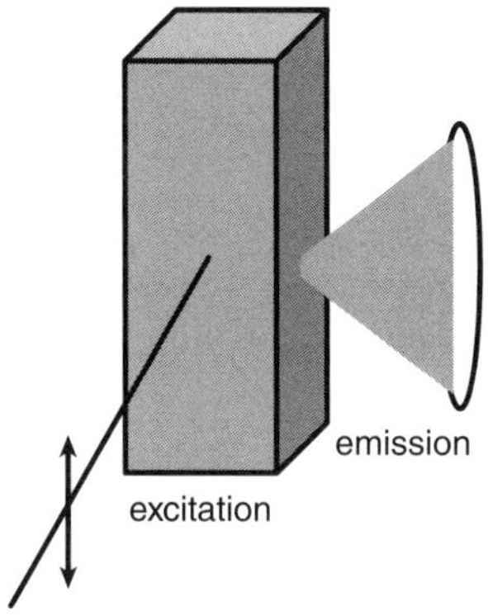

Figure 9.6. A common configuration for a photoexcited emission experiment. The excitation light, often from a laser, enters through the front of a cell, often with a well-defined polarization (shown here as vertical). The emission is collected with a lens system over some solid angle through a perpendicular face of the cell. The emission may or may not be polarization analyzed before it is detected.

through a large angle almost never occurs at normal temperatures, although molecules in solids can "rock" back and forth over a limited angular range. This type of motion is known as libration. The rotational motion of most molecules in liquids is well described as classical rotational diffusion—a "random walk" through angle space. This is an appropriate description when collisions with solvent molecules are sufficiently frequent that the molecule's angular momentum is interrupted and randomized before it has had a chance to rotate through any significant angle.

As shown in Figure 9.6, a common experimental configuration for a fluorescence experiment is one in which the propagation direction of the exciting light, the polarization direction of the exciting light, and the propagation direction of the emitted light are all mutually perpendicular. Let us assume that the exciting light is propagating along **x** and polarized along **z**, and the emitted light is detected propagating along **y**. Since the propagation and polarization directions must be perpendicular, the emitted light may be polarized either along **z** (parallel to the excitation), or along **x** (perpendicular to the excitation). When a linearly polarized electromagnetic field of polarization $\mathbf{e}_L$ interacts with a molecule having a well-defined transition dipole direction $\boldsymbol{\mu}_{abs}$, the matrix element for the interaction is proportional to $\boldsymbol{\mu}_{abs} \cdot \mathbf{e}_L$. The probability that the molecule makes a transition is therefore proportional to $|\boldsymbol{\mu}_{abs} \cdot \mathbf{z}|^2$; that is, the probability of excitation is proportional to $\cos^2\theta$. Therefore, immediately after a very short light pulsc has passed, the sample is no longer isotropic; there are more excited states oriented parallel to the laser polarization than perpendicular to it. As time goes on, the molecules begin to rotate away from their initial orientations and the degree of anisotropy goes down. The emission may arise from the same transition that was excited or a different one, if some relaxation process occurred prior to emission. The probability of emission is proportional to $|\boldsymbol{\mu}_{em} \cdot \mathbf{e}_{em}|^2$, where $\mathbf{e}_{em}$ is the detected emission polarization, either $\mathbf{z}$ or $\mathbf{x}$.

There are, unfortunately, three different conventions in common use to describe the degree of anisotropy of any optical signal that involves two fields. Raman spectroscopists normally use the depolarization ratio ρ, defined as

$$\rho = I_{\perp}/I_{\parallel}, \tag{9.28}$$

where $I_{\perp}$ and $I_{\parallel}$ are the intensities of the scattered light having polarizations perpendicular and parallel, respectively, to the incident light. In fluorescence and phosphorescence spectroscopy, workers more often use either the anisotropy, r, or the polarization, P, defined as

$$r = (I_{\parallel} - \mathrm{I}_{\perp})/(I_{\parallel} + 2\mathrm{I}_{\perp}) \tag{9.29}$$

$$P = (I_{\parallel} - \mathrm{I}_{\perp})/(I_{\parallel} + \mathrm{I}_{\perp}). \tag{9.30}$$

Note that the denominator of r is simply the total intensity emitted into all directions and with all polarizations. For this reason the anisotropy is more simply related to quantities such as the rotational diffusion constant, and is more frequently used.

A very good derivation of the emission anisotropy as a function of time is given by Tao (Tao, 1969). He shows that if the absorbing and emitting transition dipoles are parallel (e.g., they originate from the same molecular transition), then the time-dependent anisotropy as defined previously is given by

$$r(t) = \frac{2}{5} P_2[\hat{\mu}(0) \cdot \hat{\mu}(t)], \tag{9.31}$$

where P_2 is the second Legendre polynomial,

$$P_2(x) = \frac{1}{2}(3x^2 - 1),$$

and $\hat{\mu}(t)$ is a unit vector in the direction of the transition moment for the absorbing transition. The brackets represent an ensemble average over all molecules. In the case where the absorbing and emitting dipoles are not parallel, for example, when a fast internal conversion process occurs between the absorption and emission steps, then

$$r(t) = \frac{2}{5} P_2(\cos\lambda)\langle P_2[\hat{\mu}(0) \cdot \hat{\mu}(t)]\rangle, \tag{9.32}$$

where λ is the angle between the absorbing and emitting dipoles in the molecule-fixed frame. At $t = 0$, $\hat{\mu}(0) \cdot \hat{\mu}(0) = 1$ and the anisotropy has the value

$r(0) = 0.4$ for parallel absorbing and emitting dipoles, and $r(0) = -0.2$ for perpendicular dipoles.

For a spherical molecule undergoing rotational diffusion in a liquid, it can be shown that (Tao, 1969)

$$\langle P_2[\hat{\mu}(0)\cdot\hat{\mu}(t)]\rangle = \exp(-6\mathrm{D_R}t), \tag{9.33}$$

where $\mathrm{D_R}$ is the rotational diffusion constant, given by

$$D_\mathrm{R} = k_\mathrm{B}T/6\eta\mathrm{V}, \tag{9.34}$$

where η is the viscosity of the liquid, V is the volume of the sphere, $\mathrm{k_B}$ is Boltzmann's constant and T is the temperature. The general result for the time-resolved anisotropy in the case of a spherical diffuser thus becomes

$$r(t) = \frac{2}{5}P_2(\cos\lambda)\exp(-6D_\mathrm{R}t). \tag{9.35}$$

Note that the rate of reorientation by a diffusional mechanism depends only on the volume of the rotating body and not explicitly on its mass.

Viscosities are usually expressed in units of centipoise (cp) where $1\,\mathrm{cp} = 10^{-3}\,\mathrm{kg\,m^{-1}\cdot s^{-1}}$. A typical organic liquid has a viscosity of around 1 cp, and a typical medium-sized molecule of 3-Å radius has a volume of $(4\pi/3)\ (3 \times 10^{-10}\,\mathrm{m})^3 = 10^{-28}\,\mathrm{m^3}$, so its diffusion constant near room temperature would be $D_\mathrm{R} = (1.38 \times 10^{-23}\,\mathrm{kg\,m^2\cdot s^{-2}\cdot K^{-1}})(298\,\mathrm{K})/\{6(10^{-3}\,\mathrm{kg\,m^{-1}\cdot s^{-1}})(10^{-28}\,\mathrm{m^3})\} = 7 \times 10^9\,\mathrm{s^{-1}}$, giving a 1/e time for the decay of $\langle P_2[\hat{\mu}(0)\cdot\hat{\mu}(t)]\rangle$ of $1/6D_\mathrm{R} = 2.4 \times 10^{-11}$ seconds or 24 ps. Most molecules are not spherical and their rotational diffusion should more properly be represented by different diffusion coefficients around different axes. Such expressions have been derived for a variety of molecular shapes (Tao, 1969).

It is sometimes important to recognize that in an experimental configuration such as Figure 9.6, the detected emission intensity is influenced by rotational motion even if the polarization of the emitted light is not analyzed. If the light is detected propagating along **y** as shown, it must be polarized in the xz plane. For parallel absorbing and emitting dipoles, at $t = 0$, there is a net orientation of the emission dipoles along z, so the molecule cannot emit light along z. As rotational diffusion occurs, however, that orientation of the transition dipoles is lost, and once the sample has become isotropic, all directions for the emitted light become equally likely. Therefore, even if no population is lost from the excited state, the fluorescence intensity observed along y will decrease with time as rotational diffusion occurs. This can be a problem if we are interested in the excited-state population dynamics and not the rotational motion.

There is a way to get around this problem. Let us consider what happens if the emission is polarized at some arbitrary angle χ relative to the excitation. Let $\mathbf{e}_\mathrm{em} = \cos\chi\mathbf{z} + \sin\chi\mathbf{x}$. Then,

$$|\boldsymbol{\mu}_{em} \cdot \mathbf{e}_{em}|^2 = |\cos\chi(\boldsymbol{\mu}_{em} \cdot \mathbf{z}) + \sin\chi(\boldsymbol{\mu}_{em} \cdot \mathbf{x})|^2$$
$$= \cos^2\chi|\boldsymbol{\mu}_{em} \cdot \mathbf{z}|^2 + \sin^2\chi|\boldsymbol{\mu}_{em} \cdot \mathbf{x}|^2 + 2\cos\chi\sin\chi(\boldsymbol{\mu}_{em} \cdot \mathbf{z})(\boldsymbol{\mu}_{em} \cdot \mathbf{x}).$$

The last term vanishes when the ensemble average is taken since $\boldsymbol{\mu}_{em}$ is equally likely to be oriented along $+z$ and $-z$ and same for x, and so

$$I_\chi(t) = \cos^2\chi I_z(t) + \sin^2\chi I_x(t).$$

Now consider what happens if $\cos^2\chi = 1/3$, which occurs when $\chi = 54.7°$. Then $\sin^2\chi = 2/3$ and $I_\chi(t) = [I_z(t) + 2I_x(t)]/3$. Since $I_z(t) + 2I_x(t)$ is independent of the angular distribution of dipoles as discussed previously, the time dependence of the observed signal has no contribution from rotational motion. This angle is known as the magic angle. Detection with polarization at the magic angle is very useful when one wants to observe other kinetics, such as population decay, without any interference from rotation.

REFERENCES AND FURTHER READING

M. Klessinger and J. Michl, *Excited States and Photochemistry of Organic Molecules* (VCH Publishers, Inc., New York, 1995).

A. J. Stone, *The Theory of Intermolecular Forces* (Clarendon Press, Oxford, 1996).

L. Stryer, Ann. Rev. Biochem. **47**, 819 (1978).

T. Tao, Biopolymers **8**, 609 (1969).

PROBLEMS

1. *Trans*-stilbene is stable in its ground electronic state but is readily converted to *cis*-stilbene by absorbing UV light. The mechanism is thought to involve rotation of the electronically excited molecule about the central C=C bond to form a twisted "perpendicular" state that then undergoes rapid internal conversion to form ground-state trans and ground-state cis with equal probabilities. Prior to twisting, excited *trans*-stilbene can also emit fluorescence. The processes can be summarized as follows:

(1) trans + hν → trans* (light absorption)

(2) trans* → trans + hν′ (fluorescence, rate constant k_f)

(3) trans* → perpendicular* (twisting in excited state, rate constant k_{twist})

(4) perpendicular* → trans (internal conversion back to trans, rate constant k_{ic})

(5) perpendicular* → cis (internal conversion to cis, rate constant k_{ic}).

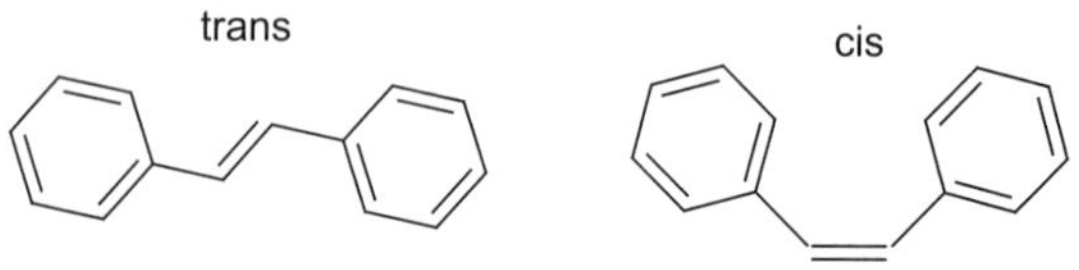

(a) In the solid state, k_{twist} is approximately zero because of steric constraints. What should be the fluorescence quantum yield in the solid state?

(b) The natural radiative lifetime is 4 ns. If the fluorescence quantum yield in hexane solution is 0.1, what is k_{twist}? Under these conditions, what is the quantum yield for formation of the cis isomer?

2. (a) If the vibrations are treated classically, the vibrational density of states as a function of energy in a molecule having n normal modes is

$$\rho(E) = \frac{E^{n-1}}{(n-1)!\prod_i \hbar\omega_i},$$

where the product is over all n normal modes having frequencies ω_i. Calculate the classical vibrational density of states at energy $E = 8000\,\text{cm}^{-1}$ for CO_2 (vibrational frequencies of 1390, 2280, and $667\,\text{cm}^{-1}$, the last one doubly degenerate) and for ethylene (frequencies of 825, 943, 950, 995, 1050, 1342, 1443, 1623, 2990, 3019, 3106, and $3272\,\text{cm}^{-1}$, all nondegenerate). Comment on the comparison between the densities of states of these two molecules.

(b) The previous expression for $\rho(E)$ is accurate only when E is large compared with the zero-point energy E_{ZP}. A more accurate expression for the vibrational density of states is the semiclassical expression

$$\rho(E) = \frac{(E + E_{ZP})^{n-1}}{(n-1)!\prod_i \hbar\omega_i}.$$

Calculate the zero-point energies for CO_2 and for ethylene and discuss whether they are large or small compared with $8000\,\text{cm}^{-1}$. Then use the semiclassical expression to recalculate $\rho(E)$ at $8000\,\text{cm}^{-1}$ for CO_2 and for ethylene.

3. Phenanthrene ($C_{14}H_{10}$) belongs to point group C_{2v}. Its ground electronic state (S_0) and the first excited singlet state (S_1, zero-zero transition at $29{,}330\,\text{cm}^{-1}$) are both of $^1A_1$ symmetry, while the second excited singlet state (S_2, zero-zero at $35{,}390\,\text{cm}^{-1}$) is $^1B_2$.

(a) Consider the internal conversion processes $S_2 \rightarrow S_0$ and $S_2 \rightarrow S_1$. What does the symmetry of a normal mode have to be for it to be a promoting mode for the internal conversion?

(b) Assuming the equilibrium geometries of these three electronic states are not greatly different, which of the previous two internal conversion processes would you expect to be faster, and why?

(c) In view of your answer for (b), what can you say about the probable quantum yield for $S_2 \rightarrow S_0$ fluorescence compared with the yield for $S_1 \rightarrow S_0$ fluorescence?

4. The rate constants for both intersystem crossing and internal conversion depend on the Franck–Condon factors for the accepting modes. In organic molecules with large energy gaps between the coupled electronic states, these accepting modes are usually C-H stretches because these are the highest frequency modes and most able to accept large amounts of energy. The intersystem crossing rates at 77 K for normal and perdeuterated structures (all hydrogens replaced by deuterium) have been measured for a number of aromatic molecules. For naphthalene ($C_{10}H_8$), the rate constant is 0.40 s^{-1} for the all-H species and 0.038 s^{-1} for the all-D. The singlet-triplet zero-zero energy gap is 21,300 cm^{-1}.

(a) An average C-H stretching frequency for an aromatic molecule is 2950 cm^{-1}. If a single C-H stretching mode is the accepting mode and the process begins from the ground vibrational level of the triplet state, to what level would the C-H stretch have to be excited to approximately conserve energy?

(b) Answer the same question for the C-D stretch if the molecule is deuterated. Calculate the isotope effect on the vibrational frequency by considering the vibration to be a diatomic C-H (or C-D) stretch.

(c) Use the expression for the one-dimensional Franck–Condon factors for harmonic modes with equal ground and excited state frequencies,

$$|\langle v|0\rangle|^2 = \exp(-S)S^v/v! \text{ where } S = \Delta^2/2,$$

to calculate the relative rates for a C-H versus a C-D stretch. Recall that

$$\Delta = (\mu\omega/\hbar)^{1/2}(r_e - r_g),$$

where r_e and r_g are the equilibrium bond lengths in the two electronic states. Assume r_e and r_g are the same for the two isotopes. Your result should be a function of $(r_e - r_g)$. If the Franck–Condon factor for the dominant v level were the only factor determining the intersystem crossing rate, what would this equilibrium geometry difference, $r_e - r_g$, have to be to get the observed isotope effect? What other important factor also contributes to the isotope dependence of the intersystem crossing rate?

5. Following $S_0 \rightarrow S_1$ excitation of a particular polar molecule, the S_1 state is known to decay almost exclusively by two mechanisms, fluorescence and internal conversion back to the ground state.

(a) In a time-resolved fluorescence experiment carried out in a weakly polar solvent (tetrahydrofuran, refractive index = 1.41), the $S_1 \rightarrow S_0$ fluorescence is centered at 500 nm and is found to decay exponentially with a time constant of 2.0 ns. The fluorescence quantum yield is also measured and is found to be 0.40. What is the natural radiative lifetime? Use this information to estimate the $S_1 \rightarrow S_0$ transition dipole moment.

(b) The time-resolved fluorescence measurement is now carried out in a strongly polar solvent (acetonitrile, refractive index 1.34), where the fluorescence is centered at 600 nm. The fluorescence decay time is found to be 1.5 ns. Assuming the transition dipole moment does not depend on the solvent, determine the fluorescence quantum yield in acetonitrile.

(c) What do you conclude about the dependence of the internal conversion rate on solvent? Discuss with reference to the energy gap law.

6. For a unimolecular reaction initiated by one-photon absorption, $A + h\nu \rightarrow B$, the rate is given by

$$-d[A]/dt = d[B]/dt = \varphi B \rho_{\text{rad}}[A],$$

where φ is the photochemical quantum yield, B is the Einstein B coefficient for absorption, and ρ_{rad} is the energy density of the radiation. Stilbene (1,2-diphenylethylene) comes in cis and trans isomers that may be interconverted by absorbing light:

$$cis\text{-stilbene} + h\nu \rightarrow trans\text{-stilbene}, \ \varphi = 0.35$$

$$trans\text{-stilbene} + h\nu \rightarrow cis\text{-stilbene}, \ \varphi = 0.45$$

If a sample of stilbenes is exposed to monochromatic radiation of constant intensity, eventually the ratio of cis to trans will reach a constant value (the photostationary steady-state). Calculate this ratio for a wavelength of 300 nm. At this wavelength, the Einstein B coefficient for *trans*-stilbene is four times larger than the Einstein B coefficient for *cis*-stilbene.

7. The dipole–dipole interaction energy, on whose square the Förster energy transfer rate depends, is given by Equation (9.24),

$$\mathbb{H}_{\text{int}} = \frac{\mu_A \cdot \mu_D}{4\pi\varepsilon_0 n^2 R_{AD}^3} - \frac{3(\mu_A \cdot \mathbf{R}_{AD})(\mu_D \cdot \mathbf{R}_{AD})}{4\pi\varepsilon_0 n^2 R_{AD}^5} = \kappa\mu_A\mu_D / 4\pi\varepsilon_0 n^2 R_{AD}^3,$$

where μ_A and μ_D are the transition dipole moments, $\mathbf{R}_{AD}$ is a vector that runs from A to D, and κ contains all the orientation dependence. Calculate the orientation factor, κ, for the following four arrangements of the transition dipoles:

$$\text{(a)} \rightarrow\rightarrow \qquad \text{(b)} \leftarrow\rightarrow \qquad \text{(c)} \uparrow\uparrow \qquad \text{(d)} \uparrow\downarrow$$

8. For molecular reorientation described by isotropic rotational diffusion, the fluorescence anisotropy given by Equations (9.32) and (9.33) is

$$r(t) = \frac{2}{5}\langle P_2[\hat{\mu}(0)\cdot\hat{\mu}(t)]\rangle = \frac{2}{5}\exp(-6D_R t), \qquad (1)$$

when the absorbing and emitting transition dipoles have the same orientation in the molecule-fixed frame. Following z-polarized excitation, the fluorescence intensity as a function of emission polarization is given by

$$I_z(t) = A\left\{\frac{1}{3} + \frac{4}{15}\langle P_2[\hat{\mu}(0)\cdot\hat{\mu}(t)]\rangle\right\}e^{-t/\tau_f} \tag{2}$$

$$I_x(t) = A\left\{\frac{1}{3} - \frac{2}{15}\langle P_2[\hat{\mu}(0)\cdot\hat{\mu}(t)]\rangle\right\}e^{-t/\tau_f}, \tag{3}$$

where τ_f is the lifetime of the probed state and A is an overall constant.

(a) Show that Equations (2) and (3) lead to Equation (1) for $r(t)$ in terms of $P_2[\hat{\mu}(0)\cdot\hat{\mu}(t)]$.

(b) Assume that the detection system is set up such that only one emitted polarization can be detected, and that the experimenter does not know about the benefits of the magic angle. Estimate the apparent lifetimes that would be measured for z- and for x-polarized emission—that is, the time it takes for $I_z(t)$ and $I_x(t)$ to decay to $1/e$ of their initial values—if the excited state lifetime τ_f is 200 ps and the rotational diffusion constant D_R is $0.1\ \text{ns}^{-1}$.

(c) Answer the same questions as (b) for $D_R = 1\ \text{ns}^{-1}$, assuming the same τ_f.

CHAPTER 10

LIGHT SCATTERING

Light scattering is a process in which an electromagnetic field interacts with a sample to produce another electromagnetic field, which may differ from the incident one in its frequency, direction, and/or polarization. To be considered a scattering process rather than absorption followed by emission the material system must evolve coherently in a quantum mechanical sense throughout the process, such that no dephasing process intervenes between destruction of the incident photon and creation of the scattered photon. The distinction is often clear-cut, but gray areas do exist for electronically resonant scattering processes.

10.1. RAYLEIGH SCATTERING FROM PARTICLES

Many aspects of light scattering can be treated using purely classical concepts. We begin our treatment of scattering by considering elastic scattering (no change in frequency between incident and scattered light) from a spherical particle. We will assume that the particle is much smaller than the wavelength of the light but much larger than a single molecule, such that it has a well-defined bulk dielectric constant. Many of the concepts developed in this section are, however, equally applicable to scattering by individual molecules.

First we define several terms. Consider a collimated beam of electromagnetic radiation impinging on a sample (Fig. 10.1). The beam prior to entering

Condensed-Phase Molecular Spectroscopy and Photophysics, First Edition. Anne Myers Kelley.

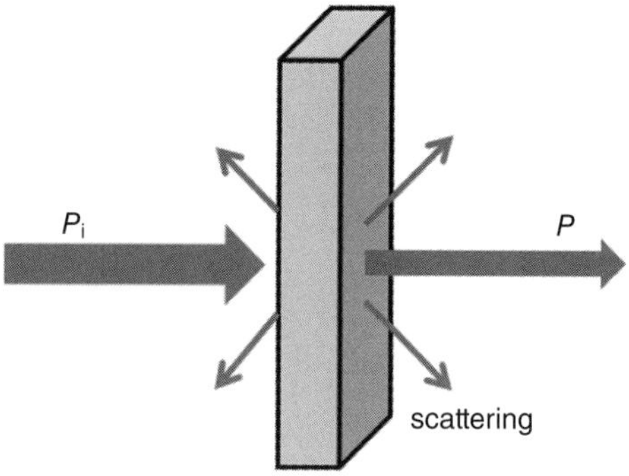

Figure 10.1. An extinction experiment measures the ratio between the power incident on a sample, P_i, and the power transmitted through the sample, P. The difference may be lost through either scattering or absorption of the radiant energy by the sample.

the sample has power P_i (units of energy per unit time, e.g., $J \cdot s^{-1} = W$) and intensity I_i (units of energy per unit time per unit area, e.g., $J \cdot s^{-1} \cdot m^{-2} = W \cdot m^{-2}$). A power detector is placed behind the sample, and it measures the transmitted power P. If $P < P_i$, we say that extinction of the incident beam has occurred. Extinction may be caused either by absorption of some of the light energy by the sample or by scattering of the incident beam into other directions that do not reach the detector. The rate at which energy is removed from the sample, W, can thus be written as $W_{ext} = W_{abs} + W_{sca}$. W has the same units as power, for example, $J \cdot s^{-1}$. If we divide the rate of energy removal by the incident intensity, we obtain $C_{ext} = C_{abs} + C_{sca}$, where $C_{ext} = W_{ext}/I_i$, $C_{abs} = W_{abs}/I_i$, and $C_{sca} = W_{sca}/I_i$. The quantities C have units of area and are referred to as the cross sections for extinction, absorption, and scattering, respectively. Note that the absorption cross section for a molecule was previously defined in Section 4.2. These are useful definitions because the fraction of the light that is removed from the beam by a single particle of extinction cross section C_{ext} is simply the ratio of C_{ext} to the cross-sectional area of the beam. The incident beam "sees" the particle as an opaque disk of area C_{ext}, which does not necessarily have any relationship to the actual physical dimensions of the particle. We can define a cross section for any process that removes energy from an incident beam of light, such as Raman scattering as discussed below.

Mie theory is the general term for the classical description of light absorption and elastic scattering from spherical particles. The term "Mie scattering" is usually applied to the situation where the scattering particles are neither very large nor very small compared with the wavelength of the light. In this case, there is no simple, closed-form solution for the absorption and scattering intensities, although for homogeneous spherical particles, the results can be expressed in terms of an infinite series in the spherical Bessel and spherical Hankel functions (Bohren and Huffman, 2004). Because of the complexity of even this "simple" geometry we will not discuss the solutions here, but will instead proceed directly to the even simpler case where the scattering particle is much smaller than the wavelength of the light. Elastic scattering in this regime is referred to as Rayleigh scattering.

We assume a homogeneous, spherical particle having radius a and complex dielectric function $\varepsilon_1(\omega)$ (Chapter 2). The particle is immersed in a medium having dielectric function $\varepsilon(\omega)$. A collimated electromagnetic wave of frequency ω_0 (vacuum wavelength λ) is assumed to be incident on the sample, and Maxwell's equations are solved both inside and outside the particle with appropriate boundary conditions to find the absorption and scattering cross sections. In the limit where the particle is very small compared with the wavelength, it is a good approximation to replace the sphere by an ideal dipole, and the following simple result is obtained for the scattering cross section (Bohren and Huffman, 2004):

$$C_{\text{sca}}(\omega_0) = \frac{8\pi k^4}{3} |\alpha_v(\omega_0)|^2, \tag{10.1}$$

where k is the wavevector of the light in the surrounding medium ($k = 2\pi n/\lambda$, where n is the real refractive index of the medium) and α_v is the polarizability volume of the sphere, defined as

$$\alpha_v(\omega) = a^3 \frac{\varepsilon_1(\omega) - \varepsilon(\omega)}{\varepsilon_1(\omega) + 2\varepsilon(\omega)}. \tag{10.2}$$

$\alpha_v(\omega)$, which has units of volume, is often referred to as simply the polarizability, but we reserve that term for the quantity defined in Equation (10.3) below.

The scattering cross section scales as the sixth power of the particle radius. As long as the dielectric functions do not have strong frequency dependences, the scattering cross section also scales as the inverse fourth power of the wavelength. Thus, the Rayleigh scattering intensity scales as the fourth power of the frequency when the frequency is far from any resonances.

10.2. CLASSICAL TREATMENT OF MOLECULAR RAMAN AND RAYLEIGH SCATTERING

We now turn to elastic and inelastic scattering from molecules. Although a complete description of light scattering by molecules requires a quantum mechanical treatment, certain aspects are more easily seen through a classical picture.

When a molecule with polarizability α (for now we neglect the frequency dependence and the tensor aspects of α) is subjected to a time-varying electric field $E(t)$, the induced time-dependent dipole moment $\mu(t)$ is given by

$$\mu(t) = \alpha E(t). \tag{10.3a}$$

The polarizability α is related to the polarizability volume α_v of Eq. (10.2) by

$$\alpha(\omega) = 4\pi\varepsilon(\omega)\alpha_v(\omega), \tag{10.3b}$$

where the frequency dependences are made explicit, and $\varepsilon(\omega)$ is the dielectric function of the medium. Since $\varepsilon(\omega)$ has the same units as ε_0, the units of α are $C^2 \cdot J^{-1} \cdot m^2$, and the units of μ are C m as correct for a dipole moment.

The molecular polarizability varies as the molecule vibrates. Considering for simplicity a single vibrational coordinate and only small amplitude vibrations, we can write $\alpha = \alpha_0 + (\partial\alpha/\partial Q)_0 Q$, where Q is the value of the vibrational coordinate. For a classical harmonic oscillator, $Q = Q_0\cos(\omega_v t + \delta)$, where Q_0 is the vibrational amplitude, ω_v the vibrational frequency, and δ an arbitrary phase of the vibration. If the incident electric field is monochromatic with frequency ω, $E(t) = E_0\cos(\omega t)$, the induced dipole moment is

$$\mu(t) = \mathrm{E}_0 \cos(\omega \mathrm{t})\{\alpha_0 + \alpha' \mathrm{Q}_0 \cos(\omega_v \mathrm{t} + \delta)\} \tag{10.4}$$

where $\alpha' = (\partial\alpha/\partial \mathrm{Q})_0$. Then using the trigonometric identity cosAcosB = ½{cos (A + B) + cos(A − B)}, we obtain

$$\begin{aligned}\mu(t) &= E_0\alpha_0 \cos(\omega t) + \left(\frac{E_0}{2}\right)\alpha' Q_0 \{\cos[(\omega+\omega_v)t+\delta] + \cos[(\omega-\omega_v)t-\delta]\} \\ &= E_0\alpha_0 \cos(\omega t) + E_0\alpha_{\mathrm{Ram}} \{\cos(\Omega_{AS}t+\delta) + \cos(\Omega_S t - \delta)\},\end{aligned} \tag{10.5}$$

where we define the Raman polarizability as

$$\alpha_{\mathrm{Ram}} = \frac{1}{2}\alpha' Q_0.$$

This induced dipole has three frequency components and will reradiate electromagnetic radiation at each frequency. The first term corresponds to scattered radiation that has the same frequency and phase as the incident light; this is referred to as elastic or Rayleigh scattering. The second and third terms correspond to radiation that is shifted up or down in frequency, respectively, by the molecular vibrational frequency. These correspond to anti-Stokes ($\Omega_{AS} = \omega + \omega_v$) and Stokes ($\Omega_S = \omega - \omega_v$) Raman scattering, respectively.

If there is more than one vibrational mode, the expansion of α as a Taylor series in the vibrational coordinates has more than one term, and Raman scattering can occur at many different frequencies. Plotting the scattered intensity as a function of Raman shift, $\omega - \Omega_{scatt}$, gives a series of peaks at positive (Stokes) and negative (anti-Stokes) frequencies corresponding to the various vibrational frequencies of the molecule. Note, however, that a frequency will appear in the Raman spectrum only if $(\partial\alpha/\partial Q)_0 \neq 0$. That is, there must be a change in the molecular polarizability along the vibrational coordinate for that vibrational mode to be Raman active.

In Stokes scattering, the radiation field transfers energy to the molecule, the molecule ending up in an excited vibrational state. In anti-Stokes scattering, the molecule transfers vibrational energy to the radiation field. The classical picture predicts that the anti-Stokes and Stokes scattering in the same vibrational mode should have equal intensities. In reality, the Stokes scattering is usually much stronger. We know from quantum mechanics that only certain vibrational energy levels are allowed, and near room temperature, most vibrational modes of most molecules are in the $v = 0$ state. These molecules do not have any vibrational energy to give up so they cannot generate anti-Stokes Raman scattering, only Stokes.

There is one other important difference between Raman and Rayleigh processes when ensembles of scatterers are involved. When we carry out an experiment on a collection of molecules, we need to consider how the contributions to the signal from different molecules add up—that is, whether the electric fields generated by different scatterers exhibit constructive or destructive interference. Note that Rayleigh scattering has a well-defined phase relationship to the incident radiation. If the positions of different scatterers are perfectly correlated, as in a perfect crystal at zero temperature, the scattered intensity over a sufficiently large volume will average to zero. Only if there are fluctuations in the positions of the scatterers will this exact cancellation not occur. Raman scattering, in contrast, does not have a well-defined phase relative to that of the driving radiation because the phase of the vibration differs randomly among different scatterers. Thus, whether or not the positions of the different scatterers are well ordered, there is no macroscopic coherence to spontaneous Raman scattering, and the scattered power increases linearly with the number of scatterers.

10.3. QUANTUM MECHANICAL TREATMENT OF MOLECULAR RAMAN AND RAYLEIGH SCATTERING

In a fully quantum mechanical description, spontaneous light scattering (Raman or Rayleigh) is a two-photon process in which the radiation field loses a photon from one mode while gaining a photon in another, previously unoccupied mode. Recall that the radiation–matter interaction produces a time-dependent perturbation term in the Hamiltonian given by Equation (3.34),

$$\mathbb{W}(\mathbf{r},\mathbf{t}) = -(e/m)\mathbb{A}\cdot\mathbb{p} + (e^2/2m)\mathbb{A}\cdot\mathbb{A}.$$

For spontaneous light scattering, the matter makes a transition from $|i\rangle$ to $|f\rangle$ while the radiation goes from $|n_{k1\alpha1}0_{k2\alpha2}\rangle$ to $|(n-1)_{k1\alpha1}1_{k2\alpha2}\rangle$ (shorthand: $|n_1\, 0_2\rangle \rightarrow |(n-1)_1\, 1_2\rangle$). Remember that the $\mathbb{A}\cdot\mathbb{p}$ term in the interaction Hamiltonian, to first order, can only change the total number of photons by one, since $\mathbb{A}$ is just a sum of creation and annihilation operators. Thus, to describe a two-photon scattering event, we must consider either the $\mathbb{A}\cdot\mathbb{A}$ term to first order or the $\mathbb{A}\cdot\mathbb{p}$ term to second order.

First, examine the $\mathbb{A}\cdot\mathbb{A}$ term. The relevant matrix element is

$$\langle f;(n-1)_1 1_2|\mathbb{W}(t)|i;n_1 0_2\rangle = (\mathrm{e}^2/2m)\langle f;(n-1)_1 1_2|\mathbb{A}\cdot\mathbb{A}|i;n_1 0_2\rangle,$$

where, referring to Eq. (2.24) and making the electric dipole approximation,

$$\mathbb{A}\cdot\mathbb{A} = \frac{\hbar}{2\varepsilon V}\sum_{p,q}\frac{1}{\sqrt{\omega_p\omega_q}}\{\mathbb{a}_p e^{-i\omega_p t} + \mathbb{a}_p^\dagger e^{i\omega_p t}\}\{\mathbb{a}_q e^{-i\omega_q t} + \mathbb{a}_q^\dagger e^{i\omega_q t}\}\boldsymbol{e}_p\cdot\boldsymbol{e}_q. \quad (10.6)$$

Here, we use the indices p and q to denote both the frequency and the polarization of a particular mode of the field. In order to get a nonzero matrix element, the operator must annihilate a photon in mode 1 and create a photon in mode 2, because states with different numbers of photons in any mode are orthogonal. Thus, we must have a product of $\mathbb{a}_1$ (the annihilation operator for mode 1) and $\mathbb{a}_2^\dagger$ (the creation operator for mode 2), and only two of the terms in $\mathbb{A}\cdot\mathbb{A}$ contain this. Furthermore, these two terms are identical because $\mathbb{a}_1$ and $\mathbb{a}_2^\dagger$ commute (they refer to different radiation modes). So the matrix element becomes

$$\langle f;(n-1)_1 1_2|\mathbb{W}(t)|i;n_1 0_2\rangle$$

$$= \frac{\hbar e^2}{2\varepsilon m V\sqrt{\omega_1\omega_2}}\langle f;(n-1)_1 1_2|\mathbb{a}_1\mathbb{a}_2^\dagger \exp[i(\omega_2-\omega_1)t]|i;n_1 0_2\rangle \boldsymbol{e}_1\cdot\boldsymbol{e}_2 \quad (10.7a)$$

$$= \frac{\hbar e^2}{2\varepsilon m V\sqrt{\omega_1\omega_2}}\sqrt{n_1}\exp[i(\omega_2-\omega_1)t]\langle f|i\rangle \boldsymbol{e}_1\cdot\boldsymbol{e}_2 \quad (10.7b)$$

$$= \frac{\hbar e^2}{2\varepsilon m V\sqrt{\omega_1\omega_2}}\sqrt{n_1}\exp[i(\omega_2-\omega_1)t]\boldsymbol{e}_1\cdot\boldsymbol{e}_2\delta_{\mathrm{fi}}. \quad (10.7c)$$

The initial and final states of the matter must be the same, because $\mathbb{A}\cdot\mathbb{A}$ evaluated in the dipole approximation contains no operators on the states of matter. It can change the radiation only. Therefore this term can contribute to Rayleigh scattering, but not to Raman scattering.

We can also get a net change of two photons by letting $\mathbb{A}\cdot\mathbb{p}$ act twice. This requires going to second order in time-dependent perturbation theory. This was not discussed in Chapter 3, so it will be developed below. A net change of two photons can occur through either one of two distinct pathways:

(1) Annihilation of a mode 1 photon followed by creation of a mode 2 photon:

$$|i;n_1 0_2\rangle \to |v;(n-1)_1 0_2\rangle \to |f;(n-1)_1 1_2\rangle.$$

(2) Creation of a mode 2 photon followed by annihilation of a mode 1 photon:

$$|i;n_1 0_2\rangle \rightarrow |w;n_1 1_2\rangle \rightarrow |f;(n-1)_1 1_2\rangle.$$

Here $|v\rangle$ and $|w\rangle$ refer to otherwise unspecified intermediate states of the matter. In Chapter 3, we calculated the matrix elements of the $\mathbb{A}\cdot\mathbb{p}$ term for creation and for annihilation of one photon, Equations (3.45) and (3.52). The ones needed are

$$\langle v;(n-1)_1 0_2|\mathbb{W}(t)|i;n_1 0_2\rangle = -(e/m)\langle v;(n-1)_1 0_2|\mathbb{A}(t)\cdot\mathbb{p}|i;n_1 0_2\rangle$$
$$= -ie\omega_{\text{vi}}(\hbar n_1/2\omega_1\varepsilon V)^{1/2}\mathbf{e}_1\cdot\langle v|\mathbb{r}|i\rangle\exp(-i\omega_1 t) \tag{10.8a}$$

$$\langle w;n_1 1_2|\mathbb{W}(t)|i;n_1 0_2\rangle = -ie\omega_{\text{wi}}(\hbar/2\omega_2\varepsilon V)^{1/2}\mathbf{e}_2\cdot\langle w|\mathbb{r}|i\rangle\exp(i\omega_2 t) \tag{10.8b}$$

$$\langle f;(n-1)_1 1_2|\mathbb{W}(t)|v;(n-1)_1 0_2\rangle = -ie\omega_{\text{fv}}(\hbar/2\omega_2\varepsilon V)^{1/2}\mathbf{e}_2\cdot\langle f|\mathbb{r}|v\rangle\exp(i\omega_2 t) \tag{10.8c}$$

$$\langle f;(n-1)_1 1_2|\mathbb{W}(t)|w;n_1 1_2\rangle = -ie\omega_{\text{fw}}(\hbar n_1/2\omega_1\varepsilon V)^{1/2}\mathbf{e}_1\cdot\langle f|\mathbb{r}|w\rangle\exp(-i\omega_1 t). \tag{10.8d}$$

In these equations, V is the volume of the box in which the field is quantized, while $|v\rangle$ is a state of the matter. Also note that we have retained the time-dependent part of $\mathbb{A}(t)$ in these equations while Equations (3.45) and (3.52) gave only the time-independent parts of the matrix elements.

With these matrix elements we can derive the second order amplitude for the system to be in the state $|f;\,(n-1)_1 1_2\rangle$ at time t assuming it started in $|i;\,n_1 0_2\rangle$ at time zero.

Starting from $|i\rangle$ (for brevity, we will label the states only by the state of *matter*), we can go in first order either to $|v\rangle$ by losing a photon or to $|w\rangle$ by gaining one. The first-order coefficients are found in the usual way, through Equation (3.17):

$$b_{\text{v}}^{(1)}(t) = \frac{1}{i\hbar}\int_0^t d\tau \exp(i\omega_{\text{vi}}\tau)W_{\text{vi}}(\tau),$$

and inserting the matrix element from Equation (10.8a) gives

$$\begin{aligned} b_{\text{v}}^{(1)}(t) &= \frac{1}{i\hbar}\int_0^t d\tau \exp(i\omega_{\text{vi}}\tau)W_{\text{vi}}(\tau) \\ &= e\omega_{\text{vi}}\sqrt{\frac{n_1}{2\hbar\omega_1\varepsilon V}}\boldsymbol{e}_1\cdot\mathbf{M}_{\text{vi}}\int_0^t d\tau\exp[i(\omega_{\text{vi}}-\omega_1)\tau] \\ &= e\omega_{\text{vi}}\sqrt{\frac{n_1}{2\hbar\omega_1\varepsilon V}}\boldsymbol{e}_1\cdot\mathbf{M}_{\text{vi}}\frac{1-\exp[i(\omega_{\text{vi}}-\omega_1)t]}{i(\omega_{\text{vi}}-\omega_1)}, \end{aligned} \tag{10.9}$$

where Equation (10.8a) is used to go from the first to the second line, and $\mathbf{M}_{vi} = \langle v|\mathbf{r}|i\rangle$, the transition length between matter states i and v. Similarly,

$$b_{w}^{(1)}(t) = e\omega_{wi}\sqrt{\frac{1}{2\hbar\omega_2 \varepsilon V}}\boldsymbol{e}_2 \cdot \mathbf{M}_{wi}\frac{1-\exp[i(\omega_{wi}+\omega_2)t]}{i(\omega_{wi}+\omega_2)}. \tag{10.10}$$

Now we need the coefficient of state $|f\rangle$ by using second-order perturbation theory. Plugging in the first-order coefficients to the right-hand side of the general differential equation for the coefficients, Equation (3.15), and integrating gives

$$\begin{aligned}
b_f^{(2)}(t) &= -\frac{i}{\hbar}\int_0^t d\tau\left\{\sum_v b_v^{(1)}(\tau)\exp(i\omega_{fv}\tau)W_{fv}(\tau)+\sum_w b_w^{(1)}(\tau)\exp(i\omega_{fw}\tau)W_{fw}(\tau)\right\}\\
&= -\sum_v\left\{\frac{e^2\omega_{fv}\omega_{vi}\boldsymbol{e}_1\cdot\mathbf{M}_{vi}\ \boldsymbol{e}_2\cdot\mathbf{M}_{fv}\sqrt{n_1}}{2\hbar V\varepsilon\sqrt{\omega_1\omega_2}}\frac{1}{i(\omega_{vi}-\omega_1)}\int_0^t d\tau\left(\exp[i(\omega_{fv}+\omega_2)\tau]\right.\right.\\
&\quad\left.\left. -\exp[i(\omega_{fv}+\omega_2+\omega_{vi}-\omega_1)\tau]\right)\right\}\\
&\quad -\sum_w\left\{\frac{e^2\omega_{fw}\omega_{wi}\boldsymbol{e}_2\cdot\mathbf{M}_{wi}\ \boldsymbol{e}_1\cdot\mathbf{M}_{fw}\sqrt{n_1}}{2\hbar V\varepsilon\sqrt{\omega_1\omega_2}}\frac{1}{i(\omega_{wi}+\omega_2)}\right.\\
&\quad\left.\cdot\int_0^t d\tau(\exp[i(\omega_{fw}-\omega_1)\tau]-\exp[i(\omega_{fw}-\omega_1+\omega_{wi}+\omega_2)\tau])\right\}\\
&= \frac{-e^2\sqrt{n_1}}{2\hbar\varepsilon V\sqrt{\omega_1\omega_2}}\left\{\sum_v\frac{\omega_{fv}\omega_{vi}\boldsymbol{e}_1\cdot\mathbf{M}_{vi}\ \boldsymbol{e}_2\cdot\mathbf{M}_{fv}}{i(\omega_{vi}-\omega_1)}\right.\\
&\quad\cdot\left[\frac{1}{i(\omega_{fv}+\omega_2)}(1-\exp[i(\omega_{fv}+\omega_2)t])\right.\\
&\quad\left.-\frac{1}{i(\omega_{fi}+\omega_2-\omega_1)}(1-\exp[i(\omega_{fi}+\omega_2-\omega_1)t])\right]\\
&\quad+\sum_w\frac{\omega_{fw}\omega_{wi}\boldsymbol{e}_2\cdot\mathbf{M}_{wi}\ \boldsymbol{e}_1\cdot\mathbf{M}_{fw}}{i(\omega_{wi}+\omega_2)}\left[\frac{1}{i(\omega_{fw}-\omega_1)}\right.\\
&\quad\left.\left.\cdot(1-\exp[i(\omega_{fw}-\omega_1)t])-\frac{1}{i(\omega_{fi}+\omega_2-\omega_1)}(1-\exp[i(\omega_{fi}+\omega_2-\omega_1)t])\right]\right\}.
\end{aligned} \tag{10.11}$$

The algebra uses the fact that $\omega_{fv} + \omega_{vi} = \omega_f - \omega_v + \omega_v - \omega_i = \omega_{fi}$. There are four terms. The first and third (denominators $\omega_{fv} + \omega_2$ and $\omega_{fw} - \omega_1$) arise from the artificial sudden turning on of the perturbation at $t = 0$; for realistic perturbations that are turned on more slowly, these terms become negligible. Keeping only the other two terms, and collapsing the sums over v and w into a sum over a single index we will call v (here $|v\rangle$ can be any state of matter), leaves

$$b_{\mathrm{f}}^{(2)}(t)=\frac{-e^2\sqrt{n_1}}{2\hbar\varepsilon V(\omega_{\mathrm{fi}}+\omega_2-\omega_1)\sqrt{\omega_1\omega_2}}\{1-\exp[i(\omega_{\mathrm{fi}}+\omega_2-\omega_1)t]\}$$
$$\cdot\sum_{\mathrm{v}}\left\{\frac{\omega_{\mathrm{fv}}\omega_{\mathrm{vi}}\boldsymbol{e}_1\cdot\mathbf{M}_{\mathrm{vi}}\ \boldsymbol{e}_2\cdot\mathbf{M}_{\mathrm{fv}}}{(\omega_{\mathrm{vi}}-\omega_1)}+\frac{\omega_{\mathrm{fv}}\omega_{\mathrm{vi}}\boldsymbol{e}_2\cdot\mathbf{M}_{\mathrm{vi}}\ \boldsymbol{e}_1\cdot\mathbf{M}_{\mathrm{fv}}}{(\omega_{vi}+\omega_2)}\right\}. \tag{10.12}$$

Remember that we got this result by using the $\mathbb{A}\cdot\mathbb{p}$ term in the radiation–matter interaction to second order. This means this contribution to the coefficient depends on the second power of $\mathbb{A}$. The $\mathbb{A}\cdot\mathbb{A}$ term to *first* order that we examined previously also depends on the second power of $\mathbb{A}$, so to be consistent (and to make use of a mathematical trick below), we should also include this term in calculating the second order coefficient of state $|f\rangle$. This contribution is

$$b_{\mathrm{f}}^{(1)}(t)=\frac{1}{i\hbar}\int_0^t d\tau\exp(i\omega_{\mathrm{fi}}\tau)W_{\mathrm{fi}}(\tau),$$

and we evaluated the matrix element earlier; plugging it in gives

$$\begin{aligned}b_{\mathrm{f}}^{(1)}(t)&=-\frac{e^2\sqrt{n_1}\boldsymbol{e}_1\cdot\boldsymbol{e}_2\delta_{\mathrm{fi}}}{2\varepsilon mV\sqrt{\omega_1\omega_2}}\int_0^t \mathrm{d}\tau\exp[i(\omega_{\mathrm{fi}}+\omega_2-\omega_1)\tau]\\&=-\frac{e^2\sqrt{n_1}\boldsymbol{e}_1\cdot\boldsymbol{e}_2\delta_{\mathrm{fi}}}{2\varepsilon mV\sqrt{\omega_1\omega_2}}\frac{1-\exp[i(\omega_{\mathrm{fi}}+\omega_2-\omega_1)t]}{\omega_{\mathrm{fi}}+\omega_2-\omega_1}.\end{aligned} \tag{10.13}$$

The total coefficient of $b_{\mathrm{f}}(t)$ is the sum of the first- and second-order contributions:

$$b_{\mathrm{f}}^{(2)}(t)=\frac{-e^2\sqrt{n_1}}{2\hbar\varepsilon V\sqrt{\omega_1\omega_2}}\frac{1-\exp[i(\omega_{\mathrm{fi}}+\omega_2-\omega_1)t]}{(\omega_{\mathrm{fi}}+\omega_2-\omega_1)}\left(\frac{\hbar}{m}\boldsymbol{e}_1\cdot\boldsymbol{e}_2\delta_{\mathrm{fi}}\right.$$
$$\left.+\sum_{\mathrm{v}}\left\{\frac{\omega_{\mathrm{fv}}\omega_{\mathrm{vi}}\boldsymbol{e}_1\cdot\mathbf{M}_{\mathrm{vi}}\ \boldsymbol{e}_2\cdot\mathbf{M}_{\mathrm{fv}}}{(\omega_{\mathrm{vi}}-\omega_1)}+\frac{\omega_{\mathrm{fv}}\omega_{\mathrm{vi}}\boldsymbol{e}_2\cdot\mathbf{M}_{\mathrm{vi}}\ \boldsymbol{e}_1\cdot\mathbf{M}_{\mathrm{fv}}}{(\omega_{\mathrm{vi}}+\omega_2)}\right\}\right). \tag{10.14}$$

We find the transition rate in the long time limit as before by taking the limit as $\mathrm{t}\to\infty$ of $|b_{\mathrm{f}}(t)|^2/t$. Since

$$\frac{1-e^{ixt}}{x}=\frac{e^{ixt/2}\left(e^{-ixt/2}-e^{ixt/2}\right)}{x}=\frac{-2i\,e^{ixt/2}\sin(xt/2)}{x},$$

taking the limit of the time-dependent part of Equation (10.14) gives

$$\begin{aligned}\lim_{t\to\infty}\left|\frac{1-\exp[i(\omega_{\mathrm{fi}}+\omega_2-\omega_1)t]}{\omega_{\mathrm{fi}}+\omega_2-\omega_1}\right|^2&=\lim_{t\to\infty}\frac{\sin^2[(\omega_{\mathrm{fi}}+\omega_2-\omega_1)t/2]}{[(\omega_{\mathrm{fi}}+\omega_2-\omega_1)/2]^2}\\&=2\pi t\delta(\omega_{\mathrm{fi}}+\omega_2-\omega_1),\end{aligned}$$

where the result that led to Equation (3.23) is used. The long-time limit of $|b_f(t)|^2 / t$ thus becomes

$$R_{if}(\omega_1, \omega_2) = \frac{\pi e^4 n_1}{2\varepsilon^2 \hbar^2 V^2 \omega_1 \omega_2} \left| \frac{\hbar}{m} \boldsymbol{e}_1 \cdot \boldsymbol{e}_2 \delta_{fi} + \sum_v \left\{ \frac{\omega_{fv}\omega_{vi} \boldsymbol{e}_1 \cdot \mathbf{M}_{vi}\ \boldsymbol{e}_2 \cdot \mathbf{M}_{fv}}{(\omega_{vi} - \omega_1)} + \frac{\omega_{fv}\omega_{vi} \boldsymbol{e}_2 \cdot \mathbf{M}_{vi}\ \boldsymbol{e}_1 \cdot \mathbf{M}_{fv}}{(\omega_{vi} + \omega_2)} \right\} \right|^2 \cdot \delta(\omega_{fi} + \omega_2 - \omega_1). \tag{10.15}$$

To continue, we make use of some algebraic manipulations to combine the $\mathbb{A}\cdot\mathbb{A}$ and $\mathbb{A}\cdot\mathbb{p}$ terms. First, note that

$$\frac{1}{m}(\mathbb{r}\cdot\boldsymbol{e}_2 \mathbb{p}\cdot\boldsymbol{e}_1 - \mathbb{p}\cdot\boldsymbol{e}_1 \mathbb{r}\cdot\boldsymbol{e}_2) = \frac{1}{m}[\mathbb{r}\cdot\boldsymbol{e}_2, \mathbb{p}\cdot\boldsymbol{e}_1] = \frac{i\hbar}{m}\boldsymbol{e}_1\cdot\boldsymbol{e}_2.$$

Therefore,

$$\frac{1}{m}\langle \mathrm{f}|\mathbb{r}\cdot\boldsymbol{e}_2 \mathbb{p}\cdot\boldsymbol{e}_1 - \mathbb{p}\cdot\boldsymbol{e}_1 \mathbb{r}\cdot\boldsymbol{e}_2|\mathrm{i}\rangle = \frac{i\hbar}{m}\boldsymbol{e}_1\cdot\boldsymbol{e}_2\delta_{fi}.$$

Inserting closure gives

$$\frac{1}{m}\sum_v \{\langle \mathrm{f}|\mathbb{r}\cdot\boldsymbol{e}_2|v\rangle\langle v|\mathbb{p}\cdot\boldsymbol{e}_1|\mathrm{i}\rangle - \langle \mathrm{f}|\mathbb{p}\cdot\boldsymbol{e}_1|v\rangle\langle v|\mathbb{r}\cdot\boldsymbol{e}_2|\mathrm{i}\rangle\} = \frac{i\hbar}{m}\boldsymbol{e}_1\cdot\boldsymbol{e}_2\delta_{fi},$$

and since we know from the substitution used to obtain Equation (3.36b) that $\langle v|\mathbb{p}|i\rangle = im\omega_{vi}\langle v|\mathbb{r}|i\rangle = im\omega_{vi}\mathbf{M}_{vi}$,

$$\frac{i\hbar}{m}\boldsymbol{e}_1\cdot\boldsymbol{e}_2\delta_{fi} = i\sum_v \{\omega_{vi}\mathbf{M}_{fv}\cdot\boldsymbol{e}_2\mathbf{M}_{vi}\cdot\boldsymbol{e}_1 - \omega_{fv}\mathbf{M}_{fv}\cdot\boldsymbol{e}_1\mathbf{M}_{vi}\cdot\boldsymbol{e}_2\}. \tag{10.16}$$

We also have the identity

$$\boldsymbol{r}\cdot\boldsymbol{e}_2\ \boldsymbol{r}\cdot\boldsymbol{e}_1 - \boldsymbol{r}\cdot\boldsymbol{e}_1\ \boldsymbol{r}\cdot\boldsymbol{e}_2 = 0$$

leading to

$$\sum_v \{\mathbf{M}_{fv}\cdot\boldsymbol{e}_2\ \mathbf{M}_{vi}\cdot\boldsymbol{e}_1 - \mathbf{M}_{fv}\cdot\boldsymbol{e}_1\ \mathbf{M}_{vi}\cdot\boldsymbol{e}_2\} = 0.$$

Multiplying by ω_2 and adding to Eq. (10.16) gives

$$\frac{\hbar}{m}\boldsymbol{e}_1\cdot\boldsymbol{e}_2\delta_{fi} = \sum_v \{(\omega_2 + \omega_{vi})\boldsymbol{M}_{fv}\cdot\boldsymbol{e}_2\ \mathbf{M}_{vi}\cdot\boldsymbol{e}_1 - (\omega_2 + \omega_{fv})\mathbf{M}_{fv}\cdot\boldsymbol{e}_1\ \mathbf{M}_{vi}\cdot\boldsymbol{e}_2\}. \tag{10.17}$$

We now substitute this into Equation (10.15), leaving

$$R_{\text{if}}(\omega_1,\omega_2)=\frac{\pi e^4 n_1}{2\varepsilon^2\hbar^2V^2\omega_1\omega_2}\left|\sum_{\text{v}}\left\{\boldsymbol{e}_1\cdot\boldsymbol{M}_{\text{vi}}\ \boldsymbol{e}_2\cdot\boldsymbol{M}_{\text{fv}}\left[\omega_2+\omega_{\text{vi}}+\frac{\omega_{\text{fv}}\omega_{\text{vi}}}{(\omega_{\text{vi}}-\omega_1)}\right]\right.\right.$$
$$\left.\left.+\boldsymbol{e}_2\cdot\mathbf{M}_{\text{vi}}\ \boldsymbol{e}_1\cdot\mathbf{M}_{\text{fv}}\left[-\omega_2-\omega_{\text{fv}}+\frac{\omega_{\text{fv}}\omega_{\text{vi}}}{(\omega_{\text{vi}}+\omega_2)}\right]\right\}\right|^2\delta(\omega_{\text{fi}}+\omega_2-\omega_1) \tag{10.18}$$

$$R_{\text{if}}(\omega_1,\omega_2)=\frac{\pi e^4 n_1\omega_1\omega_2}{2\varepsilon^2\hbar^2V^2}\left|\sum_{\text{v}}\left\{\frac{\boldsymbol{e}_1\cdot\mathbf{M}_{\text{vi}}\ \boldsymbol{e}_2\cdot\mathbf{M}_{\text{fv}}}{\omega_{\text{vi}}-\omega_1}+\frac{\boldsymbol{e}_2\cdot\mathbf{M}_{\text{vi}}\ \boldsymbol{e}_1\cdot\mathbf{M}_{\text{fv}}}{\omega_{\text{vi}}+\omega_2}\right\}\right|^2\delta(\omega_{\text{fi}}+\omega_2-\omega_1). \tag{10.19}$$

In the final step, the delta function is used to set $\omega_{\text{fv}} = \omega_1 - \omega_2 - \omega_{\text{vi}}$.

This gives the rate for emission of a photon of a particular mode given an incident field having n_1 photons in a single mode. Realistically, the incident photons are usually distributed over many modes of the radiation field, but as long as they all have very similar frequencies and the same polarization, this can be ignored for the purpose of calculating the rate. But for the emitted photons, we must realize that spontaneous emission can occur to any one of a continuum of modes, so we must replace the delta function by a density of states as was done for spontaneous emission in Chapter 3. Furthermore, when discussing Raman processes, it is conventional to use not the total cross section, as defined in Section 10.1, but the differential cross section, which is the cross section per unit solid angle Ω around a particular direction. In Chapter 3, when considering emission into all directions ($d\Omega$ integrated over all angles = 4π), we showed in Equation (3.55) that the spatially integrated density of states, converted from states per energy unit to states per frequency, is

$$\delta(\omega_{\text{fi}}+\omega_2-\omega_1)\rightarrow\rho(\omega)=(\mu\varepsilon)^{3/2}\,\omega_2^2V/\pi^2.$$

Thus, if we want only the density of states for emitted photons around a particular $d\Omega$, we need to multiply that expression by $d\Omega/4\pi$. We also need to divide it by 2 because that expression contained both polarizations of the emitted light, and here we have specified a particular polarization. Also, we replace $(\mu\varepsilon)^{3/2}$ by $(\mu_0\varepsilon_0)^{3/2}\varepsilon_r^{3/2} = n^3/c^3$, where n is the refractive index (do not confuse this n with the number of photons!) So this leaves

$$R_{\text{if}}(\omega_1,\omega_2)=\frac{n_1e^4\omega_1\omega_2^3d\Omega}{16\pi^2\varepsilon_0^2\hbar^2Vnc^3}\left|\sum_{\text{v}}\left\{\frac{\boldsymbol{e}_1\cdot\mathbf{M}_{\text{vi}}\ \boldsymbol{e}_2\cdot\mathbf{M}_{\text{fv}}}{\omega_{\text{vi}}-\omega_1}+\frac{\boldsymbol{e}_2\cdot\mathbf{M}_{\text{vi}}\ \boldsymbol{e}_1\cdot\mathbf{M}_{\text{fv}}}{\omega_{\text{vi}}+\omega_2}\right\}\right|^2. \tag{10.20}$$

The cross section is the transition rate divided by the incident photon flux in photons area^{-1}·s^{-1}. The photon density is n_1/V, so the photon flux is $(n_1/V)(c/n)$. The cross section per unit solid angle, called the differential cross section, $d\sigma/d\Omega$, is then

$$\frac{d\sigma_{\text{if}}}{d\Omega} = \frac{R_{\text{if}}(\omega_1, \omega_2)}{\left(\frac{n_1 c}{nV}\right) d\Omega} = \frac{\omega_1 \omega_2^3}{16\pi^2 \varepsilon_0^2 \hbar^2 c^4} \left| \sum_{\text{v}} \left\{ \frac{\boldsymbol{e}_1 \cdot \boldsymbol{\mu}_{\text{vi}}\, \boldsymbol{e}_2 \cdot \boldsymbol{\mu}_{\text{fv}}}{\omega_{\text{vi}} - \omega_1} + \frac{\boldsymbol{e}_2 \cdot \boldsymbol{\mu}_{\text{vi}}\, \boldsymbol{e}_1 \cdot \boldsymbol{\mu}_{\text{fv}}}{\omega_{\text{vi}} + \omega_2} \right\} \right|^2, \tag{10.21}$$

where we have converted from transition lengths to transition dipole moments for compactness. As always, it is a good idea to check that the result has the correct units. The units of ω are s^{-1}, ε_0 is in $C^2 \cdot J^{-1} \cdot m^{-1}$, $\hbar$ is in J·s or $kg \cdot m^2 \cdot s^{-1}$, c is in $m \cdot s^{-1}$, and μ is in C·m. The differential cross section is thus seen to have units of m^2, the correct units for a cross section (cm^2 or $Å^2$ are more typically used). The differential cross section is related to what is actually measured in the laboratory by

$$P_{\text{scatt}} = \int d\Omega (d\sigma / d\Omega) F_{\text{inc}}, \tag{10.22}$$

where P_{scatt} is the Raman-scattered power in photons s^{-1}, and F_{inc} is the incident photon flux in photons $m^{-2} \cdot s^{-1}$.

Note that although we have replaced the delta function by the density of states, it is understood that the density of states is evaluated at $\omega_2 = \omega_1 - \omega_{\text{fi}}$; that is, the process cannot occur unless the difference between incident and scattered frequencies exactly equals a material resonance. This just means that at long times energy must be conserved. On the other hand, there is no requirement for conservation of energy at the intermediate step; ω_1 need not equal any material resonance frequency ω_{vi} because the system does not stay in state $|v\rangle$ for very long. Conservation of energy holds only at long times!

Assuming $|i\rangle$ belongs to the ground electronic state and ω_1 is an optical frequency, $\omega_{\text{vi}} + \omega_2$ is always larger than $\omega_{\text{vi}} - \omega_1$ for most or all states $|v\rangle$. Close to resonance, where $\omega_{\text{vi}} \approx \omega_1$, it becomes a good approximation to drop the term with the denominator $\omega_{\text{vi}} + \omega_2$ (which arises from second-order paths in which the creation of the scattered photon precedes annihilation of the incident one) and to keep only a limited number of near-resonant $|v\rangle$ states in the sum.

Exactly on resonance, the term with denominator $\omega_{\text{vi}} - \omega_1$ blows up. This is caused by the neglect of the finite lifetime of the intermediate states. We can fix this by going back to time-dependent perturbation theory and assuming that each $|v\rangle$ state decays exponentially with rate constant $\gamma/2$. Then the first-order amplitude in $|v\rangle$, Eq. (10.9), becomes instead

$$\begin{aligned} b_{\text{v}}^{(1)}(t) &= \frac{1}{i\hbar} \int_0^t d\tau \exp(i\omega_{\text{vi}}\tau) W_{\text{vi}}(\tau) \exp[-\gamma(t-\tau)/2] \\ &= e\omega_{\text{vi}} \sqrt{\frac{n_1}{2\hbar\omega_1 \varepsilon V}}\, \boldsymbol{e}_1 \cdot \mathbf{M}_{\text{vi}} e^{-\gamma t/2} \int_0^t d\tau \exp[i(\omega_{\text{vi}} - \omega_1)\tau] e^{\gamma\tau/2} \\ &= -e\omega_{\text{vi}} \sqrt{\frac{n_1}{2\hbar\omega_1 \varepsilon V}}\, \boldsymbol{e}_1 \cdot \mathbf{M}_{\text{vi}} \frac{\exp[-\gamma t/2] - \exp[i(\omega_{\text{vi}} - \omega_1)t]}{i(\omega_{\text{vi}} - \omega_1) + \gamma/2}, \end{aligned} \tag{10.23}$$

with an analogous result for $b_w^{(1)}(t)$. Carrying through the rest of the development as done before finally yields a result that is the same as before except for the presence of an additional term in the denominators:

$$\frac{d\sigma_{if}}{d\Omega} = \frac{\omega_1\omega_2^3}{16\pi^2\varepsilon_0^2\hbar^2c^4}\left|\sum_v\left\{\frac{\boldsymbol{e}_1\cdot\boldsymbol{\mu}_{vi}\;\boldsymbol{e}_2\cdot\boldsymbol{\mu}_{fv}}{\omega_{vi}-\omega_1-i\gamma/2}+\frac{\boldsymbol{e}_2\cdot\boldsymbol{\mu}_{vi}\;\boldsymbol{e}_1\cdot\boldsymbol{\mu}_{fv}}{\omega_{vi}+\omega_2-i\gamma/2}\right\}\right|^2. \quad (10.24)$$

Normally, the linewidths γ are included only when $\omega_{vi} - \omega_1$ is small, and then the second term is negligible compared with the first and is dropped.

This two-photon process involves three types of states of the material system: the initial and final states $|i\rangle$ and $|f\rangle$, and a collection of intermediate states $\{|v\rangle\}$. The process in general is called Raman scattering if $|i\rangle$ and $|f\rangle$ have different energies (so ω_1 and ω_2 are different), and Rayleigh scattering if they have the same energy ($\omega_1 = \omega_2$). For Raman processes, $|i\rangle$ and $|f\rangle$ are usually different vibrational sublevels of the same electronic state (vibrational Raman scattering), although electronic Raman processes where $|i\rangle$ and $|f\rangle$ belong to different electronic states are also known. Usually, Raman is done with visible or UV light, and the intermediate states $\{|v\rangle\}$ that make the biggest contribution (large transition moments $\boldsymbol{\mu}_{vi}$ and $\boldsymbol{\mu}_{fv}$, and reasonably small denominators $\omega_{vi} - \omega_1$) belong to higher excited electronic states. In resonance Raman, ω_1 is chosen to be very close to the ω_{vi} for the vibrational levels of one particular electronic state, and then not only can the term with the $\omega_{vi} + \omega_2$ denominator be neglected, but the sum over v can reasonably be limited to vibrational sublevels of the single resonant electronic state.

An important point about all of these scattering processes is that different intermediate states exhibit *interferences* with one another because their contributions are added up at the amplitude level and then modulus squared to get the physically observable cross section. The cross section for a system with two important intermediate states is not just the sum of the cross sections for each of the intermediate states taken separately. The cross section itself is always a real positive number, but each individual contribution inside the $|\Sigma_v\cdots|^2$ can be either positive or negative (due to both the matrix element of $\boldsymbol{\mu}\cdot\mathbf{e}$ and the energy denominator $\omega_{vi} - \omega_1$) and becomes complex as resonance is approached (adding $-i\gamma/2$ to the denominators). The scattering cross section can have interesting variations with frequency due to these interference terms.

Most applications of Raman scattering involve vibrational transitions where the initial and final states are different vibrational states of the ground-state electronic surface. Thus, the relevant matrix elements can be written as

$$\boldsymbol{\mu}_{fv}\cdot\mathbf{e}_2 = \langle F|\langle\Psi_0|\boldsymbol{\mu}\cdot\mathbf{e}_2|\Psi_v\rangle|V\rangle = \langle F|\boldsymbol{\mu}_{0v}(\mathbf{Q})\cdot\mathbf{e}_2|V\rangle, \quad (10.25a)$$

and

$$\boldsymbol{\mu}_{vi}\cdot\mathbf{e}_1 = \langle V|\langle\Psi_v|\boldsymbol{\mu}\cdot\mathbf{e}_1|\Psi_0\rangle|I\rangle = \langle V|\boldsymbol{\mu}_{v0}(\mathbf{Q})\cdot\mathbf{e}_1|\mathrm{I}\rangle, \quad (10.25b)$$

where $|\Psi_0\rangle$ is the ground electronic state, $|I\rangle$ and $|F\rangle$ are vibrational levels of this ground electronic state, $|\Psi_v\rangle$ is some higher electronic state, and $|V\rangle$ is a vibrational level of this state. We can further split up the denominators as

$$\omega_{vi} - \omega_1 = \omega_{v0} + \omega_V - \omega_I - \omega_1, \tag{10.26a}$$

and

$$\omega_{vi} + \omega_2 = \omega_{v0} + \omega_V - \omega_I + \omega_2. \tag{10.26b}$$

where $\hbar\omega_{v0}$ is the purely electronic zero–zero energy difference between states $|\Psi_0\rangle$ and $|\Psi_v\rangle$, and $\hbar\omega_V$ and $\hbar\omega_I$ are the energies above the zero-point of vibrational states $|V\rangle$ and $|I\rangle$, respectively. With these substitutions, the cross-section expression becomes

$$\frac{d\sigma_{IF}}{d\Omega} \sim \omega_1\omega_2^3 \left| \sum_v \sum_V \frac{\langle F|\mu_{0v}(Q)\cdot e_2|V\rangle\langle V|\mu_{v0}(Q)\cdot e_1|I\rangle}{\hbar(\omega_{v0}+\omega_V-\omega_I-\omega_1-i\gamma/2)} + \frac{\langle F|\mu_{0v}(Q)\cdot e_1|V\rangle\langle V|\mu_{v0}(Q)\cdot e_2|I\rangle}{\hbar(\omega_{v0}+\omega_V-\omega_I+\omega_2-i\gamma/2)} \right|^2. \tag{10.27}$$

Further simplification depends on whether or not the excitation frequency is near an electronic resonance.

10.4. NONRESONANT RAMAN SCATTERING

In ordinary or nonresonant Raman scattering, the incident (laser) frequency ω_1 is not very close to any of the electronically resonant frequencies ω_{v0}, so ω_{v0}-ω_1 and $\omega_{v0} + \omega_2$ are both large quantities compared with the purely vibrational frequency difference $\omega_V - \omega_I$, and are also large compared with $\gamma/2$. Therefore, it is reasonable to neglect the vibrational part of the denominator relative to the electronic part, to neglect the difference between ω_1 and ω_2 (let $\omega_1 = \omega_2 = \omega$), and to ignore the imaginary part of the denominator (Fig. 10.2). But now there is no longer any reference to the vibrational states $|V\rangle$ in the denominator, and we can use closure to remove them from the numerators, $\Sigma_V|V\rangle\langle V| = 1$. We now have

$$\frac{d\sigma}{d\Omega} \sim \omega^4 \left| \sum_v \frac{\langle F|\mu_{0v}(Q)\cdot e_2\,\mu_{v0}(Q)\cdot e_1|I\rangle}{\hbar(\omega_{v0}-\omega)} + \frac{\langle F|\mu_{0v}(Q)\cdot e_1\,\mu_{v0}(Q)\cdot e_2|I\rangle}{\hbar(\omega_{v0}+\omega)} \right|^2 \tag{10.28a}$$

$$= \omega^4 \left| \langle F|e_2 \cdot \left\{ \sum_v \frac{\mu_{0v}(Q)\mu_{v0}(Q)}{\hbar(\omega_{v0}-\omega)} + \frac{\mu_{0v}(Q)\mu_{v0}(Q)}{\hbar(\omega_{v0}+\omega)} \right\} \cdot e_1|I\rangle \right|^2 \tag{10.28b}$$

$$= \omega^4 \left|\langle F|\mathbf{e}_2 \cdot \boldsymbol{\alpha}(\omega, \mathbf{Q}) \cdot \mathbf{e}_1|I\rangle\right|^2, \tag{10.28c}$$

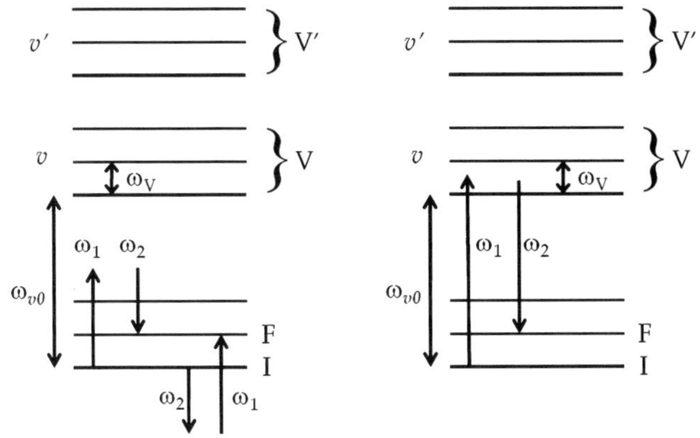

Figure 10.2. In nonresonant Raman (left), neither $\omega_{v0} + \omega_V - \omega_I - \omega_1$ nor $\omega_{v0} + \omega_V - \omega_I + \omega_2$ is small for the vibrational levels V or V' of any electronic state v or v'. In resonance Raman (right), there is one electronic state v for which $\omega_{v0} + \omega_V - \omega_I - \omega_1$ becomes small, so the second term in Equation (10.27) and all other electronic states may be ignored.

where the quantity in braces in Equation (10.28b) is the quantum mechanical expression for the polarizability tensor, $\boldsymbol{\alpha}$. This is a 3×3 tensor whose ij element gives the dipole moment induced in direction i by radiation polarized along direction j. The tensor properties are discussed further in Section 10.5 below; for the remainder of this section, we will treat $\boldsymbol{\alpha}$ as a scalar. The polarizability is a function of the applied radiation frequency, ω, and is also a weak function of the nuclear coordinates. Thus, it can be expanded as a Taylor series about the equilibrium geometry,

$$\alpha(\mathbf{Q}) = \alpha(\mathbf{Q}_0) + \sum_k (\partial\alpha/\partial Q_k) Q_k + \ldots.$$

For vibrational Raman scattering, $|I\rangle$ and $|F\rangle$ are orthogonal because they are different vibrational eigenstates of the same potential surface. Therefore, the matrix element $\langle F|\alpha(\mathbf{Q}_0)|I\rangle$ vanishes, and the leading term in the Raman cross section is

$$\frac{d\sigma}{d\Omega} \approx \omega^4 \left| \langle F | \sum_k \left[\frac{\partial\alpha(\omega)}{\partial Q_k} \right] Q_k | I \rangle \right|^2 = \omega^4 \left| \sum_k \left[\frac{\partial\alpha(\omega)}{\partial Q_k} \right] \langle F | Q_k | I \rangle \right|^2. \qquad (10.29)$$

For harmonic oscillators, the matrix element $\langle F|Q_k|I\rangle$ is nonzero only if $|I\rangle$ and $|F\rangle$ differ by just one quantum, so we get the selection rule that only fundamental transitions are allowed in nonresonant Raman. As with infrared absorption, this breaks down when vibrational anharmonicities are considered or when higher terms in the Taylor series expansion of the polarizability are important, so overtone and combination band transitions actually are observed, but are usually quite weak. From the other part of the matrix element, we get the selection rule that there must be a nonzero derivative of the polarizability along the vibrational coordinate. This is all in agreement with the purely classical treatment discussed in Section 10.2.

10.5. SYMMETRY CONSIDERATIONS AND DEPOLARIZATION RATIOS IN RAMAN SCATTERING

The components of the polarizability tensor transform as products of coordinates (xx, xy, etc.), and for the nonresonant Raman cross section not to vanish, the product of the symmetry species of the vibrational mode and of one of the components of the polarizability must contain the totally symmetric species. So for example, in C_{2v}, x^2, y^2, and z^2 transform as A_1, xy transforms as A_2, xz is B_1 and yz is B_2, so vibrations of all point groups transform as one of the components of the polarizability tensor, and all vibrations are Raman allowed. In the C_{2h} point group, x^2, y^2 z^2, and xy transform as A_g, while xz and yz are B_g. Vibrational normal modes belonging to the A_u or B_u point groups are Raman forbidden. However, the A_u and B_u modes are infrared allowed since z transforms as A_u and x and y transform as B_u, while the A_g and B_g modes are Raman allowed. This is a general feature of any point group having inversion as one of its symmetry elements: only the u modes are infrared allowed, while only the g modes are Raman allowed. The Raman and infrared spectra are complementary.

The polarizability tensor relates the components of the applied field to the dipole moment induced in the molecule:

$$\begin{pmatrix} \mu_x \\ \mu_y \\ \mu_z \end{pmatrix} = \begin{pmatrix} \alpha_{xx} & \alpha_{xy} & \alpha_{xz} \\ \alpha_{yx} & \alpha_{yy} & \alpha_{yz} \\ \alpha_{zx} & \alpha_{zy} & \alpha_{zz} \end{pmatrix} \begin{pmatrix} E_x \\ E_y \\ E_z \end{pmatrix}. \qquad (10.30)$$

These correspond to axes defined in a molecule-fixed coordinate system, while the experimenter has control of the electromagnetic field polarizations only in the laboratory-fixed coordinate system (see Section 7.5). In a Raman experiment, the polarization direction of the incident radiation ($\mathbf{e}_1$) can be defined by the experimenter, while a particular polarization of the scattered radiation ($\mathbf{e}_2$) can be selected by placing a polarization analyzer in the scattered beam. Therefore, in general, nine possible combinations of incident and scattered radiation can be obtained. For isotropic samples, however, there are only two distinct combinations: either the incident radiation and the scattered radiation have the same (parallel) polarization, or they have perpendicular polarizations. Experimentally, one defines the depolarization ratio as $\rho = I_{\perp}/I_{\parallel}$ [Eq. (9.28)]. However, for an isotropic sample, any polarization direction in the laboratory frame has projections onto all three molecule-fixed axes so a single component of the polarizability tensor referenced to laboratory coordinates has contributions from all nine components of the molecule-fixed polarizability tensor. The procedure for transforming from laboratory-fixed to molecule-fixed coordinate systems is messy but well defined (Mortensen and Hassing, 1980). The result for the depolarization ratio turns out to be

$$\rho = \{5\Sigma^1 + 3\Sigma^2\}/\{10\Sigma^0 + 4\Sigma^2\}, \tag{10.31}$$

where

$$\Sigma^0 = (1/3)|\alpha_{xx} + \alpha_{yy} + \alpha_{zz}|^2 \tag{10.32a}$$

$$\Sigma^1 = (1/2)\{|\alpha_{xy} - \alpha_{yx}|^2 + |\alpha_{xz} - \alpha_{zx}|^2 + |\alpha_{yz} - \alpha_{zy}|^2\} \tag{10.32b}$$

$$\begin{aligned}\Sigma^2 = {} & (1/2)\{|\alpha_{xy} + \alpha_{yx}|^2 + |\alpha_{xz} + \alpha_{zx}|^2 + |\alpha_{yz} + \alpha_{zy}|^2\} \\ & + (1/3)\{|\alpha_{xx} - \alpha_{yy}|^2 + |\alpha_{xx} - \alpha_{zz}|^2 + |\alpha_{yy} - \alpha_{zz}|^2\},\end{aligned} \tag{10.32c}$$

and the components of the polarizability tensor refer to the molecule-fixed coordinate system.

Examination of Equation (10.28) shows that the Raman tensor is symmetric for nonresonant scattering ($\alpha_{ij} = \alpha_{ji}$) and therefore $\Sigma^1 = 0$. For any nontotally symmetric mode, $\Sigma^0 = 0$ also, and therefore $\rho = 3/4$ for all nontotally symmetric modes. For totally symmetric modes, Σ^0 is nonzero and $\rho < 3/4$. Thus, depolarization ratios provide a way to experimentally distinguish totally symmetric from nontotally symmetric vibrations. For molecules with tetrahedral or octahedral symmetry, the *x*-, *y*-, and *z*-directions are equivalent and remain equivalent as the molecule vibrates along a totally symmetric coordinate, so only Σ^0 is nonzero and $\rho = 0$. In nonresonant Raman scattering, ρ cannot exceed 3/4, although it can do so in resonance Raman (anomalous polarization) since Σ^1 can be nonzero.

10.6. RESONANCE RAMAN SPECTROSCOPY

When the incident frequency is very close to or in resonance with an electronic transition of the molecule, that is when $\omega_{v0} - \omega_1$ becomes small, then the approximations made for nonresonant Raman are no longer valid. Near resonance when $\omega_{v0} - \omega_1$ becomes small, the other denominator $\omega_{v0} + \omega_2$ is necessarily large for all $|\Psi_v\rangle$. Therefore, it becomes a good approximation to neglect entirely the second term in the Raman amplitude. Furthermore, if $\omega_{v0} - \omega_1$ is small for some electronic excited state $|\Psi_v\rangle$, it will generally be much larger for all others, so it is usually reasonable to assume that only one electronic state will contribute significantly to the sum over v (Fig. 10.3). (If there are several near-degenerate electronic states, more than one state may have to be

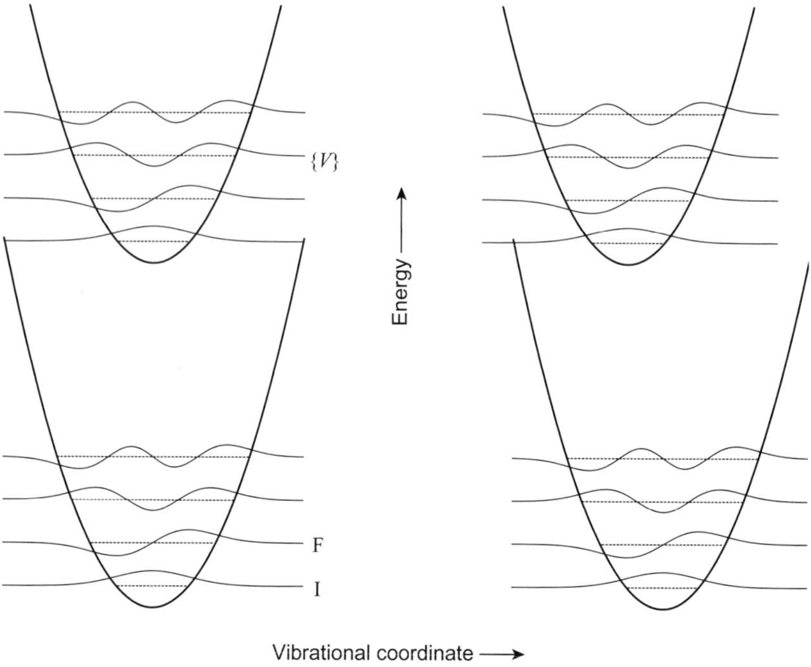

Figure 10.3. A resonance Raman transition has intensity only if one or more excited-state vibrational levels V have good overlaps with both the initial state I and the final state F. This cannot occur if the ground and excited state potential surfaces have very similar equilibrium geometries and curvatures (left), but it can if the equilibrium geometries are significantly different along the normal mode of interest (right).

considered.) On the other hand, it is no longer reasonable to neglect $\omega_V - \omega_I$ relative to $\omega_{v0} - \omega_1$, so we cannot remove the dependence of the denominator on vibrational state $|V\rangle$ and use closure as we could for nonresonant Raman. The cross-section expression valid for resonance Raman becomes

$$\frac{d\sigma_{IF}}{d\Omega} \sim \omega_1 \omega_2^3 \left| \sum_V \frac{\langle F | \boldsymbol{\mu}_{0v}(Q) \cdot \boldsymbol{e}_2 | V \rangle \langle V | \boldsymbol{\mu}_{v0}(Q) \cdot \boldsymbol{e}_1 | I \rangle}{\hbar(\omega_{v0} + \omega_V - \omega_I - \omega_1 - i\gamma/2)} \right|^2 . \tag{10.33}$$

Now we expand the electronic transition length in a Taylor series around the equilibrium nuclear geometry just as we did for electronic spectroscopy [Eq. (8.2)]:

$$\boldsymbol{\mu}_{0v}(\mathbf{Q}) = \boldsymbol{\mu}_{0v}(Q_0) + \Sigma_k [\partial \boldsymbol{\mu}_{0v} / \partial Q_k] Q_k + \ldots,$$

so the matrix elements appearing in the numerator of the cross section expression become

$$\langle F|\boldsymbol{\mu}_{0v}(\mathbf{Q})\cdot\mathbf{e}_2|V\rangle = \boldsymbol{\mu}_{0v}(Q_0)\cdot\mathbf{e}_2\langle F|V\rangle + \sum_k [\partial\boldsymbol{\mu}_{0v}/\partial Q_k]\cdot\mathbf{e}_2\langle F|Q_k|V\rangle + \ldots.$$

As discussed in Chapter 8 with regard to electronic transitions, for strongly allowed transitions the electronic transition moment itself is considerably larger than its derivatives with respect to nuclear coordinates. Furthermore, there is no reason why the vibrational overlap integral, $\langle F|V\rangle$, should be zero or small, since $|F\rangle$ and $|V\rangle$ are vibrational levels of different electronic surfaces. Therefore, we normally keep only the term in $\boldsymbol{\mu}_{0v}(Q_0)$. This is the Condon approximation as employed in electronic spectroscopy, for exactly the same reasons. So in the Condon approximation the resonance Raman cross section becomes

$$\frac{d\sigma_{IF}}{d\Omega} \sim \omega_1\omega_2^3\,|\boldsymbol{\mu}_{0v}(Q)\cdot\boldsymbol{e}_2\ \boldsymbol{\mu}_{v0}(Q)\cdot\boldsymbol{e}_1|^2\left|\sum_V \frac{\langle F|V\rangle\langle V|I\rangle}{\hbar(\omega_{v0}+\omega_V-\omega_I-\omega_1-i\gamma/2)}\right|^2. \tag{10.34}$$

For nondegenerate electronic states, the electronic transition moment $\boldsymbol{\mu}_{0v}$ has a unique direction in the molecule-fixed frame, so we can always define our molecule-fixed coordinate system such that only one of the elements of the Raman polarizability tensor, say α_{xx}, is nonzero. Then from the expression for ρ in Equations (10.31) and (10.32), we can see that $\rho = 1/3$ for Raman scattering on resonance with a single, nondegenerate electronic state, which will hold for all Raman transitions. The vibrational overlaps $\langle V|I\rangle$ and $\langle F|V\rangle$ are the same overlaps that appear in the expressions for absorption and fluorescence spectroscopy; they are $3N$–6-dimensional vibrational overlaps for a polyatomic molecule. In Raman fundamental scattering, $|I\rangle$ and $|F\rangle$ differ by one quantum in one vibrational mode, usually $|I\rangle = |0\rangle$ and $|F\rangle = |1\rangle$. For the cross section to be nonzero, there must be one or more vibrational levels $|V\rangle$ of the excited electronic state that have nonzero overlaps with both $|0\rangle$ and $|1\rangle$ of the ground state, and that can happen only if the excited state has its potential minimum at a different geometry from the ground state (Fig. 10.3). This can be the case only for totally symmetric modes. Therefore, resonance Raman fundamental scattering is allowed only for totally symmetric modes, and they are the same modes that have good Franck–Condon factors in absorption and fluorescence.

Within the assumption of harmonic oscillators and no Duschinsky rotation, combination bands (those in which $|I\rangle$ and $|F\rangle$ differ by one quantum in two or more vibrational modes) can also involve only totally symmetric modes that have nonzero Franck–Condon factors. Overtone transitions ($|I\rangle$ and $|F\rangle$ differ by two or more quanta in some mode) can be allowed for nontotally symmetric modes since for an even overtone transition $|I\rangle$ and $|F\rangle$ are both even or both odd symmetry, and there can be some level $|V\rangle$ of the excited state that has good overlaps with both of them even if there is no change in equilibrium geometry in that mode upon excitation. There does, however, have to be a substantial change in frequency between ground and excited states for this to be possible. Thus, resonance Raman overtone scattering is allowed for totally symmetric modes and for nontotally symmetric modes having a large change in vibrational frequency between ground and resonant excited states (Fig. 10.4).

A Raman spectrum (Fig. 10.4) is a plot of scattered intensity as a function of scattered frequency (ω_2) at a fixed incident frequency (ω_1). It has peaks where the frequency difference $\omega_1 - \omega_2$ equals a Raman-allowed vibrational transition of the molecule, $\omega_F - \omega_I$. The explicit dependence on $\omega_F - \omega_I$ was lost from the cross-section equations above when we replaced the delta function, $\delta(\omega_{fi} + \omega_2 - \omega_1)$, by the density of states for the scattered photons. A Raman excitation profile (Fig. 10.5) is a plot of the cross section of a particular $|I\rangle$ to $|F\rangle$ transition as a function of incident frequency ω_1. It is somewhat analogous to an absorption spectrum in that it spans the same frequency axis and generally has peaks at about the same places because it has a frequency denominator that gets small in the same places. This is the source of resonance enhancement—the Raman cross sections on resonance with a strongly allowed electronic transition are larger by typically 3–6 orders of magnitude than the nonresonant cross sections. However, the structure of the Raman excitation profile is usually different from that of the absorption spectrum because in Raman scattering, the contributions from the different $|V\rangle$ states are summed at the amplitude level before squaring. This leads to interferences between the

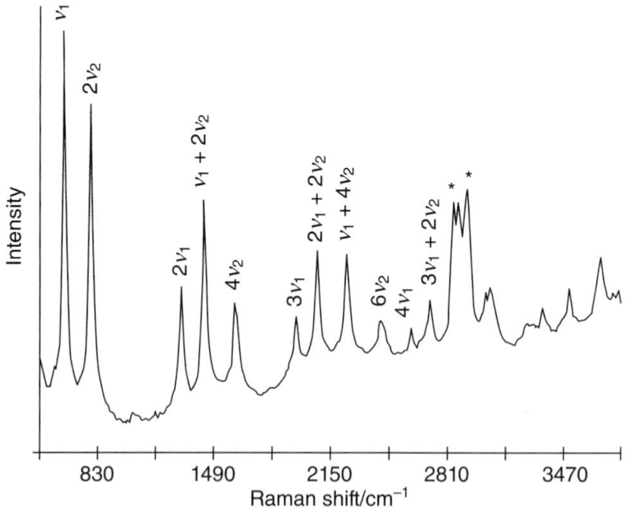

Figure 10.4. Resonance Raman spectrum (208 nm excitation) of CS_2 in cyclohexane, showing long progressions in the symmetric stretch (ν_1) and even overtones of the bend (ν_2). The antisymmetric stretch, ν_3, also appears as even overtones, but they are very weak. The asterisks mark solvent Raman lines.

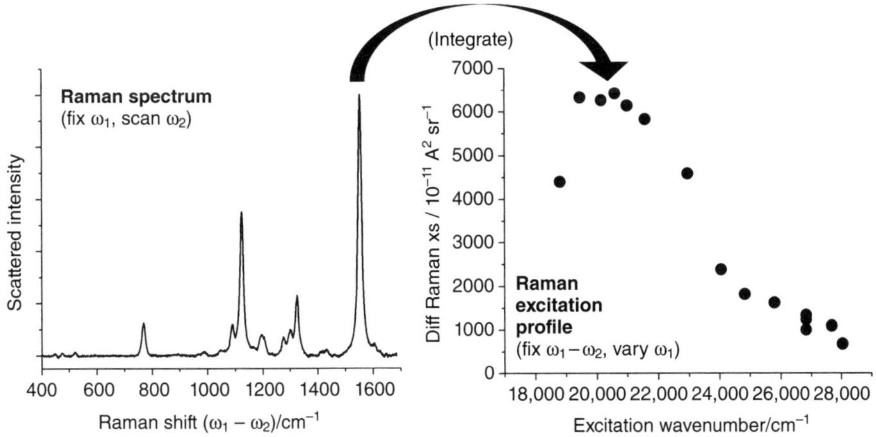

Figure 10.5. Illustration of what is measured in a Raman spectrum (left) and a Raman excitation profile (right). The excitation profile measures the Raman cross section, which is proportional to the integrated area under a particular Raman line as a function of excitation wavelength.

different $|V\rangle$ states, because in a two-photon process, we do not measure the intermediate state of the system and therefore cannot know which intermediate state the process went through. They all interfere as in a multiple-slit wave interference process.

REFERENCES AND FURTHER READING

C. F. Bohren and D. R. Huffman, *Absorption and Scattering of Light by Small Particles* (Wiley-VCH, Weinheim, 2004).

O. S. Mortensen and S. Hassing, Adv. Infrared Raman Spectrosc. **6**, 1 (1980).

A. B. Myers, in *Laser Techniques in Chemistry*, A. B. Myers and T. R. Rizzo, eds. (Wiley, New York, 1995). p. 325.

PROBLEMS

1. Zinc oxide (ZnO) is a semiconductor often used in sunscreens because of its uv absorbing capabilities. In the visible region of the spectrum, it is nonabsorbing and has a relative dielectric constant of about 8.5 that is nearly independent of wavelength. Compare the Rayleigh scattering cross section, Equation (10.1), to the physical cross section for 10-nm diameter spherical

ZnO nanoparticles at 650 nm (red light) and at 450 nm (blue light). Assume the medium is vacuum. What can you say about the scattering efficiency from such particles?

2. A system has three eigenstates, φ_a, φ_b, and φ_c, with energies $E_a < E_c < E_b$. At time $t = 0$, it is in its ground state, φ_a. At $t = 0$, an oscillatory perturbation having two frequency components is turned on:

$$\mathbb{W}(t) = 2\mathbb{W}[\cos\omega_1 t + \cos\omega_2 t].$$

The frequencies ω_1 and ω_2 are close to resonance with the a → b and b → c transitions, respectively; that is, $(\omega_{ba} - \omega_1)$ and $(\omega_{bc} - \omega_2)$ are both small but nonzero. The matrix elements W_{ab} and W_{bc} are nonzero, but $W_{ac} = 0$, so to get from φ_a to φ_c requires that the perturbation act twice.

(a) At time $t > 0$, the state of the system is a superposition of the three eigenstates:

$$|\Psi(t)\rangle = d_a(t)\exp(-i\omega_a t)\varphi_a + d_b(t)\exp(-i\omega_b t)\varphi_b + d_c(t)\exp(-i\omega_c t)\varphi_c.$$

Calculate the coefficients $d_b(t)$ and $d_c(t)$ to first order in the perturbation, keeping only the near-resonant terms.

(b) Now use these first-order corrected coefficients as the starting point to find the second-order correction to the coefficient $d_c(t)$. Again, keep only the most resonant terms.

(c) Your result should involve a difference of two terms with different frequency denominators. Assuming $(\omega_{bc} - \omega_2) >> (\omega_{ca} - \omega_1 + \omega_2)$, what is the probability of finding the system in state φ_c after time t?

(d) Find the second order steady-state transition rate into state φ_c in the limit of long times.

(e) Now consider that instead of just a single intermediate state φ_b, there are two such states, φ_{b1} and φ_{b2}, which have slightly different energies and matrix elements with φ_a and/or φ_c. Find the steady-state transition rate into state φ_c for this case. Is this rate just the sum of the rates for each of the two intermediate states considered separately? Discuss.

3. Use Equation (10.21) to make a rough estimate of the magnitude of a near-resonant Raman cross section. Assume a single vibration with a frequency of 1600 cm^{-1} (like a C=C stretch) and consider a single excited vibronic state having transition dipole moments μ_{vi} and μ_{fv} each equal to 1 Debye and neglect the vector properties (assume the polarization vectors of the radiation are parallel to the transition dipoles). Let the incident laser frequency be $\omega_1 = 40{,}000\,\text{cm}^{-1}$ and the excited-state resonance frequency be $\omega_{vi} = 45{,}000\,\text{cm}^{-1}$. Calculate the differential Raman cross section in units of Å^2. What does this imply about the fraction of the incident light that is Raman scattered?

4. Show that $\alpha(\omega,\mathbf{Q})$ as defined in Equation (10.28) has the same units as the polarizability defined in Equation (10.3).

5. Carbon tetrachloride, CCl_4, is a tetrahedral molecule (T_d) with four normal modes:

$$\nu_1(A_1),\, 459\ \text{cm}^{-1}$$
$$\nu_2(E), \approx 225\ \text{cm}^{-1}$$
$$\nu_3(T_2), \approx 315\ \text{cm}^{-1}$$
$$\nu_4(T_2), \approx 770\ \text{cm}^{-1}$$

(a) Which fundamentals will be allowed in the nonresonant Raman spectrum?

(b) Of the allowed fundamentals, what can you say about their depolarization ratios?

6. The following diagram shows four vibrational normal modes of the planar C_2H_4 molecule. Atoms labeled with the + sign are moving out of the plane of the paper, while those labeled with the – sign are moving into the plane of the paper. Which of the vibrational modes shown are expected to be Raman active?

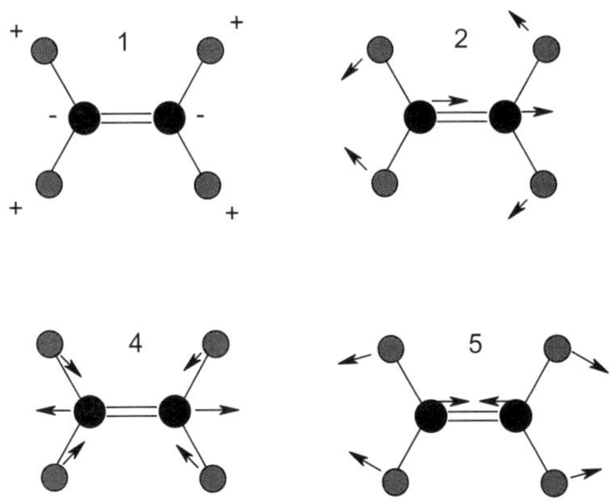

7. Consider three possible structures for ICl_3 (all three Cl atoms are bonded to the I): trigonal planar, pyramidal, or T-shaped.

(a) To which point group to each of these three structures belong?

(b) Which shape is predicted by valence shell electron pair repulsion (VSEPR) theory?

(c) ICl_3 has six vibrational normal modes, all of which are observed in both IR and Raman. Which of the possible shapes can be excluded?

8. Use Eq. (10.34) to calculate the resonance Raman spectrum for a simple case of a molecule having a resonant electronic state with $\omega_{v0} = 30{,}000\,\text{cm}^{-1}$ and only one important vibrational mode with a frequency of $1000\,\text{cm}^{-1}$. Assume that the vibrational mode has the same frequency in the ground and excited states, and that the displacement between ground and excited state potential minima in dimensionless normal coordinates is $\Delta = 1.0$. Take the homogeneous linewidth γ to be $200\,\text{cm}^{-1}$ and let the incident laser frequency be $30{,}300\,\text{cm}^{-1}$. Assume that all of the molecules start out in their vibrational ground state, $|0\rangle$. Calculate the relative Raman cross sections (i.e., neglect constants that are the same for all transitions) for transitions to the final states $|0\rangle$, $|1\rangle$, and $|2\rangle$ (the elastic Rayleigh scattering, the Raman fundamental, and the Raman overtone) and plot a stick spectrum of intensity as a function of Raman shift (the difference between incident and scattered frequency). You will have to make some reasonable assumptions about how many of the excited-state vibrational levels $|V\rangle$ to include in the sum over intermediate states. For this case of equal ground and excited state vibrational frequencies, the one-dimensional vibrational overlap integrals you will need are:

$$\langle 0|V\rangle\langle V|0\rangle = \exp(-S)S^V/V!$$

$$\langle 1|V\rangle\langle V|0\rangle = 2^{-1/2}\Delta\{\langle 0|V\rangle\langle V|0\rangle - \langle 0|V-1\rangle\langle V-1|0\rangle\}$$

$$\langle 2|V\rangle\langle V|0\rangle = 2^{-1/2}S\{\langle 0|V\rangle\langle V|0\rangle - \langle 0|V-1\rangle\langle V-1|0\rangle + \langle 0|V-2\rangle\langle V-2|0\rangle\}$$

where $|V\rangle$ is a vibrational level of the excited electronic state and $|0\rangle$, $|1\rangle$, and $|2\rangle$ are levels of the ground electronic state. Here, $S = \Delta^2/2$, as defined previously. (The second terms in the expressions for $\langle 1|V\rangle\langle V|0\rangle$ and $\langle 2|V\rangle\langle V|0\rangle$ are omitted when $V = 0$, and the last term in $\langle 2|V\rangle\langle V|0\rangle$ is omitted when $V = 1$, since states with negative V do not exist.)

9. Equation (10.34) for the resonance Raman cross section involves the product of overlap integrals $\langle F|V\rangle\langle V|I\rangle$, which can be written more explicitly as $\langle 1|V\rangle\langle V|0\rangle$ for fundamental scattering starting from the vibrational ground state. If the ground and excited states have identical vibrational frequencies, and there is no Duschinsky rotation, these overlaps have the explicit form given in Problem 8. Derive expressions for $\langle 1|V\rangle\langle V|0\rangle$ valid in the limit of small Δ. Show that only two V states contribute significantly in this limit and explain physically why this is in terms of the wavefunction overlaps.

CHAPTER 11

NONLINEAR AND PUMP–PROBE SPECTROSCOPIES

11.1. LINEAR AND NONLINEAR SUSCEPTIBILITIES

Ordinary absorption spectroscopy, whether electronic (UV-vis) or vibrational (infrared), is a linear process. It involves one photon, is described by first order time-dependent perturbation theory, and its rate depends on the first power of the intensity of the radiation (square of the electric field). Ordinary spontaneous and stimulated emission are also described by first-order perturbation theory, although in spontaneous emission, the transition rate has no dependence on any applied field. Raman spectroscopy involves two photons and requires taking time-dependent perturbation theory to second order, but since one of the photons is spontaneously emitted, it is still considered a linear spectroscopy from an experimental point of view in that the signal depends linearly on the incident light intensity.

There are, however, many processes that are nonlinear—they depend on higher powers of the intensity of the incident light, or on products of the intensities of several applied fields. The proper quantum mechanical description of such processes involves second-order or higher time-dependent perturbation theory if a state vector approach is used, or evolving the perturbed density matrix to third order or higher if system–bath interactions are to be considered. Some nonlinear techniques can be described fairly well using a strictly classical picture in which the dipole moment induced in a molecule by

Condensed-Phase Molecular Spectroscopy and Photophysics, First Edition. Anne Myers Kelley.

an applied field is recognized to have components that depend on higher than the first power of the field. That is, the polarization $P(r,t)$ induced in a material, or the dipole moment $\mu(r,t)$ induced in a molecule, by an applied electric field $E(r,t)$ is given by

$$P(r,t) = \chi^{(1)}E(r,t) + \chi^{(2)}E^2(r,t) + \chi^{(3)}E^3(r,t) + \ldots \tag{11.1a}$$

$$\mu(r,t) = \mu_0 + \alpha E(r,t) + \beta E^2(r,t) + \gamma E^3(r,t) + \ldots \tag{11.1b}$$

(P and E are actually vectors and the susceptibilities $\chi^{(n)}$, α, β, and γ are tensors). α, β, and γ play the same role for individual molecules that $\chi^{(1)}$, $\chi^{(2)}$, and $\chi^{(3)}$ play for bulk materials. The induced polarization radiates a field whose intensity is detected as the signal, so the signal goes as the polarization squared. Thus, $\chi^{(1)}$ contributes to linear optical responses, $\chi^{(2)}$ to quadratic responses, and so on.

11.2. MULTIPHOTON ABSORPTION

The simplest nonlinear spectroscopy is the "simultaneous" absorption of two photons. Although the experiment can be configured such that the material absorbs one photon from each of two different light sources, most often the two photons are absorbed from the same light source. For developing the quantum mechanical treatment of two-photon absorption, we will assume that both absorbed photons have the same wavevector and polarization direction (Fig. 11.1).

Quantum mechanically, two-photon absorption is described by second-order time-dependent perturbation theory. The process is very similar to Raman scattering (Section 10.3) except that both photons are absorbed instead of one being absorbed and one emitted. Absorption of two identical photons of wavevector **k** and polarization α involves the sequence of steps $|i; n_{\mathbf{k},\alpha}\rangle \rightarrow |v; (n-1)_{\mathbf{k},\alpha}\rangle \rightarrow |f; (n-2)_{\mathbf{k},\alpha}\rangle$. Because there is no spontaneous emission step, the transition rate may be derived using either the classical electromagnetic field or the quantized field; the former is considerably simpler. Following the same steps used to generate the Raman transition rate in Chapter 10, we find for the rate of absorption of two identical photons

$$R_{if}(\omega) = \frac{\pi u_{avg}(\omega)\rho_{\mathrm{rad}}(\omega)}{2\hbar^4\varepsilon_0^2 n^4}\left|\sum_{\mathrm{v}} \frac{\mu_{\mathrm{fv}}\cdot \boldsymbol{e}\, \mu_{\mathrm{vi}}\cdot \boldsymbol{e}}{\omega_{\mathrm{vi}} - \omega}\right|^2. \tag{11.2a}$$

Here, u_{avg} is the energy density of the applied field and ρ_{rad} is the energy density per unit frequency, both evaluated at the energy-conserving frequency $\omega = \omega_{\mathrm{fi}}/2$, and n is the refractive index of the medium. The quantity $u_{\mathrm{avg}}(\omega)$ is defined in Equation (2.10) and $\rho_{\mathrm{rad}}(\omega) = u_{\mathrm{avg}}(\omega)\rho(\omega)$. Alternatively, we may rewrite this expression in terms of the incident light intensity $I(\omega) = u_{\mathrm{avg}}(c/n)$:

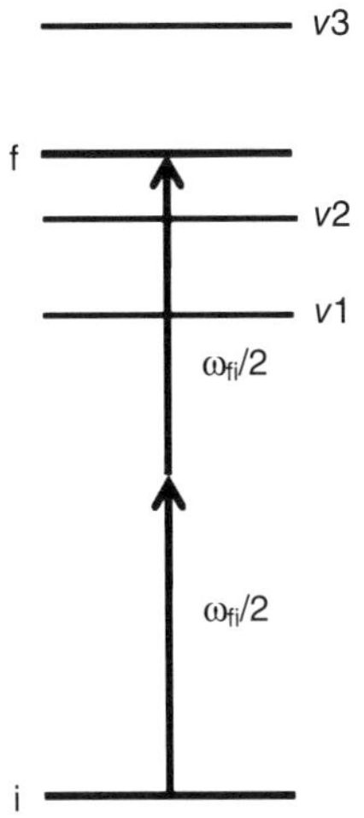

Figure 11.1. Two-photon absorption. The system makes a transition from initial state i, usually the ground state, to a higher-energy final state f by absorbing two photons of frequency $\omega_{fi}/2$. Quantum mechanically, the process proceeds through all possible states $v1$, $v2$, $v3$, . . . as intermediate states; these states may be either higher or lower in energy than f.

$$R_{if}(\omega) = \frac{\pi I^2(\omega)\rho(\omega)}{2\hbar^4\varepsilon_0^2 n^2 c^2}\left|\sum_v \frac{\mu_{fv}\cdot e\,\mu_{vi}\cdot e}{\omega_{vi}-\omega}\right|^2. \tag{11.2b}$$

Equation (11.2) gives the transition rate for a single absorber that has particular orientations of its transition dipole moments, μ_{fv} and μ_{vi}, relative to the polarization vector of the light in the laboratory-fixed coordinate system, **e**. When we dealt with one-photon absorption in Chapter 3, a random distribution of molecular orientations was assumed to convert the initial expression for the transition rate, Equation (3.49), to the orientationally averaged result, Equation (3.51). The corresponding averaging is more complicated in the two-photon case because it involves multiple pairs of transition moments. This problem has been worked out in detail (Monson and McClain, 1970) and will not be further discussed here.

The energy denominator expresses the fact that the process becomes more probable when some $|i\rangle \rightarrow |v\rangle$ transition occurs near the radiation frequency ω. When this denominator is small, the finite lifetime of the intermediate states must be considered just as it was for resonance Raman, the result being that an extra term $-i\gamma/2$ is added to the denominator. However, when the applied light frequency approaches resonance with an allowed one-photon absorption transition, the process consisting of two sequential one-photon absorptions becomes more probable than the coupled two-photon absorption described by Equation (11.2). The distinction is that in a true two-photon absorption process, there is no dephasing or energy relaxation of the intermediate state. The energies and phases of the states $|i\rangle$, $|v\rangle$, and $|f\rangle$ are well defined through the entire process. If instead the system absorbs the first photon to make a transition to state $|v\rangle$, that state can then lose energy through any of the relaxation processes described in Chapter 9, or it can undergo an interruption of its phase (dephasing) as described in Chapter 5. At that point, there is no longer a well defined phase relationship between states $|i\rangle$ and $|v\rangle$; the system has "forgotten" how it got to state $|v\rangle$. Subsequent absorption of a photon by state $|v\rangle$ to go to state $|f\rangle$ is then described as a second one-photon absorption process. The distinction is the same one as between resonance Raman scattering (a two-photon process) and absorption followed by fluorescence (two one-photon processes).

Two- or higher-photon absorption experiments are usually performed with pulsed lasers because only a pulsed laser provides a high enough peak light intensity. Linear absorption depends only on the average intensity (photons or joules per second per area) integrated over the duration of the experiment; it does not matter whether the light source is continuous or broken up into pulses. But since two-photon absorption depends on the square of the intensity, the number of transitions per second will be much higher if the source has the form of short light pulses separated by long dark periods, even if the total number of photons per second is no greater than for the corresponding continuous-wave light source. One consequence, beyond experimental feasibility considerations, is that since a short light pulse cannot have a well-defined single frequency (Section 2.3), Equation (11.2) should be integrated over the spectral bandwidth of the field or fields used.

One-photon absorption strengths can be measured in units of absorption cross section, σ, as defined in Equation (4.9). Equation (4.9) is obtained by integrating, from $z = 0$ to $z = \ell$, the following expression, which describes the attenuation of the incident light:

$$dI(\omega, z)/dz = -I(\omega, z)\sigma N, \tag{11.3}$$

where N is the number density of absorbers. The corresponding expression for absorption of two photons from the same field is

$$dI(\omega, z)/dz = -I^2(\omega, z)\delta N, \tag{11.4}$$

where δ is the two-photon cross section, and I is generally expressed in units of photons per area per second rather than energy per area per second. With I in these units and N in units of molecules/volume, δ has units of $(\text{area})^2 \cdot (\text{photon/s})^{-1} \cdot \text{molecule}^{-1}$. It is often expressed in units of marias (for Maria Goeppert-Mayer, who originally developed the theory of two-photon absorption) where 1 maria = 10^{-50} $\text{cm}^4 \cdot \text{s} \cdot \text{molecule}^{-1} \cdot \text{photon}^{-1}$, and "strong" two-photon cross sections when there is no intermediate state resonance are on the order of 10–100 marias. The two-photon cross section δ may be related to the calculated two-photon transition rate in Eq. (11.2) following the general approach outlined in Section 4.2 after carrying out the appropriate orientational averaging. Because two-photon absorption is usually quite weak, the attenuation of the incident beam is hard to measure directly. Most two-photon absorption experiments actually detect the excited state formed by some more sensitive method, often fluorescence.

Two-photon absorption involves two dot products of molecular transition dipole moments with polarization vectors for the radiation. Therefore, the symmetry of the effective two-photon absorption operator is that of products of coordinates: xx, yy, zz, xy, and so on, the same as for Raman. The selection rules for two-photon absorption are therefore also the same as for Raman, with the distinction that two-photon absorption usually connects two different electronic states, while Raman usually involves vibrational transitions within the ground electronic state. Two-photon absorption shares with Raman the selection rule that in molecules with a center of symmetry, only g $\leftrightarrow$ g and u $\leftrightarrow$ u transitions are allowed, while only g$\leftrightarrow$u are one-photon allowed. There is complete complementarity between one-photon and two-photon allowed transitions in molecules with inversion symmetry.

Three- or more-photon absorption processes are derived in much the same way. The rate for absorption of n photons from the same field goes as the nth power of the incident intensity. The dependence on higher powers of the light intensity makes such processes even more strongly dependent on having a light source with high peak power.

11.3. PUMP–PROBE SPECTROSCOPY: TRANSIENT ABSORPTION AND STIMULATED EMISSION

In a generic pump–probe experiment, a sample is exposed to two light pulses separated by a time delay that is usually variable. The two pulses may come from splitting a single laser pulse and thus have identical spectral and temporal properties, or they may be from different sources. The "pump" pulse promotes molecules from their initial Boltzmann distribution to an excited state or mixture of excited states, and the subsequent transmission of the probe pulse through the sample is monitored. As the probe passes through the sample, it can be absorbed, reducing its intensity, or it can stimulate emission from the sample, increasing its intensity. The balance between these two processes determines whether the probe emerges with higher or lower intensity than it had when it entered (Fig. 11.2).

The transmitted probe intensity depends on the time delay between pump and probe for a number of reasons: vibrational relaxation and solvent reorganization within the initially prepared excited electronic state, radiative and/or nonradiative decay of the excited state back to the ground state, radiationless transitions to other excited electronic states (such as singlet to triplet), chemical reactions of the photoexcited molecules, and any other photoinduced

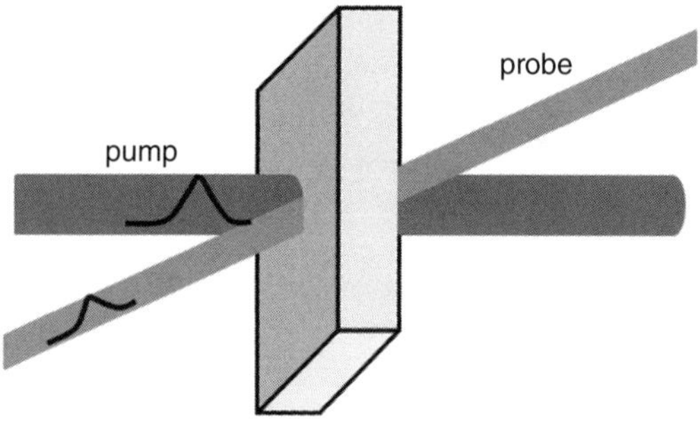

Figure 11.2. Schematic of a pump–probe experiment. The time delay between pump and probe is varied by changing the distance the probe beam travels before reaching the sample. The angle between pump and probe is normally much smaller than drawn in this cartoon.

processes. The transmitted intensity also depends on the wavelengths of both pump and probe since the former determines the nature of the excited states formed, and the absorption and stimulated emission coefficients of those states depend on probe wavelength. Finally, the probe transmission also depends on the relative polarization of pump and probe and the dynamics of rotational motion of the pumped molecule. We will not further discuss polarization effects in pump–probe experiments. As long as both the pumped and the probed transitions involve electric dipole interactions, which is usually the case, the sensitivity of the signal to rotational motion can be eliminated by choosing the angle between the linear polarizations of pump and probe to be 54.7° (the "magic angle") as discussed in Section 9.7.

Usually, the experiment is set up such that the pump and probe beams cross within the sample at a small angle, thus allowing them to be spatially separated outside the sample. Alternatively, if pump and probe are at sufficiently different wavelengths, they may pass through the sample collinearly and a dichroic beamsplitter which reflects only one of the wavelengths may be used to separate the pump and probe after they have passed through the sample. In either case, one can measure the intensity of the probe beam alone. In that case, the signal is usually expressed as the pump-induced change in transmittance or absorbance. That is, one measures $T_{ref} = I_{inc}/I_{trans}$ or, equivalently, converts to absorbance units by forming $A_{ref} = \log(T_{ref})$ for the probe beam passing through the sample in the absence of the pump, and then measures the same thing, T_{sig} or A_{sig}, in the presence of the pump as a function of pump–probe delay τ. The quantity reported is usually either

$$\Delta T(\lambda_{pump}, \lambda_{probe}, \tau) = (T_{sig} - T_{ref})/T_{ref}, \tag{11.5a}$$

or

$$\Delta A(\lambda_{pump}, \lambda_{probe}, \tau) = (A_{sig} - A_{ref})/A_{ref}, \tag{11.5b}$$

which expresses the change in transmittance or absorbance of the sample at wavelength λ_{probe} measured at a time τ after pumping at wavelength λ_{pump}. If there is net absorption, then ΔT will be negative and ΔA positive; the opposite holds when stimulated emission dominates the signal.

Most experiments vary only one of the three parameters λ_{pump}, λ_{probe}, and τ at a time. Usually either the time delay τ is varied at fixed λ_{pump} and λ_{probe} (kinetics at fixed wavelength) or λ_{probe} is varied at fixed λ_{pump} and τ (spectrum at fixed delay). In some experiments, a probe pulse is generated that contains a very broad spectrum of wavelengths ("white light"), either by using a pulsed lamp that produces intrinsically broadband radiation or by starting with a short laser pulse and converting it to a broadband frequency distribution through the optical phenomenon of self-phase modulation. Here the entire broadband probe pulse is passed through the sample, then separated into its frequency components with a grating or prism spectrograph and sent onto a multichannel detector. In this way, the entire spectrum can be obtained simultaneously at each pump–probe time delay (Fig. 11.3).

How one thinks about, models, and interprets a pump–probe experiment depends on whether the experiment can be adequately described as two sequential one-photon processes or needs to be discussed as a coherent two-photon process. The former holds when the off-diagonal density matrix terms induced by the pump pulse, which are initially zero for a system at thermal equilibrium, have decayed back to zero by the time the probe pulse arrives. Then the description of the process is conceptually simple: the pump pulse creates populations in the various excited states, populations that are different from the initial Boltzmann distribution and evolve with time, but the system retains no explicit "memory" of the details of the pump pulse that

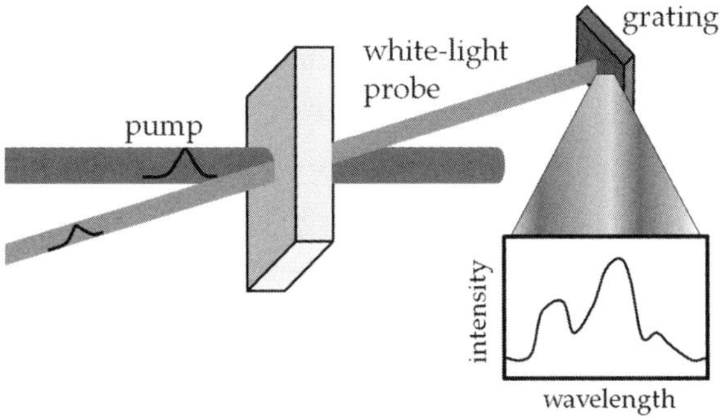

Figure 11.3. Highly schematic diagram of a time-resolved absorption experiment using a "white-light" probe. The probe is variably delayed relative to the pump, passes through the sample, and is spectrally dispersed onto a detector. A shutter is used to block and unblock the pump beam and the spectra of the probe with and without the pump are compared.

excited it. The probe pulse comes along at a later time and interacts with whatever distribution of populations exists at that time delay. We can then describe the process as one-photon absorption of the pump pulse followed by some intervening kinetics, followed by a combination of one-photon absorption and one-photon stimulated emission from the probe pulse. This situation always holds at sufficiently long pump–probe separations. But when the pulses are short and arrive close together, the coherences in the density matrix may also be important at the time the probe starts to interact with the sample, and then the whole pump–probe sequence has to be described as a coupled two-photon process. Also, regardless of how fast the dephasing is, if the pump and probe pulses overlap in time, it is not possible to know that the "pump" interacted with the sample before the "probe" did; all possible time orderings of the interactions of the molecule with the electric fields of pump and probe must be considered. Usually, this also requires a coupled two-photon description.

When the pump and probe pulses are well separated in time and the dephasing is fast enough that all pump-induced coherences have decayed by the time the probe arrives, the pump–probe experiment can be described using a kinetic scheme. This can be as complicated or as simple as the experimenter wants to consider. Assume that in the pumping step, only two states are important, the ground state $|g\rangle$ and some excited state $|e\rangle$. The rate of change of population in the upper state is given in general by

$$dN_e/dt = W_{ge} - W_{eg},$$

where N_g and N_e are the number densities (molecules per unit volume) in each of the two states and W_{ge} and W_{eg} are the rates of transitions per unit volume from g to e and from e to g, respectively. These rates are given by the Einstein coefficients [Eq. (4.1)]:

$$\begin{aligned} W_{ge} &= N_g B(\omega)\rho_{rad}(\omega) \\ W_{eg} &= N_e B(\omega)\rho_{rad}(\omega) + \mathrm{N_e A} \end{aligned}$$

where $B(\omega)$ is the Einstein coefficient for either absorption or stimulated emission (recall these coefficients are equal, but depend on wavelength), A is the Einstein coefficient for spontaneous emission, and $\rho_{rad}(\omega)$ is the energy density (energy per unit volume per unit frequency) of the applied pump field, which is related to the intensity $I(\omega)$ of the pump pulse by $\rho_{rad}(\omega) = I(\omega)(n/c)\rho(\omega)$, where n is the refractive index, and $\rho(\omega)$ is the frequency distribution of the radiation. In the limit where the fraction of molecules pumped is small, we can neglect emission and assume N_g is approximately constant with time, leaving

$$dN_e = [N_g BI(\omega)(n/c)\rho(\omega)]dt, \tag{11.6a}$$

or using Equation (4.14), $B\rho(\omega) = (c/\hbar n)\sigma(\omega)/\omega$, where σ is the absorption cross section,

$$dN_e = [N_g I(\omega)\sigma(\omega)/\hbar\omega]dt. \tag{11.6b}$$

Assuming the population of excited molecules in the initial thermal equilibrium was negligible, this integrates to

$$N_e(t) = \frac{N_g\sigma(\omega)}{\hbar\omega}\int_0^t d\tau I(\omega,\tau). \tag{11.7}$$

That is, the number of excited states formed is simply proportional to the absorption cross section and the integrated intensity of the pumping pulse up until the observation time. In the limit of well-separated pump and probe pulses, we make no observations until the pump pulse has completely passed through the sample, and then it is okay to replace the upper limit of the time integral by ∞:

$$N_e(t > t_{\text{pulse}})/N_g = \frac{\sigma(\omega)}{\hbar\omega}\int_0^\infty d\tau I(\omega,\tau). \tag{11.8}$$

This gives the fraction of molecules in the initial sample that are excited by the pump pulse, in the limit that the pump does not significantly deplete the ground state population.

The initially prepared excited state $|e\rangle$ may then evolve with time in a variety of ways. For interpreting pump–probe experiments, one usually does not try to keep track of every individual quantum state, but rather groups of states are lumped together as "species" that have distinguishable spectral and/or kinetic properties. Let us just call all those species $\{j\}$, and ask what happens to a probe pulse on passing through a sample that consists of some mixture of a number of different species that have different absorption coefficients at different frequencies. On passing through an infinitesimal path length dz, the intensity of the probe pulse $I(\omega,t)$ will be changed by

$$\begin{aligned} dI(\omega,t) &= \hbar\omega\sum_j\sum_{k<j}[W_{jk}(\omega,t) - W_{kj}(\omega,t)]dz \\ &= \hbar\omega\sum_j\sum_{k<j}[N_j(t)B_{jk}(\omega)\rho_{\text{rad}}(\omega,t) + N_j(t)A_{jk}(\omega) \\ &\quad - N_k(t)B_{kj}(\omega)\rho_{\text{rad}}(\omega,t)]dz. \end{aligned} \tag{11.9}$$

which accounts for gain due to emission (transitions from states j to lower states k) and loss due to absorption (transitions from states k to higher states j).

Normally, spontaneous emission makes a negligible contribution to the change of intensity of the probe beam, because the number of photons emitted per second is small and because only a tiny fraction of them have the same wavevector as the probe beam. Therefore, we can drop the N_jA_{jk} term in Eq. (11.9). Substituting $\rho_{rad}(\omega,t) = I(\omega,t)(n/c)\rho(\omega)$ leaves

$$dI(\omega,t) = I(\omega,t)(\hbar\omega n/c)\sum_j \sum_{k<j} [N_j(t)B_{jk}(\omega)\rho(\omega) - N_k(t)B_{kj}(\omega)\rho(\omega)]dz, \tag{11.10}$$

and then using $B\rho(\omega) = (c/\hbar n)\sigma(\omega)/\omega$ gives

$$dI(\omega,t)/I(\omega,t) = \sum_j \sum_{k<j} [N_j(t) - N_k(t)]\sigma_{kj}(\omega)dz, \tag{11.11}$$

which, when integrated over the path length from $z = 0$ to $z = \ell$, becomes

$$I_\ell(\omega,t)/I_0(\omega,t) = T(\omega,t) = \exp\left\{\sum_j \sum_{k<j} [N_j(t) - N_k(t)]\sigma_{kj}(\omega)\ell\right\} \tag{11.12}$$

The transmission $T(\omega,t)$ depends on time through the populations of the various states and on frequency through the absorption cross section of each species. In principle, one rarely measures $T(\omega,t)$ directly, but rather the integrated intensity of the probe pulse which has a finite spectral bandwidth and a finite temporal duration. So in principle, one has to integrate the above equation over the temporal envelope of the pulse and over its spectral distribution, but this is often neglected when the kinetics are slow compared with the pulse duration and the spectra are broad.

Consider a simple but typical kinetic scheme:

$$A \xrightarrow{h\nu} B^* \xrightarrow{k_{vr}} B \xrightarrow{k_{ic}} A^* \xrightarrow{k'_{vr}} A. \tag{11.13}$$

Here, A represents the thermally equilibrated ground state, B^* is an initially excited electronic state reached by light absorption, B is this state after vibrational and solvent relaxation, and A* is the vibrationally excited ground state reached from B through internal conversion. Usually, one would choose the pump wavelength λ_1 to be near the maximum of the $A \rightarrow B^*$ electronic absorption spectrum. In a ground-state recovery experiment, one would choose the probe wavelength λ_2 to be identical or close to that of the pump ("one-color" experiments, where pump and probe are formed by splitting a single laser pulse, tend to be technically easier than two-color experiments). At very short pump–probe time delays, there will be contributions to the signal from $B^* \rightarrow A$ stimulated emission, but usually $B^* \rightarrow B$ is quite fast, and unless B or A^* just happens to also absorb strongly at λ_2, the net result will be that the probe will be less strongly absorbed because the concentration of A is reduced. That is, the transmission will be increased (transient bleaching). It will return to its initial value only after all of the A has been regenerated. This time depends on all three rate constants k_{vr}, k_{ic}, and k'_{vr}, with k_{ic} often but not always being rate-limiting. If on the other hand one wants specifically to measure the lifetime of the excited state B, one would try to choose a probe wavelength λ_3 that is not absorbed by A but is on resonance with some electronic transition $B \rightarrow C$ (i.e., if B is the first excited singlet state, C might be some higher singlet state). Then, the transmission at λ_3 is near unity in the absence of the pump, is reduced at short positive pump–probe delays where there is a high concentration of B (transient absorption), and decays with rate constant k_{ic}.

11.4. VIBRATIONAL OSCILLATIONS AND IMPULSIVE STIMULATED SCATTERING

If the light pulses used are short compared with vibrational frequencies, pump–probe or time-resolved fluorescence signals may have components due to molecular vibrations (or collective phonons in crystals—see Chapter 15). These processes can be interpreted simply, although crudely, using a classical-like model for the vibrational motion. It is easiest to interpret the case of oscillations in fluorescence or transient absorption. Consider a molecule having a ground state $|g\rangle$ and two excited states $|e1\rangle$ and $|e2\rangle$ whose potential energy surfaces all have different shapes along one or more vibrational coordinates (Fig. 11.4). If the molecule is excited from $|g\rangle$ to $|e1\rangle$ using an electronically resonant light pulse that is much shorter than the vibrational period, then the initially excited molecules have essentially the same distribution of vibrational coordinates as the equilibrated ground state did. But since the excited state has a different potential minimum than the ground state, this distribution of coordinates will then evolve with time, undergoing oscillatory motion on the excited state potential surface. Because all three surfaces have different shapes, the energy difference between $|g\rangle$ and $|e1\rangle$ and between $|e1\rangle$ and $|e2\rangle$ is a function of vibrational coordinate. Therefore, as the nuclei move on the $|e1\rangle$ state surface, the $|e1\rangle \rightarrow |e2\rangle$ absorption spectrum and $|e1\rangle \rightarrow |g\rangle$ emission spectrum will vary with time. If one fixes the probe frequency or the frequency of the detected spontaneous emission, the intensity of the signal at that frequency will oscillate in time at the excited state vibrational frequency or frequencies, giving information about the vibrations coupled to the electronic transition in much the same manner as a resonance Raman spectrum does.

Vibrations on the ground-state potential energy surface can be excited by a similar mechanism through a process known as impulsive stimulated Raman scattering (ISRS). Here, the interaction with a short pump pulse, instead of transfering the ground-state vibrational wavefunction to an excited-state potential energy surface, instead creates a coherent superposition of one or more ground-state vibrational wavefunctions. This occurs through a resonant or nonresonant Raman process (Chapter 10), which takes part of the $v = 0$ wavefunction and promotes it to $v = 1$, and, potentially, other higher v states.

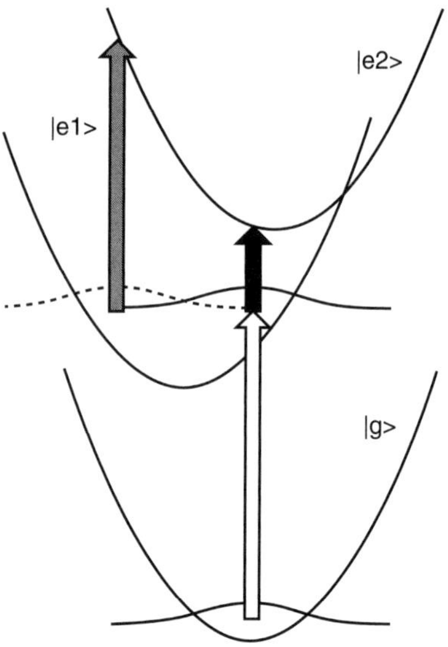

Figure 11.4. Origin of vibrational oscillations in transient absorption (and stimulated emission) experiments. Optical excitation with a pulse short compared with the vibrational period (white arrow) places the ground-state vibrational coordinate distribution (solid curve) onto the potential energy surface of excited state |e1⟩, where it becomes a moving wavepacket oscillating between the distributions shown as solid and dashed curves. If absorption to a higher excited state |e2⟩ is probed by a time-delayed pulse, the peak absorption frequency oscillates between the values shown by the black and gray arrows.

Note that this can only happen if the pump pulse is spectrally broad compared with the vibrational frequency, such that the condition $\omega_1 - \omega_2 = \omega_{vib}$ can be satisfied with ω_1 and ω_2 both coming from the same pump pulse. Recalling the Fourier transform relationship between frequency and time (Section 2.3), this corresponds to the requirement that the pump pulse be temporally short compared with the vibrational period.

This coherent superposition of ground-state vibrational wavefunctions corresponds to a wavepacket whose position oscillates back and forth on the ground-state potential energy surface. To see this, consider the simplest possible case where the pump pulse is a delta function in time and transfers exactly half of the wavefunction from v = 0 to $v = 1$. At $t = 0$, the total vibrational wavefunction is then

$$\Psi(q,0) = \frac{1}{\sqrt{2}}\{\psi_0(q) + \psi_1(q)\}. \tag{11.14}$$

At a later time, it is given by

$$\begin{aligned}\Psi(q,t) &= \frac{1}{\sqrt{2}}\{e^{-iHt/\hbar}\psi_0(q)+e^{-iHt/\hbar}\psi_1(q)\}\\ &= \frac{1}{\sqrt{2}}\{\psi_0(q)+\exp(-i\omega_{\text{vib}}t)\psi_1(q)\},\end{aligned} \tag{11.15}$$

where the v = 0 state is taken to be the zero of energy. The expectation value of the vibrational coordinate as a function of time is given by

$$\begin{aligned}\langle q\rangle_t &= \langle\Psi(t)|q|\Psi(t)\rangle\\ &= \frac{1}{2}\{\langle\psi_0|q|\psi_0\rangle+\langle\psi_1|q|\psi_1\rangle+\exp(-i\omega_{\text{vib}}t)\langle\psi_0|q|\psi_1\rangle+\exp(i\omega_{\text{vib}}t)\langle\psi_1|q|\psi_0\rangle\}\\ &= \frac{1}{2}\left\{(2\cos\omega_{\text{vib}}t)\frac{1}{\sqrt{2}}\right\} = \frac{1}{\sqrt{2}}\cos(\omega_{\text{vib}}t).\end{aligned} \tag{11.16}$$

Thus the average value of the vibrational coordinate oscillates between $+(1/\sqrt{2})$ and $-(1/\sqrt{2})$ at the ground-state vibrational frequency. As the molecule oscillates along the vibrational coordinate, other optical properties also change; in particular, the wavelength of maximum absorption to an excited state will be modulated, so in a pump–probe absorption experiment at a fixed probe wavelength, the signal will oscillate at the vibrational frequency. When the pump pulse is electronically resonant, generally, both an excited-state wavepacket (Fig. 11.4) and a ground-state one (Fig. 11.5) are created. While they oscillate at different frequencies if the ground and excited electronic states have different curvatures, the two contributions can be difficult to distinguish.

11.5. SECOND HARMONIC AND SUM FREQUENCY GENERATION

A related optical process that depends on higher powers of the incident intensity is sum frequency generation. In sum frequency generation, two beams characterized by wavevectors $\mathbf{k}_1$ and $\mathbf{k}_2$, and frequencies ω_1 and ω_2, are directed into a material, and light of a new frequency $\omega_{\text{sum}} = \omega_1 + \omega_2$ and wavevector $\mathbf{k}_{\text{sum}} = \mathbf{k}_1 + \mathbf{k}_2$ comes out. Unlike two-photon absorption, there is no net change in state of the material and no net transfer of energy or momentum between radiation and matter. The matter acts as a "catalyst" to change the radiation field. Second harmonic generation is a special case where the two input frequencies are the same. In a quantum mechanical description, sum frequency generation can be considered to proceed through the matter transitions

$$|\text{i}\rangle \to |\text{v}_1\rangle \to |\text{v}_2\rangle \to |\text{i}\rangle, \tag{11.17}$$

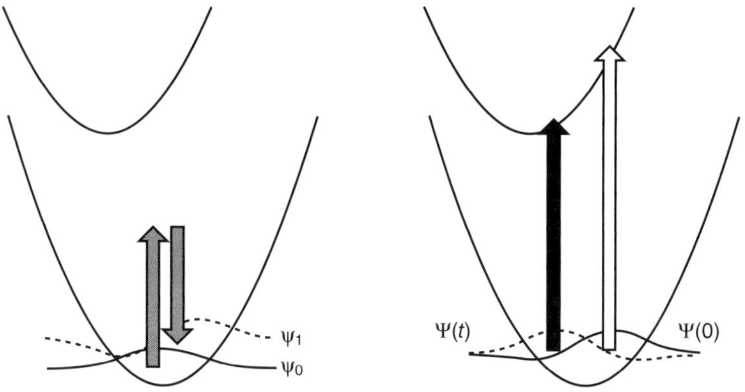

Figure 11.5. Creation of a ground-state wavepacket through impulsive stimulated scattering. Left: a very short light pulse interacts with a system initially in the vibrational ground state, ψ_0, and through a Raman process (gray arrows) transfers part of the amplitude to the first excited vibrational state, ψ_1. Right: the coherent superposition of these two vibrational states at $t = 0$ or $t = 2n\pi/\omega_{vib}$ (black curve) and at $t = (2n + 1)\pi/\omega_{vib}$ (dashed curve). Transient absorption to a higher electronic state measured at different wavelengths (white and black arrows) oscillates at ω_{vib} with different phases.

where each of the first two steps step annihilates one photon and the final step creates a photon of frequency ω_{sum}.

In second harmonic generation, a field of frequency ω and wavevector **k** is directed into a nonlinear material (usually a crystal for reasons discussed below), and a new field of frequency 2ω and wavevector 2**k** is generated. This is a second-order nonlinear process, and the harmonic field is generated by a polarization given by [Eq. (11.1a)]

$$P^{(2)}(t) = \chi^{(2)}|E(t)|^2. \tag{11.18}$$

If the driving field is monochromatic, $E(t) = E_0\cos(\omega t)$ (we neglect the vector properties of the field for now), then

$$|E(t)|^2 = E_0^2\cos^2(\omega t) = (E_0^2/2)[\cos(2\omega t) + 1], \tag{11.19}$$

and we see that the polarization induced through the second-order nonlinearity has both a constant component and a component oscillating at twice the applied frequency. This polarization will radiate a field at frequency 2ω, the second harmonic.

Second harmonic generation can occur only in media that are not centrosymmetric. To see why, realize that both the electric field E and the polarization $P^{(2)}$ are vector quantities. If we invert the coordinate system, $(x, y, z) \rightarrow (-x, -y, -z)$, then the electric field changes sign. If the medium is centrosymmetric, that is, inversion is one of its symmetry operations, then the polarization must also have the same magnitude and a reversed sign. This says

$$\mathbf{P}^{(2)}(x, y, z) = \chi^{(2)}|\mathbf{E}(x, y, z)|^2,$$

and

$$\mathbf{P}^{(2)}(-x, -y, -z) = -\mathbf{P}^{(2)}(x, y, z) = \chi^{(2)}|\mathbf{E}(-x, -y, -z)|^2 = \chi^{(2)}|\mathbf{E}(x, y, z)|^2,$$

thus we have

$$\chi^{(2)}|\mathbf{E}(x, y, z)|^2 = -\chi^{(2)}|\mathbf{E}(x, y, z)|^2,$$

which is satisfied only if $\chi^{(2)} = 0$. Thus, only noncentrosymmetric media can have nonvanishing $\chi^{(2)}$. This means second harmonic generation can occur in certain crystals and at interfaces between unlike media, but not in isotropic gases or liquids.

The requirement that both energy and momentum must be conserved (since there is no net change in the state of the matter) presents some complications. In a medium, $|\mathbf{k}| = \mathrm{k} = \mathrm{n}\omega/\mathrm{c}$, where n is the refractive index. Therefore, the wavevector matching condition, $\mathrm{k}(2\omega) = 2\mathrm{k}(\omega)$, becomes

$$n_{2\omega}2\omega/\mathrm{c} = 2n_{\omega}\omega/\mathrm{c} \text{ or } n_{2\omega} = n_{\omega}. \tag{11.20}$$

The physical basis for this requirement is that as the fundamental field at ω propagates through the sample, it continually generates the second harmonic field at 2ω. If the wavelength at 2ω is not exactly half the wavelength at ω, then the different components of the new field generated at different positions cannot all interfere constructively, and the amplitude of the 2ω field will be greatly reduced (Fig. 11.6). This does not matter much if the radiation–matter interaction occurs only over a very short distance (on the order of the wavelength of the light), but the greater the interaction length, the more severe the penalty for any refractive index mismatch. One can show that the second harmonic intensity depends on interaction length L and wavevector mismatch, $\Delta k = 2k(\omega) - k(2\omega)$, as (Yariv, 1975)

$$I(2\omega) \sim |\chi^{(2)}(-2\omega; \omega, \omega)|^2 I^2(\omega)L^2[\sin(\Delta kL/2)/(\Delta kL/2)]^2, \tag{11.21}$$

so if $\Delta k = 0$, the intensity grows quadratically with path length, but if $\Delta k \neq 0$, then the second harmonic intensity "saturates" at some short interaction length and the total second harmonic intensity cannot be very large. In most media (those with normal dispersion), $n_{2\omega} > n_{\omega}$ so $\Delta k = 0$ cannot be satisfied. However, it can be in certain classes of anisotropic crystals in which the refractive index is different for light polarized along different crystal axes. By carefully adjusting either the angle between the crystal axes and the fundamental beam (angle tuning) or the temperature of the crystal (temperature tuning),

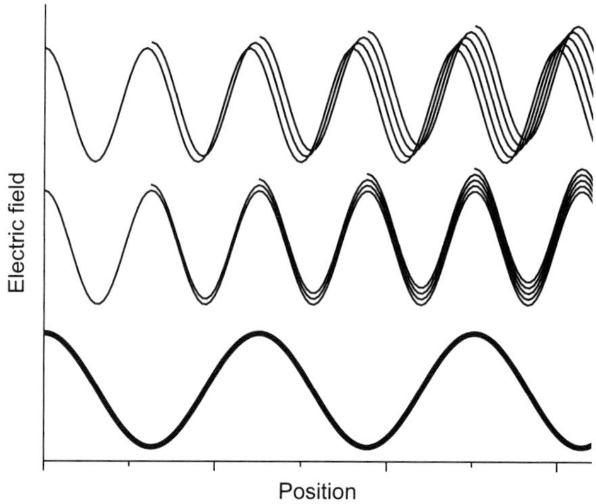

Figure 11.6. Illustration of phase matching in second harmonic generation. As the fundamental (thick curve) propagates through the medium, it launches second harmonic waves with a well defined phase relationship. If the refractive index is the same at ω and 2ω (center), all waves add constructively to form an intense second harmonic beam. If the refractive indices differ (top), the second harmonic waves get out of phase.

the condition $n_{2\omega} = n_{\omega}$ can be met, and useful output power can be generated over a large interaction distance.

Second harmonic generation is widely used as a technique for converting the frequency of a laser from infrared to visible or from visible to UV. It can also be used to some degree as a spectroscopic tool because it undergoes enhancement as resonance with either the fundamental or the harmonic transition is approached. In particular, it is often used for probing both the spectra and the orientation (through the polarization dependence of the signal) of molecules at interfaces. Since isotropic materials do not contribute to $\chi^{(2)}$, the few molecules at an interface between two materials can be probed without interference from a much larger quantity of material in the bulk.

Sum frequency generation is just like second harmonic generation except that two input fields at different frequencies are involved, and $\omega_{sum} = \omega_1 + \omega_2$. The phase matching condition $k(\omega_{sum}) = k(\omega_1) + k(\omega_2)$ requires $n_{sum} = n_1 + n_2$, which can be achieved through angle or temperature tuning, or ignored for spectroscopic applications at interfaces where the effective interaction length is very short. Sum frequency generation is widely used as an optical technique for shifting laser output to shorter wavelengths. For example, the "third harmonic" of the Nd:YAG laser at 1064 nm is generated by first using second harmonic generation to convert part of the fundamental to 532 nm, and then summing the 1064 nm with the 532 nm in another crystal to create 355 nm. It is also the principle behind many optical gating methods for time-resolving spontaneous emission. A short pulse of light at ω_1, usually in the visible or near-UV, is used to excite a sample, which emits long-lived fluorescence at slightly lower frequencies ω_2. A part of the original excitation beam at ω_1 is variably delayed and focused into a nonlinear crystal along with the fluorescence, generating a UV signal at ω_{sum}. The intensity of the sum frequency radiation is proportional to the product of the ω_1 and ω_2 intensities during the time interval when the two beams overlap in the crystal, thereby allowing the time dependence of the fluorescence to be mapped out.

Sum frequency generation is also a useful spectroscopic tool, particularly for vibrational spectroscopy of surfaces and interfaces. These applications use ω_1 as an infrared beam tunable around some vibrational resonance, while ω_2 is in the visible and far from any material resonances. ω_2 is held fixed while ω_1 is tuned over a narrow region, and the intensity of light generated at $\omega_{sum} = \omega_1 + \omega_2$ is monitored. When ω_1 is resonant with a $|0\rangle \rightarrow |V\rangle$ vibrational transition, the intensity at ω_{sum} increases. In order for enhancement to be observed, the vibrational transition must be both infrared and Raman allowed, as the first step (one interaction with ω_1 to promote the system from $|0\rangle \rightarrow |V\rangle$) is like an infrared absorption, while the second pair of steps (interaction with ω_2 and then ω_{sum} to take the system from $|V\rangle \rightarrow |e\rangle \rightarrow |0\rangle$) is like a Raman transition (Fig. 11.7).

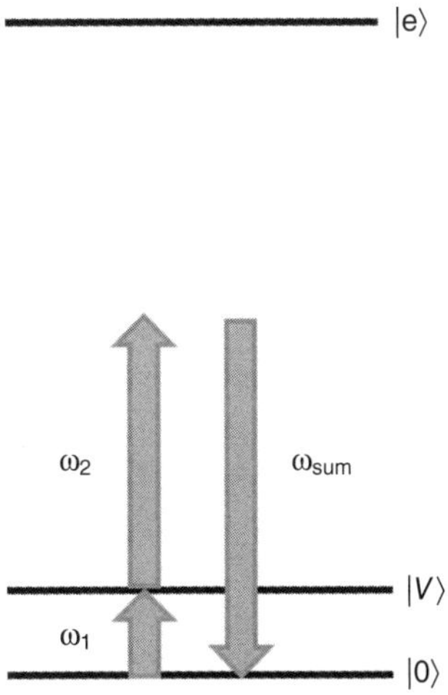

Figure 11.7. Sum frequency generation spectroscopy. ω_2, usually in the visible, is normally held fixed while ω_1, an infrared wavelength, is tuned, and signal is detected at ω_{sum}. When ω_1 comes into resonance with a vibrational transition that is both IR allowed ($|0\rangle \rightarrow |V\rangle$) and Raman allowed ($|V\rangle \rightarrow |e\rangle \rightarrow |0\rangle$), the signal strength greatly increases. The sensitivity may be further increased by selecting ω_2 such that $\omega_1 + \omega_2$ is in or near resonance with excited state $|e\rangle$.

11.6. FOUR-WAVE MIXING

Sum frequency and second harmonic generation are examples of three-wave mixing processes in which two input fields, ω_1 and ω_2 or ω_1 and ω_1, interact with a material system to generate a third field at ω_{sum}. Considerably more common are four-wave mixing processes, which can occur in isotropic media, such as liquids.

Four-wave mixing includes such phenomena as third harmonic generation, where three photons at ω_1 are destroyed while creating one at $3\omega_1$. It also includes other processes in which there is no net transfer of energy or momentum between radiation and matter but the state of the radiation is changed. For example, in coherent antistokes Raman scattering (CARS), there are two input fields at frequencies ω_1 and ω_2. Two photons are lost from the ω_1 field while one is gained at ω_2 by stimulated emission, and the resulting polarization generates a fourth field at frequency $\omega_{CARS} = 2\omega_1 - \omega_2$. Since $2\omega_1 - \omega_2 - \omega_{CARS} = 0$, there is no change in the state of the matter. When $\omega_1 - \omega_2$ equals a ground-state vibrational frequency of the material, the four-wave mixing signal is greatly enhanced by the intermediate state resonance and the signal at ω_{CARS} becomes large, so this is a way to measure Raman spectra (Fig. 11.8). Since momentum must also be conserved in this process, $\mathbf{k}_{CARS} = 2\mathbf{k}_1 - \mathbf{k}_2$ and the signal comes out as a laser-like beam with a well defined direction, unlike spontaneous Raman scattering. CARS is therefore particularly useful when ordinary spontaneous Raman scattering is hidden under large fluorescence backgrounds.

11.7. PHOTON ECHOES

The photon echo (Fig. 11.9), analogous to the spin echo in NMR, is a type of four-wave mixing that can measure dephasing rates for electronic states of

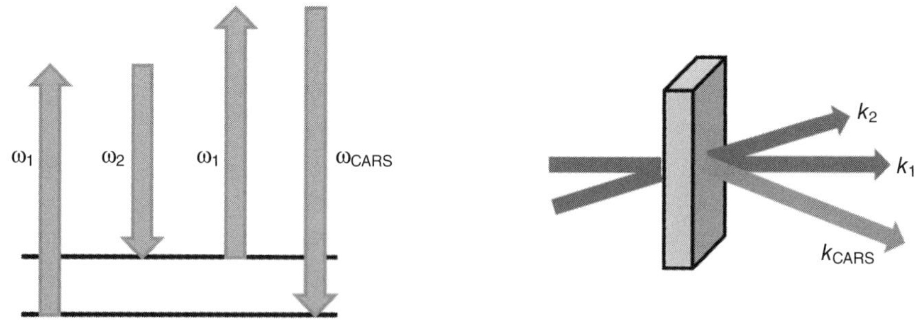

Figure 11.8. Coherent anti-Stokes Raman scattering (CARS) as an example of a four-wave mixing process. The material system interacts with two photons at ω_1 and one photon at ω_2 to produce a signal at a new frequency $\omega_{CARS} = 2\omega_1 - \omega_2$, undergoing no net change in its state. Resonances occur whenever $\omega_1 - \omega_2$ equals the frequency of a Raman-allowed transition. The new field is produced with a well-defined wavevector $\mathbf{k}_{CARS} = 2\mathbf{k}_1 - \mathbf{k}_2$.

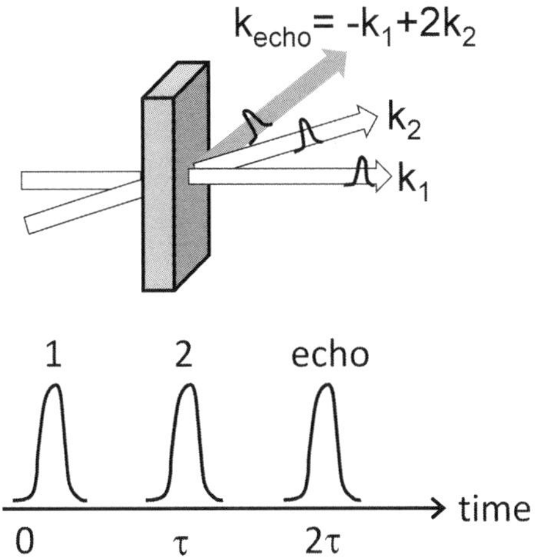

Figure 11.9. Two-pulse photon echo. The first-order coherence created by pulse 1 decays during time τ at the pure dephasing rate Γ_e. The remaining coherence is converted to a third-order coherence by pulse 2, which also reverses the sign of the dephasing due to inhomogeneous broadening. The different oscillator frequencies rephase at time 2τ to form an "echo" in the wavevector-matched direction.

molecules, often in condensed phases where these dephasing processes are subpicosecond. The photon echo has a specific ability to separate the homogeneous and inhomogeneous contributions to the electronic spectral breadth. The photon echo experiment can be described formally using density matrix methods as discussed in Chapter 5, but the description requires evolving the density matrix to third order in the applied field, and the mathematics becomes somewhat involved, so only an outline is presented here.

Consider a material system having just two important energy levels, $|1>$ and $|2>$, which can be coupled by a transition dipole μ. Coherences between these two levels undergo dephasing due to the solvent environment with a rate constant Γ_e. There may also be inhomogeneous broadening, such that for any given molecule, the energy separation between states 2 and 1 is $\hbar(\omega_{21} + \delta)$, where δ is the shift from the average transition frequency experienced by a molecule in a particular environment. In the simplest form of the photon echo experiment, two short pulses resonant with the $|1\rangle \rightarrow |2\rangle$ transition arrive at the sample separated by a time interval τ. The system interacts once with the first field to put its density matrix into a coherence, $\rho_{12}^{(1)}(t)$, as described in Section 5.3. This coherence evolves freely during the interpulse time delay τ, decaying with its dephasing rate as $\exp(-\Gamma_e t)$. When the second pulse arrives at time τ, the system interacts with it twice to produce a second-order population, $\rho_{11}^{(2)}(t)$ or $\rho_{22}^{(2)}(t)$, and then a third-order coherence, $\rho_{12}^{(3)}(t)$. This coherence represents an induced polarization that radiates a fourth field whose intensity is detected as the signal. The four-wave mixing process can occur only if the coherence generated by the first pulse is still present when the second pulse arrives, so the magnitude of the generated signal will depend on pump–probe delay as $\exp(-2\Gamma_e\tau)$ (the factor of two comes from the fact that it is the square of the generated field that is detected). Therefore, one can deduce the homogeneous linewidth Γ_e from this experiment.

The photon echo experiment gives the dephasing time of the ground to excited state coherence even if the lineshape of the $|1\rangle \rightarrow |2\rangle$ transition is dominated by the inhomogeneous broadening, because the inhomogeneous width does not contribute to the time dependence of the echo. A macroscopic picture of this is that the different oscillators with different frequencies ω_{12} get out of phase with one another as the coherences evolve during the time delay τ, but application of the second and third pulses produces a sort of time reversal that causes all the different oscillators to again come back into phase at $t = 2\tau$, producing a signal that is an "echo" of the first pulse. There are many variations on the basic photon echo experiment that use more than two pulses (the most common three-pulse photon echo is also referred to as a stimulated photon echo) and that time- and/or frequency-resolve the echo pulse itself rather than merely detecting its integrated intensity as in the basic experiment.

REFERENCES AND FURTHER READING

P. R. Monson and W. M. McClain, J. Chem. Phys. **53**, 29 (1970).

W. L. Peticolas, Annu. Rev. Phys. Chem. **18**, 233 (1967).

A. M. Weiner, S. De Silvestri, and E. P. Ippen, J. Opt. Soc. Am. **2**, 654 (1985).

A. Yariv, *Quantum Electronics*, 2nd edition (John Wiley & Sons, New York, 1975).

PROBLEMS

1. Consider a molecule with an electronic transition centered at 25,000 cm^{-1} which is allowed in both one-photon and two-photon absorption. The one-photon absorption cross section at 400 nm is 10^{-16} $cm^2 \cdot molecule^{-1}$ and the cross section for absorption of two identical photons at 800 nm is 10^{-48} cm^4 s $photon^{-1}$ $molecule^{-1}$. Assume you have available two different lasers, each of which can produce an average power of 2 W at 800 nm. Laser 1 is a continuous-wave laser, for which frequency doubling is very inefficient; although it produces 2 W at 800 nm, when doubled to 400 nm, the power is only 2 μW. Laser 2 is a pulsed laser that produces its 2 W at 800 nm in the form of 1 ps duration pulses at a 100 MHz repetition rate. It can be frequency doubled much more efficiently, and its average power at 400 nm is 200 mW. Both lasers can be focused to a circular spot of diameter 10 μm.

(a) For each laser source, assuming the same concentration of molecules, calculate the ratio of one-photon absorption at 400 nm to two-photon absorption at 800 nm when each laser is putting out its maximum power at each wavelength. Assume the path length is so short that the change in beam intensity as it passes through the sample can be neglected. You may treat the pulses from the pulsed laser as having a rectangular shape, that is, the light is completely on for 1 ps every 10 ns, and completely off the rest of the time. Neglect any explicit consideration of the finite spectral bandwidth of the pulsed laser.

(b) Is this a surprising result? What important factor, apart from the different spectral bandwidths of the pulsed and cw light, has been left out of this calculation?

2. In the text, it was shown that second harmonic generation is possible only in bulk noncentrosymmetric media. Is the same true for sum frequency generation?

3. Refer to Equations (11.3) and (11.4) for the attenuation of the incident light as a function of the path length, z.

(a) Assume that in a linear absorption process, the light intensity transmitted through a path length of 5 mm is half the initial intensity. What will be the transmitted intensity if the path length is increased to 10 mm? Then answer the same question for a two-photon absorption process.

(b) In a linear absorption experiment with a fixed concentration and path length, when the initial light intensity is I_0, half the light is transmitted through the sample. What fraction of the light is transmitted if the initial intensity is increased to $2I_0$? Then answer the same question for a two-photon absorption process.

4. Azulene ($C_{10}H_8$) is famous for its anomalous photophysics. The zero–zero transition between the ground state and the first excited singlet state ($S_0 \rightarrow S_1$) is at about 14,300 cm^{-1}, while the $S_0 \rightarrow S_2$ is at almost twice that energy (~28,600 cm^{-1}). Thus, the $S_0 \rightarrow S_1$ and $S_1 \rightarrow S_2$ energy gaps are almost the same. Furthermore, it turns out that the Franck–Condon factors for $S_1 \rightarrow S_0$ internal conversion are much more favorable than those for $S_2 \rightarrow S_1$ because the change in geometry is much larger for the former transition. As a result, the S_2 state of azulene has a much longer lifetime (1.3 ns) than the S_1 state (1.1 ps). The $S_0 \rightarrow S_1$, $S_0 \rightarrow S_2$, and $S_1 \rightarrow S_2$ transitions are all electric dipole allowed.

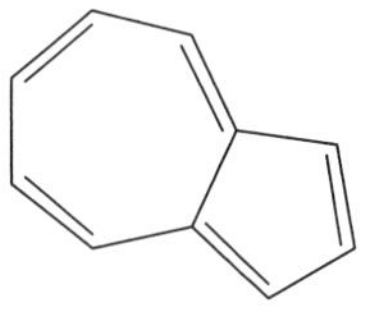

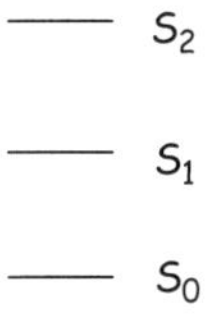

The 1.1 ps lifetime of the S_1 state is determined by carrying out a pump–probe experiment. Two very short light pulses centered at 700 nm, of equal intensity and having a relative time delay τ, are overlapped in the sample, and the time-integrated fluorescence between 350 and 400 nm is detected.

(a) What transition gives rise to this fluorescence?

(b) Draw a rough sketch of how the fluorescence signal should vary with time delay τ. You may assume, for the purpose of working out the time dependence, that the two pulses are very short compared with the lifetime of the S_1 state. Consider both positive and negative times.

(c) How else could you measure the S_1 state lifetime?

5. Eq. (11.8) gives the fraction of ground-state molecules that will be pumped into the excited state by a laser pulse under the assumption that the ground-state population is not significantly depleted, that is, N_g is not changed much from its initial equilibrium value. A frequency-doubled picosecond mode-locked Ti:sapphire laser produces light at 430 nm in the form of pulses with a duration of about 10 ps and at a repetition rate of 82 MHz. Assume this laser is used to pump a sample containing molecules having a moderately large absorption cross section of 10^{-16} cm^2 at 430 nm.

(a) If the average power of the laser at the sample is 10 mW, how large must the beam diameter be to insure that no more than 5% of the molecules in the sample are excited during a single pulse? Note that as long as only linear absorption occurs, you do not need to make any assumption about the pulse shape, that is, the functional form of $I(\omega,\tau)$.

(b) Would your answer be any different if the pulse duration were 50 fs instead of 10 ps? What other considerations might come into play if you were doing the experiment with 50 fs pulses?

6. The high overtone region of the infrared absorption spectrum of a small polyatomic molecule in the gas phase shows two weak lines attributed to absorption from the ground vibrational state to excited states, which we will call Ψ_L and Ψ_H (L and H stand for low and high). The transition to Ψ_L is 50 cm^{-1} lower than the transition to Ψ_H. If the harmonic oscillator approximation holds, it is expected that there should be only one state in this region, φ_B (B for bright), that has much intensity. The presence of two lines is attributed to anharmonic coupling between φ_B and another state close in frequency, φ_D (D for dark). That is, the true eigenstates are given in terms of the harmonic oscillator basis states by

$$\Psi_L = \alpha\varphi_B - \beta\varphi_D$$

$$\Psi_H = \beta\varphi_B + \alpha\varphi_D,$$

where α and β are real numerical coefficients.

(a) The infrared absorption strength of the higher frequency band is 4/3 that of the lower frequency band. Assuming only the φ_B basis state carries any intensity, determine the values of the coefficients α and β. Assume φ_B and φ_D are orthonormal.

(b) Using the values of α and β found in (a) and the 50 cm^{-1} splitting between the eigenstates, calculate the matrix element of the perturbation that couples the two basis states, $\langle\varphi_B|V|\varphi_D\rangle$, in cm^{-1}.

(c) A very short light pulse is used to excite the molecule (i.e., to prepare φ_B) at time t = 0, and then a second short light pulse is used to probe the remaining population in φ_B at later times through stimulated emission back to the ground state. Give the frequency, in inverse picoseconds (ps^{-1}), of the resulting oscillations in the stimulated emission. In order to solve this problem, first write φ_B as a linear combination of the eigenstates Ψ_L and Ψ_H, then use the time-dependent Schrödinger equation to write the wavefunction at time *t* and look at its time-dependent overlap with φ_B.

7. Quartz (SiO_2) has refractive indices along its "ordinary" crystallographic axis of 1.5341 at 1064 nm and 1.54687 at 532 nm. For second harmonic generation of 1064 nm light in quartz, plot the relative second harmonic intensity from Eq. (11.21) as a function of interaction length for $L = 0$ to 1 mm.

8. Consider a four-wave mixing process carried out with nearly degenerate input beams having frequencies ω_1 and $\omega_2 = \omega_1 + \delta$. The signal beam has frequency $\omega_{signal} = 2\omega_1 - \omega_2$ and wavevector $\mathbf{k}_{signal} = 2\mathbf{k}_1 - \mathbf{k}_2$. Assume δ is small enough that we do not have to consider the dependence of the refractive index on frequency. If $\mathbf{k}_1$ and $\mathbf{k}_2$ cross in the sample at an angle θ, what is the angle φ between $\mathbf{k}_1$ and $\mathbf{k}_{signal}$?

CHAPTER 12

ELECTRON TRANSFER PROCESSES

12.1. CHARGE–TRANSFER TRANSITIONS

Any electronic transition in a molecule necessarily involves some change in the molecule's charge distribution. An "electron transfer" transition is an extreme example of a spectroscopically induced charge redistribution in which essentially a full electron charge is transferred from one molecule or part of a molecule to another. In the most classic example of an electron transfer process, a neutral "donor" molecule D and a neutral "acceptor" molecule A, in close physical proximity, absorb light to reach a final state that consists of a donor cation and an acceptor anion: $DA \rightarrow D^+A^-$. Intermediate cases are also common, for example, where D and A are different functional groups on a single covalently bound molecule, and the amount of charge transferred from D to A is something less than a full electron. Alternatively, electron transfer may occur subsequent to electronic excitation of either the donor or the acceptor, for example, $DA \rightarrow D^*A \rightarrow D^+A^-$. Electron transfer processes, both direct and indirect, are of great importance in many biological and technological processes including photosynthesis, photography, xerography, and artificial solar energy capture.

A simple picture of an electron transfer process can be indicated with a molecular orbital model as shown in Figure 12.1. Both the donor and the acceptor individually can undergo transitions that promote an electron from

Condensed-Phase Molecular Spectroscopy and Photophysics, First Edition. Anne Myers Kelley.

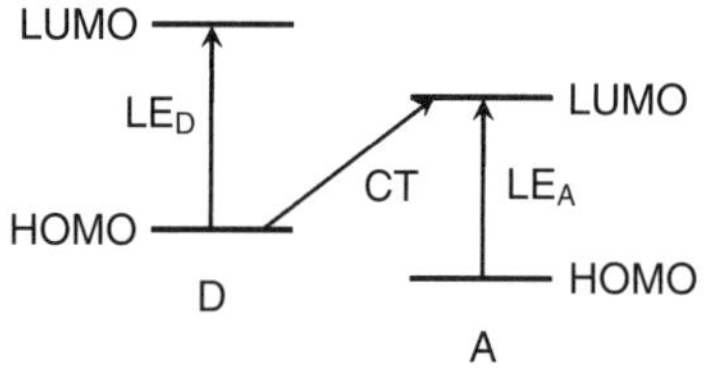

Figure 12.1. Energy level diagram for a charge-transfer transition. The charge-transfer transition (CT) occurs at a lower energy than the locally excited transition (LE) on either the donor (D) or the acceptor (A).

the highest occupied molecular orbital (HOMO) to the lowest unoccupied molecular orbital (LUMO), known in this context as "locally excited" transitions. These are the type of transitions discussed in Chapter 8. However, if D and A are in sufficiently close physical proximity that there is some overlap between their electronic wavefunctions, it is also possible to promote an electron directly from the HOMO of D to the LUMO of A, producing the ion pair D^+A^-. This charge-transfer transition can occur at a lower energy than either of the locally excited transitions. The energy of the charge-transfer transition can be approximated as $E_{CT} = E_{LUMO} - E_{HOMO} - C$, where C, here defined as a positive number, is the electrostatic interaction between D^+ and A^-. It can be seen that it is also energetically favorable for the locally excited state of the donor, D*, to transfer an electron to the LUMO of the acceptor, producing the same ion pair D^+A^- through a two-step process. Alternatively, the locally excited state of the acceptor, A*, can receive an electron from the HOMO of the donor resulting in the same ion pair.

Consider an optically induced direct charge-transfer transition, $DA \xrightarrow{h\nu} D^+A^-$. The oscillator strength for this transition depends on the square of the matrix element $\langle \Psi_{DA}|r|\Psi_{D+A-}\rangle$. In a simple MO picture, DA and D^+A^- differ only in the occupations of the HOMO and LUMO, so the interesting part of the matrix element may be abbreviated as $\langle HOMO|r|LUMO\rangle$. If the HOMO and the LUMO occupy completely different regions of space, for example, the donor and acceptor are far apart, then the matrix element is zero and the transition has no intensity. If the HOMO and LUMO have considerable spatial overlap, the matrix element may be large, but then it is not a true "charge transfer" transition because the electron has not been moved completely. The conclusion is that a true optically induced electron transfer process is not possible; there has to be some overlap between the initial and final electron distributions in order for the process to have intensity. Nevertheless, the concept of an optical charge transfer transition is a useful one.

The intensities and band shapes of charge transfer transitions can be described with the general formalism for electronic transitions given in Chapter 8. However, some qualitative aspects particular to charge-transfer spectra should be pointed out. Charge-transfer spectra are often measured in polar solvents, which stabilize the ionic products, and when they are, the optimum arrangement of solvent dipoles is often very different for DA than for D^+A^- (Fig. 12.2). Thus the nuclear reorganization that accompanies electron transfer is often dominated by the solvent rather than by internal vibrations. The contribution of solvent reorganization to the shape and width of a charge-transfer absorption spectrum can be developed by reference to Figure 12.3. Here the energies of

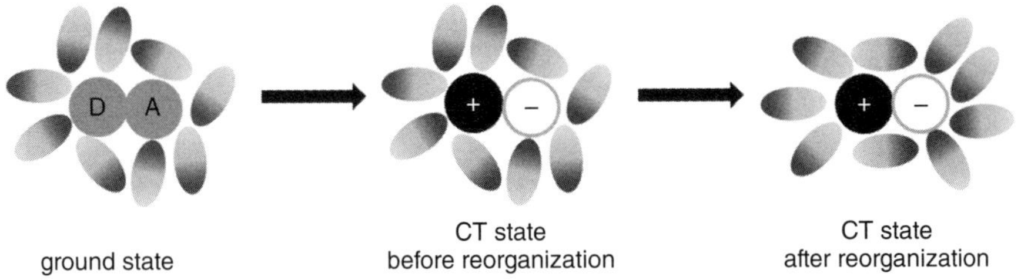

Figure 12.2. Solvent reorganization in a charge-transfer transition. The arrangement of polar solvent molecules around the neutral ground state is essentially random (left). Optical excitation of a charge-transfer transition creates an ion pair that is not optimally solvated by the ground-state solvent distribution (middle). The solvent molecules reorganize by rotation and translation to minimize the energy of the ion pair state (right).

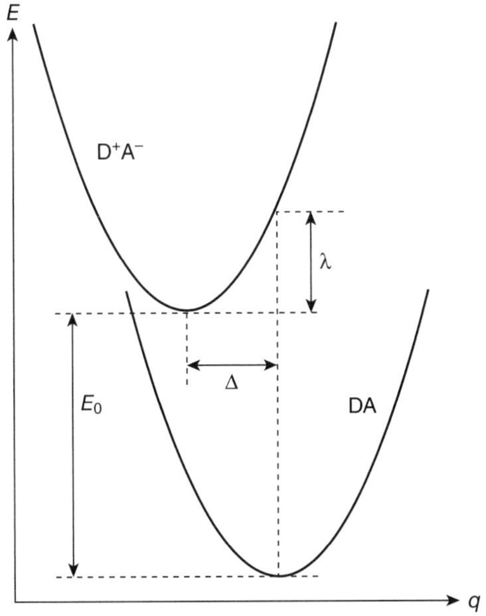

Figure 12.3. Quantities defining the charge-transfer bandshape.

the ground state (DA) and the excited state (D^+A^-) are plotted with respect to a generic solvation coordinate that represents the collective reorientation of solvent dipoles diagrammed in Figure 12.2. It is assumed that the energy varies quadratically with solvation coordinate and that the curvature is the same for both ground and excited states, but the two states have their minimum energies at different values of the solvation coordinate. Thus the ground-state energy is given by $E_g = kq^2$, where q is the solvation coordinate and k plays the role of a force constant. The excited-state energy is given by $E_e = k(q - \Delta)^2 + E_0$, where Δ is the displacement of the two minima along the solvation coordinate, and E_0 is the displacement of the two curves along the energy coordinate. At any given value of the solvation coordinate, the energy separation between the two surfaces, which corresponds to the energy of the optical transition, is

$$E = E_e - E_g = E_0 - 2k\Delta q + k\Delta^2 = E_0 - 2k\Delta q + \lambda, \tag{12.1}$$

where the reorganization energy, λ, is identified as $k\Delta^2$. The relative probability of finding the system originally at coordinate q at thermal equilibrium is given by a Boltzmann distribution,

$$P(q) \sim \exp(-E_g / k_B T) = \exp(-kq^2 / k_B T). \tag{12.2}$$

Solving Equation (12.1) for q and substituting into Equation (12.2) gives the relative probability that the system will absorb light of energy E as

$$P(E) \sim \exp\{-(E_0 - E + \lambda)^2 / 4\lambda k_B T\}, \tag{12.3a}$$

or converting to frequency,

$$P(\omega) \sim \exp\{-\hbar^2 (\omega_0 - \omega + \lambda/\hbar)^2 / 4\lambda k_B T\} \tag{12.3b}$$

$$P(\omega) = \hbar(4\pi\lambda k_B T)^{-1/2} \exp(-\hbar^2 [\omega_0 + \lambda/\hbar - \omega]^2 / 4\lambda k_B T), \tag{12.3c}$$

where the prefactor in Equation (12.3c) normalizes the total probability to unity.

Equation (12.3c) describes a Gaussian frequency distribution of standard deviation $(2\lambda k_B T/\hbar^2)^{1/2}$, centered at the frequency of the vertical transition from the ground-state minimum, $\omega_0 + \lambda/\hbar$. The larger the solvent reorganization energy, the broader the band. However, charge-transfer transitions also have contributions to the absorption spectral width from the same processes that broaden other optical spectra, primarily vibronic structure. Both of these sources can be accounted for by using Equation (8.8), with the delta functions replaced by the Gaussian lineshape of Equation (12.3c). For simplicity, we will assume that all of the molecular vibrations start in their zero-point level (all $i_j = 0$), and also that the ground-state and excited-state frequencies of all of the molecular vibrations are identical. In this limit, Equation (8.8) becomes

$$\sigma_{i\to f}(\omega) \sim |M_{if}(Q_0)|^2 \sum_{v_1=0}^{\infty} \cdots \sum_{v_{3N-6}=0}^{\infty} \left(\prod_{j=1}^{3N-6} |\langle v_j | 0_j \rangle|^2 \right)$$
$$\delta\left(\omega_0 - \omega + \sum_{j=1}^{3N-6} \{\omega_{ej} v_j\}\right),$$

and then replacing the delta function with a Gaussian centered at the same frequency gives

$$\sigma_{i\to f}(\omega) \sim |M_{if}(Q_0)|^2 \sum_{v_1=0}^{\infty} \cdots \sum_{v_{3N-6}=0}^{\infty} \left(\prod_{j=1}^{3N-6} |\langle v_j | 0_j \rangle|^2 \right)$$
$$\cdot \exp\left\{ -\hbar^2 \left(\omega_0 + \lambda/\hbar - \omega + \sum_{j=1}^{3N-6} \omega_{ej} v_j \right)^2 \Big/ 4\lambda k_B T \right\}. \tag{12.4}$$

This is a superposition of Gaussian bands, each centered at the frequency of a vibronic transition and weighted by the vibrational Franck–Condon factor for that transition. When the reorganization energy is large compared with vibronic energy spacings, as is typical for charge-transfer transitions in polar solvents, the absorption spectra tend to exhibit little or no vibronic structure as illustrated previously in Figure 8.6.

Calculation of solvent reorganization energies in electron transfer processes has been an active area of research and many different levels of sophistication can be employed. If D and A are treated as spheres of radii a_1 and a_2, respectively, separated by a distance r, then the reorganization energy accompanying electron transfer between them is given by dielectric continuum theory as (Marcus and Sutin, 1985)

$$\lambda = \frac{e^2}{4\pi\varepsilon_0} \left(\frac{1}{2a_1} + \frac{1}{2a_2} - \frac{1}{r} \right) \left(\frac{1}{n^2} - \frac{1}{\varepsilon} \right), \tag{12.5}$$

where n is the refractive index of the medium (n^2 is its dielectric constant at optical frequencies), and ε is its static relative dielectric constant.

12.2. MARCUS THEORY

As discussed above, transfer of an electron over a long distance is difficult to achieve optically because the corresponding transitions tend to be quite weak. Photoinduced charge transfer is more often accomplished through the alternate route of preparing a locally excited state which then undergoes an energetically favorable electron transfer reaction. The process $D^*A \to D^+A^-$ or $DA^* \to D^+A^-$ may be treated as an internal conversion process via Fermi's Golden Rule, which says that the rate is given by Equation (9.2),

$$R_{i\to\{f\}} = \frac{2\pi}{\hbar} |W_{\text{if,avg}}|^2 \rho(E_f = E_i).$$

This describes the transition rate between a particular initial state of D^*A or DA^* and all possible isoenergetic final states of D^+A^-. We now divide the possible final states into a set of discrete energy levels arising from the vibrations of donor and acceptor, and a set of continuously distributed levels resulting from solvent nuclear motions. We also assume that the matrix element $|W_{\text{if,avg}}|^2$ may be factored into a purely electronic part and a nuclear part (the Condon approximation). For the discrete, quantum mechanical vibrations, the nuclear part is just the vibrational Franck–Condon factor, the square of the vibrational overlap integral. Note that the vibrational wavefunctions extend to nuclear coordinates that are in the classically forbidden region, and the overlap integrals show the effects of this quantum mechanical tunneling. For the continuous, classically behaved solvent modes, tunneling is assumed to be negligible, and the "wavefunctions" are assumed to be a set of delta functions in the solvation coordinate q. The "matrix element" between them is unity when the two functions occupy the same q, and zero otherwise. Accordingly, calculation of the nuclear contribution to the matrix element times the density of states amounts to simply calculating the Boltzmann probability of the initial state at the crossing point of the two curves (Fig. 12.4). This is a very similar calculation to that carried out above to describe the charge-transfer bandshape.

The parabolas describing the energies of the initial and final states are, respectively,

$$E_{\text{i}} = kq^2 \tag{12.6a}$$

$$E_{\text{f}} = k(q+\Delta)^2 - E_0. \tag{12.6b}$$

The position along the solvation coordinate (value of q) at which these two curves intersect, q^+, is found by solving $k(q^+)^2 = k(q^+ + \Delta)^2 - E_0$. The result is $q^+ = (E_0 - k\Delta^2)/2k\Delta$, and the energy of the initial state at this q value is given by $E_{\text{i}}(q^+) = k(q^+)^2 = (E_0 - k\Delta^2)^2/4k\Delta^2$. We now recognize that the quantity $k\Delta^2$

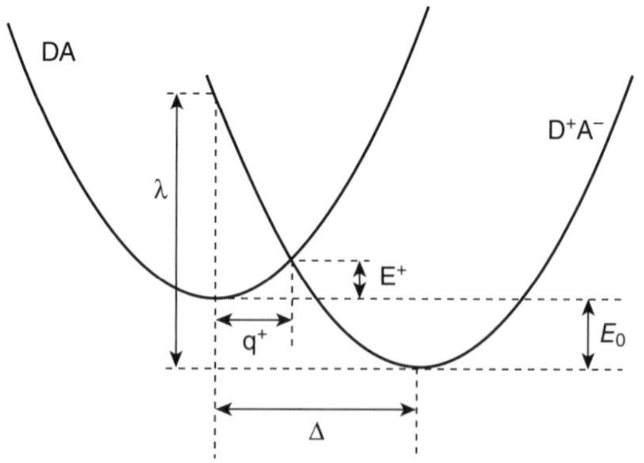

Figure 12.4. Definition of quantities used in deriving electron transfer rate (Marcus theory). The y-axis is energy (actually free energy), and the x-axis is solvation coordinate.

is the reorganization energy λ, the energy difference between the final state at the initial equilibrium geometry and the final state at the final equilibrium geometry. We also recognize that if these curves are associated with free energies rather than energies, the quantity E_0 is simply minus the free energy change associated with the reaction, $-\Delta G^0$. So the energy of the initial state at the curve crossing point is $(\Delta G^0 + \lambda)^2/4\lambda$. The Boltzmann probability of finding the system at this position is $\exp(-[\Delta G^0 + \lambda]^2/4\lambda k_B T)/(4\pi\lambda k_B T)^{1/2}$.

Putting the quantum mechanical and classical parts together, we get for the electron transfer rate

$$\begin{aligned}\text{Rate} = \frac{2\pi}{\hbar}|W_{el}|^2 \frac{1}{\sqrt{4\pi\lambda k_B T}} \sum_{v_1=0}^{\infty} \cdots \sum_{v_{3N-6}=0}^{\infty} \left(\prod_{j=1}^{3N-6} |\langle v_j | 0_j \rangle|^2 \right) \\ \cdot \exp\left\{ -\left(\Delta G^0 + \lambda + \sum_{j=1}^{3N-6} \hbar\omega_j v_j \right)^2 \Big/ 4\lambda k_B T \right\},\end{aligned} \tag{12.7}$$

where W_{el} is the electronic part of the matrix element, which depends on the overlap between the donor and acceptor orbitals. This equation was first derived by Marcus, although in his original derivation, the high-frequency quantized molecular vibrations were omitted. Let us focus on the behavior of the rate if we neglect the high-frequency modes. Note that in the figure above, we have assumed that ΔG^0 is negative—that is, the electron transfer reaction is energetically favorable. When the reorganization energy λ is greater than $-\Delta G^0$, as shown in Figures 12.4 and 12.5 (thin black to thick black curve), making the reaction more favorable (moving the free energy curve for the

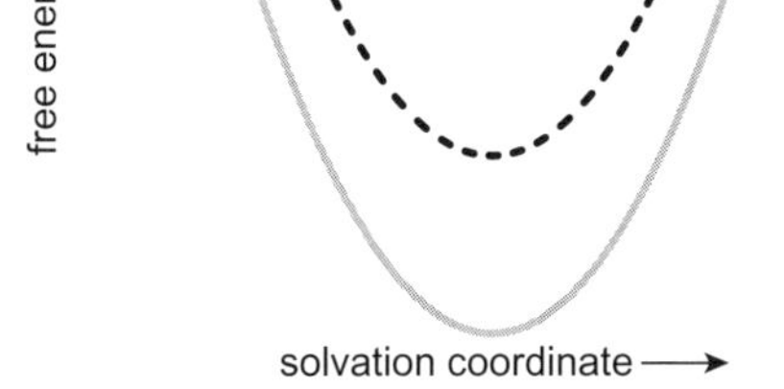

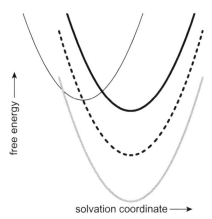

Figure 12.5. Illustration of activation energy as a function of reaction free energy for DA (thin black curve) → D^+A^- (thick curves). As the reaction becomes increasingly energetically favorable from black to dashed to gray, the activation energy for the curve crossing first decreases, reaches a minimum for the dashed curve, and then increases again.

final state down to make ΔG^0 more negative) makes the negative argument of the exponential smaller and increases the rate. This agrees with the usual expectation that the reaction should go faster as it becomes more energetically favorable. Continuing to move the final state curve down in energy eventually makes λ and ΔG^0 equal in magnitude, as shown in Figure 12.5 (dashed curve for D^+A^-), and now there is no activation energy from the solvent coordinate and the reaction rate is a maximum. Making ΔG^0 even more negative (Fig. 12.5, thick gray curve) again creates an activation barrier and makes the reaction slow down again! This decrease in the electron transfer rate with increasing reaction exoergicity is known as the Marcus inverted region.

Note again the close resemblance between Equation (12.4) for the charge-transfer absorption spectrum and Equation (12.7) for the nonphotochemical electron transfer rate. While the electronic matrix elements are different, the Franck–Condon factors and solvent reorganization enter into the equations in the same way. From the viewpoint of the nuclear degrees of freedom, thermal electron transfer behaves just like optical electron transfer performed with a zero-frequency photon.

Figure 12.4 is slightly misleading in that it shows the potential energy surfaces for DA and D^+A^- intersecting at the point q^+. Recall from basic quantum mechanics that any time two basis states are exactly degenerate, any perturbation coupling the two will cause the actual eigenstates to be linear combinations of the basis states and to become nondegenerate. Thus, the potential energy surfaces do not exactly cross, but rather split into an upper surface and a lower surface, as shown in Figure 12.6 for varying coupling strengths. Equation (12.7) is derived in the nonadiabatic limit, where the electron transfer is assumed to involve the hopping of the probability from the DA surface to the D^+A^- surface. This limit is most likely to be valid when the coupling is weak, such that the two surfaces nearly cross. In the opposite, adiabatic limit, the electron transfer process is better described as motion along a single potential

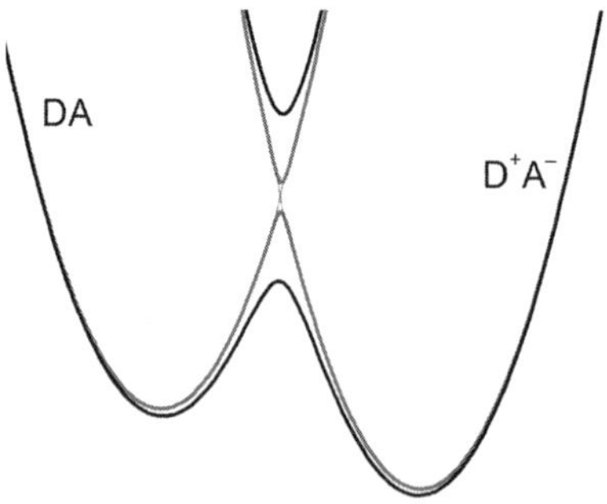

Figure 12.6. Adiabatic potential energy surfaces for weak coupling (gray curve) and strong coupling (black curve). The two curves nearly cross when the coupling is weak, but split progressively farther apart in energy near the original crossing point as the coupling strength increases.

energy surface that evolves smoothly from having mostly DA character to having mostly D^+A^- character as the system moves along the solvation coordinate. In this case, the time scale for solvent reorganization may play an important role in determining the electron transfer rate. While only the nonadiabatic theory is discussed here in detail, real electron transfer processes often fall somewhere between these limits.

12.3. SPECTROSCOPY OF ANIONS AND CATIONS

Figure 12.1 is not meant to be quantitative, but it does highlight some general features of the spectroscopy of molecular ions. If D and A are both stable, closed-shell molecules, there is usually a fairly large energy separation between the HOMO and the LUMO on both D and A, and both molecules typically have their lowest electronic transitions in the UV. As discussed above, in a charge-transfer system, the LUMO of the acceptor is lower in energy than the LUMO of the donor, so the DA $\rightarrow$ D^+A^- excitation occurs at a longer wavelength than either of the locally excited transitions, often in the visible region of the spectrum. Now consider the ions produced, D^+ and A^-. Within a given molecule, the energetic spacing between the molecular orbitals near the top of the occupied band, and between those near the bottom of the unoccupied band, is usually smaller than between the HOMO and the LUMO. D^+ has a vacancy in its HOMO, so transitions such as (HOMO–1) $\rightarrow$ HOMO, (HOMO–2) $\rightarrow$ HOMO, and so on, become possible. Some of these transitions may be forbidden for symmetry or other reasons, but if they are allowed, these often occur at much lower energies (longer wavelengths) than the lowest

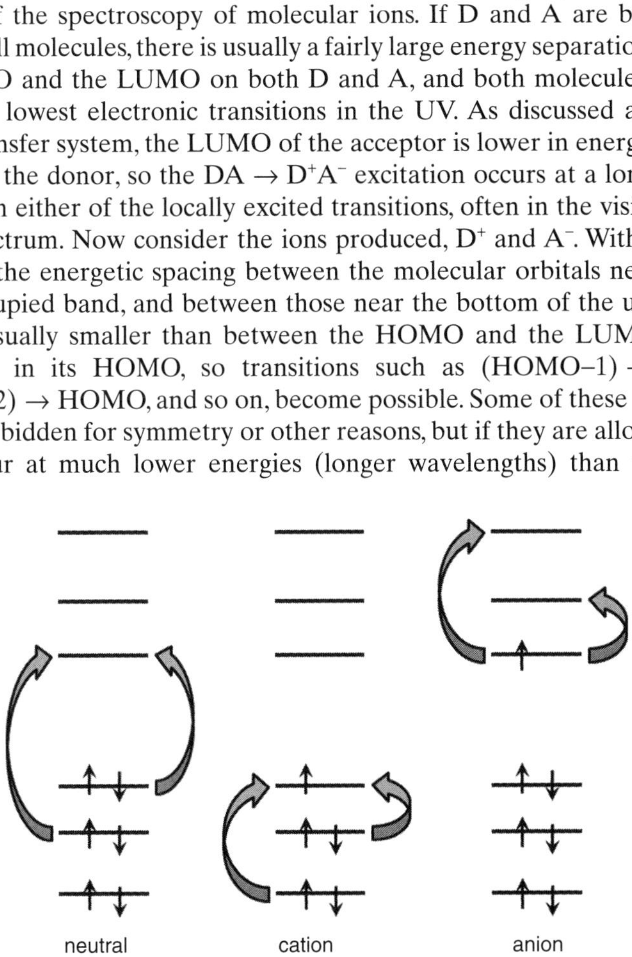

Figure 12.7. Neutral, closed-shell molecules typically have a fairly large energy gap between HOMO and LUMO, so their lowest energy transitions (arrows) occur at fairly short wavelengths. The corresponding cation or anion has lower-energy optical transitions.

locally excited transition of the neutral donor. Similarly, A^- has an extra electron in its LUMO, which can make transitions to (LUMO+1), (LUMO+2), and so on. These transitions, if allowed, may be much lower in energy than the HOMO → LUMO of the neutral A molecule.

Single-electron MO energies are only a rough approximation to the actual state to state transition energies of complex molecules, but these qualitative conclusions tend to hold. Electronic spectra of radical ions, whether cations or anions, usually have their lowest transitions considerably to the red of the absorptions of the neutral, closed-shell parent molecules (Fig. 12.7). Pump-probe experiments intended to detect short-lived radical intermediates usually involve pumping in the UV or blue and probing in the visible or near-IR. This generality does not always hold, and in particular, it may be reversed if the anion and/or cation is a closed-shell species, while the neutral is a radical.

REFERENCES AND FURTHER READING

R. A. Marcus, J. Phys. Chem. **93**, 3078 (1989).

R. A. Marcus and N. Sutin, Biochim. Biophys. Acta **811**, 265 (1985).

PROBLEMS

1. Consider an ion pair in which a full electronic charge has been transferred between the donor and the acceptor. What is the dipole moment in Debye if the distance between donor and acceptor is 5 Å? One Debye = 3.3356×10^{-30} C·m.

2. **(a)** Consider a donor-acceptor pair (donor is 1, acceptor is 2), with the energy levels and neutral ground-state electron configurations shown in diagram (a), and refer to Figure 12.1. Show the electron configuration

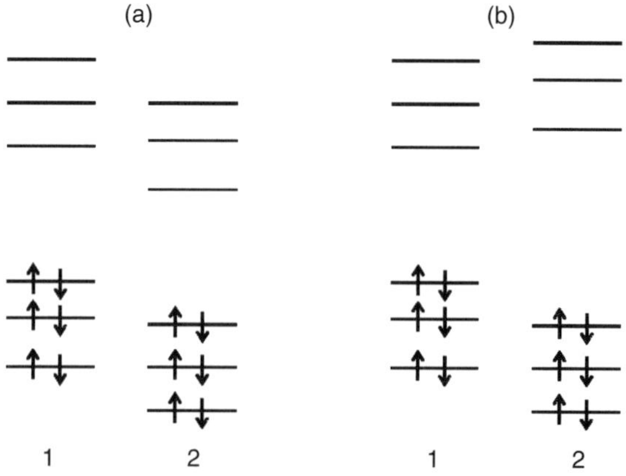

(how many electrons in each orbital) following charge-transfer excitation. Show the electron configuration following the lowest-energy locally excited transition on the donor and the lowest-energy locally excited transition on the acceptor. Explain why both of the local excitations should relax to the same electron configuration as the direct charge-transfer excitation.

(b) Now consider the energy level ordering in diagram (b). How does this change your answers to the above questions?

3. A number of "symmetric" charge transfer processes are possible, such as the ferrous-ferric exchange reaction, $Fe^{2+} + Fe^{3+} \rightarrow Fe^{3+} + Fe^{2+}$. For a reaction of this type, make a plot analogous to Figure 12.4 and discuss it. What is E_0? Why does the plot have two minima? What does the x-axis mean? Make a cartoon of the solvation shell around the two iron atoms at each of the minima. Explain why the reorganization energy is not zero for such a reaction.

4. (a) Use Equation (12.5) to estimate the reorganization energy, in cm^{-1} and eV, for electron transfer between two molecules with radii 3 and 5 Å, respectively, in direct contact (r = 8 Å). Do this calculation in a strongly polar solvent (water, $\varepsilon = 80$, $n = 1.33$) and a weakly polar solvent (diethyl ether, $\varepsilon = 4.3$, $n = 1.35$).

(b) If the free energy change for the reaction, ΔG^0, is –0.5 eV, and the effects of molecular vibrations can be ignored, use Equation (12.7) to calculate the relative electron transfer rates for the two solvents at room temperature, assuming no change in the electronic matrix element.

5. Use Equation (12.4) to calculate the bandshape (don't worry about the overall scaling) of a charge-transfer transition as a function of the solvent reorganization energy. Assume there is only one vibrational mode that matters, and that it has a frequency of 1400 cm^{-1} (typical of ring stretching modes in aromatic molecules). Calculate the vibrational overlap integral using the result from Equation (8.9), $|\langle v|0\rangle|^2 = e^{-S}\left(S^v / v!\right)$ and take $S = 0.5$. Take the electronic zero–zero frequency to be 20,000 cm^{-1}. Write a simple computer program or use a spreadsheet to calculate the spectrum at room temperature for $\lambda = 0.1$ eV and $\lambda = 1$ eV. Discuss how and why the spectrum depends on solvent reorganization.

6. (a) Use Equation (12.7) to calculate the temperature dependence of the electron transfer rate.

Assume that the molecular vibrations can be ignored, so the relevant expression is

$$\text{Rate} = \frac{2\pi}{\hbar}|W_{\text{el}}|^2 \frac{1}{\sqrt{4\pi\lambda k_B T}} \exp\left\{-\left(\Delta G^0 + \lambda\right)^2 / 4\lambda k_B T\right\}.$$

Take the reorganization energy to be $\lambda = 4000\ \text{cm}^{-1}$. Calculate the relative rates for temperatures of 280, 300, 320, and 340 K, for free energy changes of $\Delta G^0 = -0.15$ eV and -0.5 eV.

(b) How does the temperature dependence of the rate compare with the standard Arrhenius equation, $k = A\exp(-E_a/RT)$?

7. While the chapter uses the term "reorganization energy", λ, to describe only the component arising from the solvent, one can similarly define the vibrational reorganization energy, λ_v, as the energy difference between the vibrational potential energy for the charge-transfer excited state at the neutral ground-state geometry and the minimum of the charge-transfer vibrational potential function.

(a) For a single vibrational mode, find an expression for λ_v in terms of the vibrational frequency ω_v and the dimensionless displacement Δ under the assumption that the ground-state and excited-state vibrational potentials have the same curvature.

(b) Generalize this expression to multiple vibrational modes.

(c) Recall that the zero-point vibrational energy in each mode is $\hbar\omega_v/2$. Does this lead to a problem with the interpretation of λ_v for small displacements? Explain.

8. Consider the vibrational reorganization energy as defined in Problem 7. How do you expect λ_v for the charge-transfer transition, $DA \rightarrow D^+A^-$, to compare with the sum of the λ_v for the individual lowest-energy locally excited transitions, $D \rightarrow D^*$ and $A \rightarrow A^*$?

CHAPTER 13

COLLECTIONS OF MOLECULES

13.1. VAN DER WAALS MOLECULES

Van der Waals molecules are complexes among two or more molecules that are not covalently bound. Usually, one of the species is a chromophore (a molecule that has a reasonably strong transition, vibrational or usually electronic, that can be probed by light absorption and/or emission), and the other species is either an atom (e.g., Ar) or a fairly simple molecule (CO_2, H_2O) that does not have any spectroscopic transitions overlapping with those of the chromophore. The two species have a weak attractive interaction arising from dipolar or polarizability interactions; the van der Waals "bond" usually has a frequency of a few tens of cm^{-1}, and a dissociation energy of a few hundred cm^{-1}. Thus, they do not live very long except at very low temperatures, and are usually made and studied in the gas phase under conditions that produce a beam of rotationally and vibrationally cold molecules (supersonic free-jet expansion) (Smalley et al., 1977).

Van der Waals molecules are of interest for several reasons. First, they comprise a sort of intermediate system between an isolated chromophore and the same chromophore in a condensed phase (solid or liquid) environment. They allow the spectroscopic properties of a chromophore to be studied as it is "solvated" by one solvent molecule at a time. They are also interesting systems in which to study intramolecular vibrational relaxation, since the

Condensed-Phase Molecular Spectroscopy and Photophysics, First Edition. Anne Myers Kelley.

coupling between the chromophore and the van der Waals "ligand" is quite weak, and IVR between the chromophore and the ligand is often much slower than within the chromophore itself.

13.2. DIMERS AND AGGREGATES

A common and interesting problem in spectroscopy is that arising from two identical chromophores that are in sufficiently close physical proximity to have some interaction, yet far enough apart that there is little overlap between their wavefunctions. The identical (homodimer) problem is easy to solve, and the results form a useful starting point for more complex situations where the two molecules are not identical (heterodimer), where the wavefunctions do have some overlap, or where the number of interacting molecules is greater than two.

Consider two identical molecules, labeled A and B, which have the same transition frequency and oscillator strength. When A and B are far apart, they do not interact at all, and the absorption spectrum of the two molecules is simply the sum of the two individual spectra. That is, there are two degenerate excited states, A*B and AB*, each having energy E_0 above the ground state. As the two molecules are brought closer together, the transition dipole moments on A and on B begin to interact, producing a cross term in the Hamiltonian between the basis states A*B and AB*. The interaction energy between two dipole moments A and B (here, the transition dipoles associated with the transitions AB→A*B and AB→AB*, respectively) was given in the discussion of energy transfer in Chapter 9, Equation (9.24):

$$V = \frac{\boldsymbol{\mu}_A \cdot \boldsymbol{\mu}_B}{4\pi\varepsilon_0 n^2 R_{AB}^3} - \frac{3(\boldsymbol{\mu}_A \cdot \boldsymbol{R}_{AB})(\boldsymbol{\mu}_B \cdot \boldsymbol{R}_{AB})}{4\pi\varepsilon_0 n^2 R_{AB}^5}, \tag{13.1}$$

where $\boldsymbol{\mu}_A$ and $\boldsymbol{\mu}_B$ are the transition dipole moments of the two monomers (identical in magnitude for a homodimer, but potentially having different orientations in space), $\boldsymbol{R}_{AB}$ is the vector pointing from the center of A to the center of B having length R_{AB}, and n is the refractive index of the medium. The matrix of the Hamiltonian for the system of two interacting excited states, in the basis $\{|A^*B\rangle, |AB^*\rangle\}$, is

$$H = \begin{pmatrix} E_0 & V \\ V & E_0 \end{pmatrix}. \tag{13.2}$$

The resulting eigenstates of the coupled system, the actual excited states, are now linear combinations of the excitations on the two monomers:

$$\Psi^+ = (|A^*B\rangle + |AB^*\rangle)/\sqrt{2},\ E^+ = E_0 + V \tag{13.3a}$$

$$\Psi^- = (|A^*B\rangle - |AB^*\rangle)/\sqrt{2},\ E^- = E_0 - V, \tag{13.3b}$$

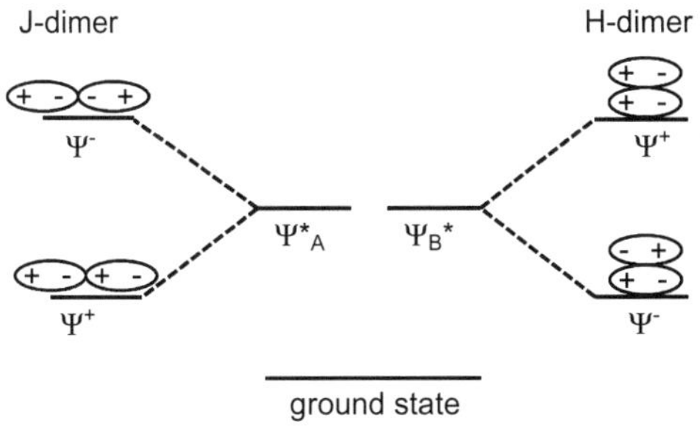

Figure 13.1. Energies of the plus and minus combinations of the two originally degenerate excited states for a "J-dimer" and an "H-dimer." The splitting of the excited-state energy levels relative to the ground to excited state energy separation is greatly exaggerated.

Which state is lower in energy depends upon the sign of V, which is determined by the relative orientations of the transition dipole moments on A and B (Fig. 13.1). If the two molecules are arranged head-to-tail as shown on the left, then the lower-energy state is the one where the two transition dipoles are in phase (Ψ^+), and Ψ^+ has twice the oscillator strength of one monomer while Ψ^- has zero oscillator strength. This is a J-dimer, where the allowed electronic transition is shifted to lower energy than the monomer. If the two monomers are arranged side-by-side as on the right, the lower-energy state is Ψ^-, which still has zero oscillator strength. This is an H-dimer, where the allowed dimer transition is shifted to higher energy than in the monomer. H-dimers normally exhibit very weak fluorescence because allowed excitation to Ψ^+ is followed by rapid internal conversion to Ψ^-, which has a very small (zero in the ideal case) radiative rate. For intermediate geometries, the oscillator strength is distributed over both transitions.

A system of N interacting identical monomers can be treated by a straightforward extension of the method used to treat dimers. The details will not be provided here. The N monomers couple to form N different excited states whose energies and transition moments depend on the geometric arrangement of the monomers. In many cases, the result can be characterized as either a J-aggregate, in which most or all of the oscillator strength piles up in the lowest-energy transition red-shifted from that of the monomer, or an H-aggregate, in which most of the oscillator strength is in a blue-shifted transition. Examples are shown in Figures 13.2 and 13.3.

13.3. LOCALIZED AND DELOCALIZED EXCITED STATES

Notice that the functional form of the wavefunctions Ψ^+ and Ψ^- for the homodimer problem is independent of the value of V. Even if the interaction

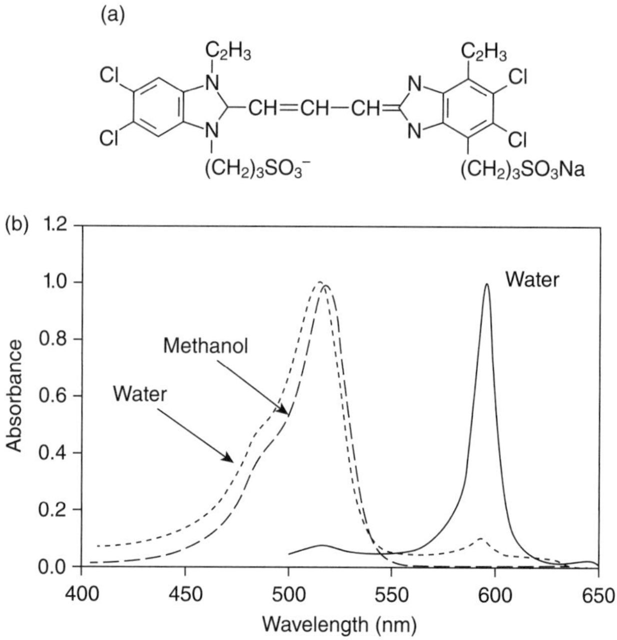

Figure 13.2. Absorption spectra of a cyanine dye in water and methanol at low concentration (mostly monomers) and in water at high concentration (mostly J-aggregates). Reprinted with permission from Ohta et al., J. Chem. Phys. **115**, 7609 (2001). Copyright 2001, American Institute of Physics.

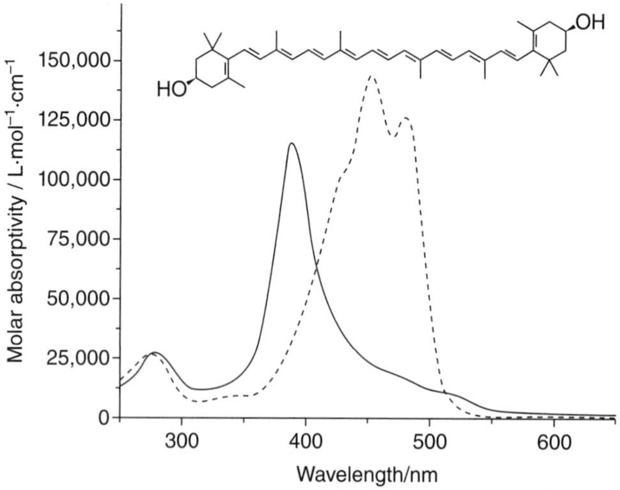

Figure 13.3. Absorption spectra of zeaxanthin in dilute ethanol solution (dashed curve, monomers) and in 80:20 v/v H_2O:ethanol (solid curve, mostly H-aggregates). Data courtesy of Michael J. Tauber, UC San Diego.

matrix element is very small, both basis states make equal contributions (apart from sign) to the total wavefunction. That is, as long as the uncoupled states are exactly degenerate, any coupling, no matter how small, is enough to completely delocalize the excited-state wavefunction over the two monomers. For a system larger than a dimer, the coefficients of each monomer in each eigenfunction are generally not identical, but symmetrically equivalent monomers (e.g., the first and last ones in a linear chain) will always have the same squared coefficient $|c_{in}|^2$ in each eigenstate ψ_n as long as the localized excitations are exactly degenerate.

Any factor that breaks the degeneracy of the basis excitations will tend to localize the wavefunctions. Consider the dimer case again, but now assume that the two basis state excitations are split by an amount 2δ, such that the localized excitations have the value $E_0 + \delta$ and $E_0 - \delta$, respectively. If the coupling strength remains V, then solution yields

$$\Psi^+ = N^+\{|\mathrm{A^*B}\rangle + [(\delta^2/V^2+1)^{1/2} - \delta/V]|\mathrm{AB^*}\rangle\},\ E^+ = E_0 + (\delta^2+V^2)^{1/2} \quad (13.4a)$$

$$\Psi^- = N^-\{|\mathrm{A^*B}\rangle + [-(\delta^2/V^2+1)^{1/2} - \delta/V]|\mathrm{AB^*}\rangle\},\ E^- = E_0 - (\delta^2+V^2)^{1/2}, \quad (13.4b)$$

where the N are normalization factors. These reduce to Equation (13.3) for $\delta = 0$. If the energy splitting between the basis states is large compared with the coupling, $\delta >> V$, then $\Psi^+ \approx |\mathrm{A^*B}\rangle$ and $\Psi^- \approx |\mathrm{AB^*}\rangle$; that is, the eigenstates are essentially the same as the localized basis states. For intermediate values, the extent to which the basis states are mixed in the eigenstates increases as the basis states become more nearly degenerate (decreasing δ) and/or the coupling becomes stronger (increasing V).

In real systems, there are a variety of effects that may cause apparently equivalent excitations to actually be nondegenerate, leading to localization of the coupled wavefunctions. Even if the interacting moieties are chemically equivalent, they often sit in slightly different environments which perturb their transition energies differently. This is usually a significant effect for molecules in disordered environments such as liquids, glasses, or proteins, and a much smaller but often non-negligible effect in crystals. In nanocrystals, there is always some heterogeneity in the size and shape of the crystals, thus making the individual monomers intrinsically different. Even effects of isotopic substitution, such as the ~1% natural abundance of ^{13}C in organic compounds, can be a large enough perturbation to matter. Finally, finite temperature is a major contributor to rendering transition frequencies nondegenerate, both by disordering the environment and by placing the chromophore itself in a different initial quantum state (low-frequency vibrations/librations).

The above are examples of static effects. They act on the ground state of the system to break the degeneracy of otherwise degenerate transitions of individual chromophores. The excited states at the ground state configuration are already localized. There are also dynamic effects, which act on some finite time scale to localize an initially delocalized excited state, usually an

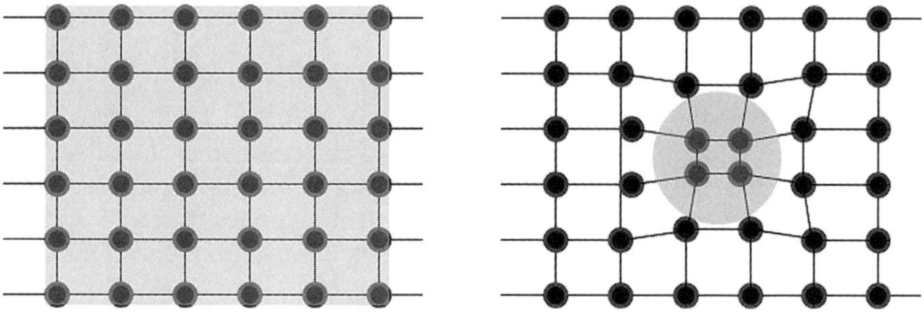

Figure 13.4. Illustration of a polaron. When the crystal lattice is unperturbed, the electronic excitation is delocalized over the whole crystal (left). Distortion of the nuclei accompanied by localization of the excitation (polaron formation, right) can lower the total energy of the system.

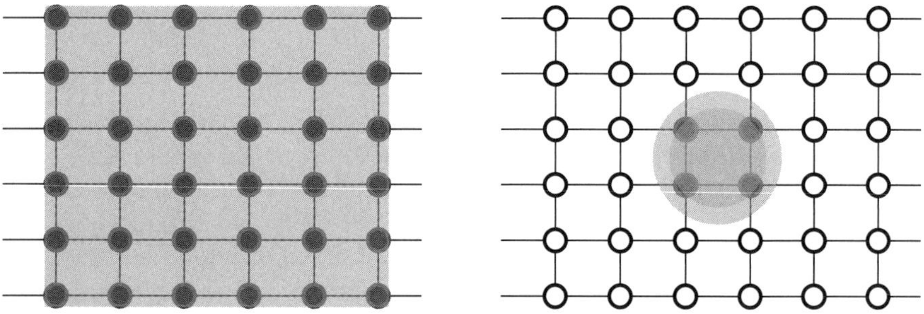

Figure 13.5. Illustration of an exciton. If the electron (gray) and hole (black) interact weakly enough, both are delocalized over the crystal (left). Stronger Coulombic interaction localizes the charges (right).

electronically excited state. In the language of solid state physics, which is often adapted to other types of collective excitations, a distortion of the nuclei that lowers the energy of a localized electronic state relative to that of the corresponding delocalized state is referred to as a polaron (Fig. 13.4). As discussed further in Chapter 16, the Coulombic interactions among the negatively charged electrons and the positively charged nuclear cores may also limit the localization of the excited state; a Coulombically bound electron–"hole" pair is referred to as an exciton (Fig. 13.5). Both polarons and excitons may move through the system, the excitation visiting different groups of monomers as a function of time.

13.4. CONJUGATED POLYMERS

Conjugated polymers are a scientifically interesting and technologically relevant class of materials that consist of large numbers of coupled chromophores.

The simplest example of a conjugated polymer is polyethylene. The linear polyenes, $C_{2n}H_{2n+2}$, form a series of molecules often used as examples for simple pi-electron theories and/or examples of one-dimensional particle in a box behavior (Section 1.4). The first few members of the series (ethylene, butadiene, hexatriene, and octatetraene) show a strongly allowed electronic transition that is well described as excitation of a single electron from the HOMO to the LUMO using the molecular orbitals obtained from simple pi-electron MO theory. (There is also a nominally forbidden transition that involves large contributions from configurations that involve two-electron excitations, but this transition is higher in energy than the strongly allowed one in all but the shortest polyenes and makes little contribution to the electronic spectroscopy.)

A model that is slightly more sophisticated than the simple particle in a box, but still mathematically very simple, is the Hückel or tight-binding model. Here, the molecular orbitals of a planar molecule containing N conjugated carbon atoms in the xy plane are taken as linear combinations of the $2p_z$ orbitals on each of the conjugated carbon atoms. The matrix elements of the pi-electron Hamiltonian are taken to be

$$H_{ii} = \langle \varphi_i | \mathbb{H} | \varphi_i \rangle = \alpha \tag{13.5a}$$

$$H_{ij} = \langle \varphi_i | \mathbb{H} | \varphi_j \rangle = \beta \text{ for } i \text{ and } j \text{ adjacent (bonded atoms)} \tag{13.5b}$$

$$H_{ij} = \langle \varphi_i | \mathbb{H} | \varphi_j \rangle = 0 \text{ for } i \text{ and } j \text{ not adjacent (not bonded atoms),} \tag{13.5c}$$

where φ_i is the $2p_z$ atomic orbital on atom i, and α and β are parameters that are determined empirically. The one-electron Hamiltonian thus has the form

$$\begin{bmatrix} \alpha & \beta & 0 & 0 & 0 & 0 & \cdots & 0 \\ \beta & \alpha & \beta & 0 & 0 & 0 & \cdots & 0 \\ 0 & \beta & \alpha & \beta & 0 & 0 & \cdots & 0 \\ 0 & 0 & \beta & \alpha & \beta & 0 & \cdots & 0 \\ 0 & 0 & 0 & \beta & \alpha & \beta & \cdots & 0 \\ 0 & 0 & 0 & 0 & \beta & \alpha & \cdots & 0 \\ \vdots & \vdots & \vdots & \vdots & \vdots & \vdots & \cdots & \vdots \\ 0 & 0 & 0 & 0 & 0 & 0 & \cdots & \alpha \end{bmatrix}. \tag{13.6}$$

Diagonalization of this $N \times N$ matrix gives the energies of the molecular orbitals as

$$E(n, N) = \alpha + 2\beta \cos\left(\frac{n\pi}{N+1}\right) \quad n = 1, 2, \ldots N. \tag{13.7}$$

In the ground state, the N pi electrons are placed spin-paired into the $N/2$ lowest energy orbitals. The energy of the lowest $\pi \rightarrow \pi^*$ transition is then taken

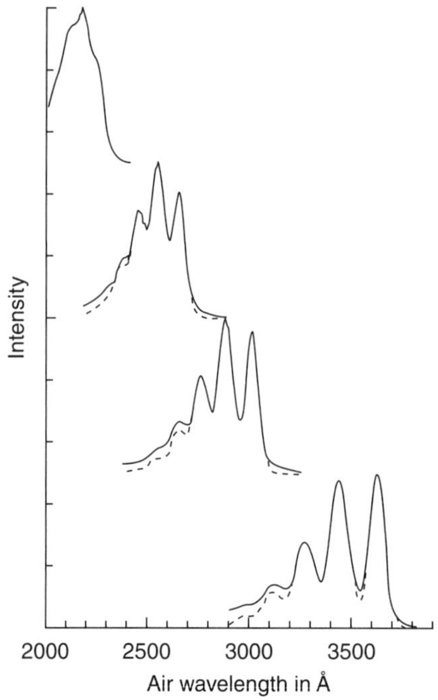

Figure 13.6. Absorption spectra (room temperature, hexane solvent) of the linear polyenes, $C_{2n}H_{2n+2}$, for (top to bottom) $n = 2, 3, 4$, and 6 (butadiene, hexatriene, octatetraene, and dodecahexaene). The dashed curves are calculated fits. As the polyene length increases, the absorption shifts to longer wavelengths and an increasingly large fraction of the total intensity is found in the electronic origin (the longest-wavelength band). Reprinted with permission from Granville et al., J. Chem. Phys. **75**, 3765 (1981). Copyright 1981, American Institute of Physics.

to be the energy difference between the highest occupied and lowest unoccupied orbitals. As N becomes large, this energy difference becomes

$$E_{\pi\to\pi^*} = 2\beta\left[\cos\left(\frac{\left(1+\frac{N}{2}\right)\pi}{N+1}\right)-\cos\left(\frac{N\pi/2}{N+1}\right)\right]$$
$$= 2\beta\left[\cos\left(\frac{N\pi/2}{N+1}\right)\cos\left(\frac{\pi}{N+1}\right)-\sin\left(\frac{N\pi/2}{N+1}\right)\sin\left(\frac{\pi}{N+1}\right)-\cos\left(\frac{N\pi/2}{N+1}\right)\right]$$
$$\approx -2\beta\left[\left(\frac{\pi}{N+1}\right)\sin\left(\frac{N\pi/2}{N+1}\right)\right]\approx\frac{-2\beta\pi}{N}\quad \text{for } N \gg 1.$$

Thus, simple Hückel theory predicts that in the limit of many conjugated units, the electronic transition energy should scale as $1/N$. Experimentally, this is not observed; the electronic transition does not red-shift as quickly as expected with increasing chain length, and the experimental energy gap reaches an approximately constant value for large N rather than approaching zero as simple pi-electron theories predict. This suggests that in the actual material, the excited states are not delocalized over the entire chain, which may contain hundreds to thousands of carbon atoms. Spectroscopic and photophysical studies on a wide variety of pi-conjugated organic polymers imply that the effective number of fully conjugated units ("conjugation length") spans a wide range in any given sample and rarely exceeds about 10 repeat units. Thus, an "infinite" polyene actually consists, from a spectroscopic point of view, of a mixture of shorter polyenes.

Simple pi-electron theories also predict that the amount of vibronic structure in the transition should decrease as delocalization of the transition over more atoms reduces the accompanying nuclear distortions. This is observed in the shorter polyenes of the series (Fig. 13.6).

REFERENCES

R. E. Smalley, L. Wharton, and D. H. Levy, Acc. Chem. Res. **10**, 139 (1977).

PROBLEMS

1. The interaction energy between nonpolar, nonbonded atoms or molecules that stick together purely through van der Waals forces is often approximated by the Lennard–Jones potential,

$$V_{LJ}(r) = 4\varepsilon\left[\left(\frac{\sigma}{r}\right)^{12} - \left(\frac{\sigma}{r}\right)^{6}\right],$$

where ε is a constant with units of energy, σ is a constant with units of distance, and r is the intermolecular separation.

(a) Find the value of r at which the potential energy is a minimum, r_0, in terms of ε and σ.

(b) Obtain an expression for the second derivative of the Lennard–Jones potential at the potential minimum, that is, the harmonic force constant associated with the van der Waals "bond."

(c) Use Lennard–Jones parameters typical of methane ($\varepsilon/k_B = 148\,\text{K}$, $\sigma = 3.81\,\text{Å}$) to calculate the expected vibrational frequency in cm^{-1} of the methane–methane van der Waals bond. Treat methane as an atom of atomic mass 16.

2. Figure 13.1 shows that when the transition dipole moments on two interacting monomers are parallel (μ_A and μ_B have identical magnitudes and directions), the dipole–dipole interaction raises the energy relative to that of the monomer if the two molecules are side by side ($\boldsymbol{R}_{AB}$ is perpendicular to both μ_A and μ_B), but lowers the energy if the two molecules are head-to-tail ($\boldsymbol{R}_{AB}$ is parallel to both μ_A and μ_B). This suggests that there should be an intermediate arrangement for which the energy does not change. Find the angle between $\boldsymbol{R}_{AB}$ and μ_A and μ_B for which the interaction energy in Equation (13.1) is zero.

3. The chapter discussed the excited-state energies and wavefunctions for a dimer of neutral monomers whose electronic excitations interact through their transition dipole moments. Now extend this treatment to a chain of N monomers, arranged head-to-tail as in the J-dimer shown in Figure 13.1. Assume that each monomer interacts only with its nearest neighbor. Calculate the energies of the resulting excited states in terms of the monomer transition energy, E_0, and the coupling strength, V. (Note that the $N \times N$ determinant to be solved has the same form as the Hückel determinant for an N-atom linear polyene.) Using physical intuition and analogy with the dimer problem, which of these states do you expect to carry most of the transition strength?

4. For the coupled homodimer problem whose solutions are Equation (13.3), calculate the magnitudes of the transition dipole moments $\left|\langle\Psi^+|\mu|\Psi_0\rangle\right|$ and $\left|\langle\Psi^-|\mu|\Psi_0\rangle\right|$ as a function of the angle between the monomer transition dipoles, μ_A and μ_B. How does the integrated absorption cross section for both components of the dimer compare with that of the monomer?

5. Estimate the energy shift V, in units of cm^{-1}, between the monomer and dimer absorptions for the J-dimer ($\boldsymbol{R}_{AB}$ parallel to both μ_A and μ_B) and the H-dimer ($\boldsymbol{R}_{AB}$ perpendicular to both μ_A and μ_B) shown in Figure 13.1. Make reasonable assumptions for the transition dipole moment appropriate for a strongly allowed electronic transition and for the distance between the centers of the two molecules when two medium-sized molecules are in direct contact. Use the refractive index appropriate for an organic solvent in the visible region of the spectrum.

6. The simplest model for the pi-electronic states of the linear polyenes (Fig. 13.6) is the one-dimensional particle in a box discussed in Chapter 1. The energy levels are given by $E_n = (h^2n^2/8ma^2)$, for $n = 1, 2, \ldots$, where m is the electron mass and a is the length of the box. For application to a linear polyene, the length of the box is taken to be the average CC bond length (about 1.4 Å) multiplied by the number of conjugated CC bonds (three for butadiene, five for hexatriene, etc.), plus half a "bond" at each end of the molecule. The N pi electrons, one for each sp^2 hybridized carbon atom, are then spin paired in the lowest $N/2$ particle in a box states, and the lowest energy electronic transition is taken to be $E_{N/2+1} - E_{N/2}$.

(a) Use the one-dimensional particle-in-a-box model to calculate the expected longest electronic absorption wavelengths of butadiene, hexatriene, octatetraene, and dodecahexaene. Compare your results with Figure 13.6.

(b) What does this model predict for the absorption wavelength of an infinite polyene, polyethylene?

7. An exciton consists of an electron and a "hole" (the net positive charge left over by the removal of an electron) that are Coulombically attracted to each other. It is highly analogous to a hydrogen atom. Recall from Chapter 1 that the energy levels of the hydrogen atom are given by $E_n = -m_e e^4/32\pi^2\varepsilon_0^2\hbar^2 n^2$ for $n = 1$ to ∞. The binding energy for the H atom is thus $E_\infty - E_1 = m_e e^4/32\pi^2\varepsilon_0^2\hbar^2 = 13.6$ eV. In a crystal, the electron mass m_e has to be replaced by an "effective mass," m^*m_e, which accounts for the interactions between the electron and the periodic lattice. In addition, ε_0 must be replaced by $\varepsilon = \varepsilon_0\varepsilon_r$ where ε_r is the relative dielectric constant. For a semiconductor (Ge) in which $m^* = 0.55$ and $\varepsilon_r = 16$, calculate the exciton binding energy. Do you expect bound excitons to exist at room temperature?

8. By diagonalizing the Hückel Hamiltonian for butadiene, find the energies and molecular orbital coefficients for the four pi orbitals (you do not have to normalize the MOs). Sketch the four MOs, paying particular attention to the nodal patterns. Discuss similarities and differences between the Hückel MOs and those obtained from the linear particle in a box model.

9. **(a)** Use the Hückel model to calculate the lowest transition energies of butadiene, hexatriene, octatetraene, and dodecahexaene in terms of the parameters α and β, assuming that the parameters do not depend on the length of the polyene.

(b) Using whatever units of energy you prefer, determine the experimental transition energy of butadiene from Figure 13.6 and use it to determine the value of the parameter β.

(c) Use this value of β to predict the absorption wavelengths of hexatriene, octatetraene, and dodecahexaene. How well do these compare with the experimental data in Figure 13.6?

10. Equation (13.7) gives the pi-electron molecular orbital energies for a linear chain of identical conjugated carbon atoms in the Hückel model. If the ends of the chain are joined (e.g., benzene instead of hexatriene), the energy levels become

$$E(n, N) = \alpha + 2\beta cos\left(\frac{2n\pi}{N}\right) \quad n = 1, 2, \ldots N \ (N \text{ even}).$$

Discuss qualitatively how the energy levels of the cyclic polyene differ from those of the corresponding linear polyene. Compare the band gaps in the limit of large N.

CHAPTER 14

METALS AND PLASMONS

Very small metal clusters (up to dozens of atoms) have discrete electronic states and electronic transitions similar to those of molecules. However, bulk metals or metal clusters with hundreds of atoms or more exhibit the defining characteristic of a metal: electronic states that are more or less continuously distributed in energy and have no energy gap between the states occupied at low temperatures and those that are unoccupied. The electrons in metals are delocalized over the whole material, leading to high electrical conductivity. The optical properties of metals are usually discussed not in terms of quantum mechanical energy levels, but through the complex dielectric function of the bulk metal.

14.1. DIELECTRIC FUNCTION OF A METAL

The optical properties of a bulk material may be described phenomenologically in terms of either the complex refractive index, Equation (2.6), $\sqrt{\varepsilon_r} = n + i\kappa$, or the complex relative dielectric function, $\varepsilon_r = \varepsilon_r' + i\varepsilon_r''$ (recall $\varepsilon_r = \varepsilon/\varepsilon_0$). For nonmagnetic materials, the relative dielectric constant is the square of the refractive index, so $\varepsilon_r' = n^2 - \kappa^2$ and $\varepsilon_r'' = 2n\kappa$.

A very simple yet flexible classical microscopic model of polarizable matter is that due to Lorentz (Bohren and Huffman, 2004). It treats the electrons as

Condensed-Phase Molecular Spectroscopy and Photophysics, First Edition. Anne Myers Kelley.

particles of mass m and charge e, bound to fixed centers through isotropic harmonic springs and subject to frictional damping. Newton's equation of motion for such an oscillator is

$$m(d^2\mathbf{x}/dt^2)+b(d\mathbf{x}/dt)+K\mathbf{x}=e\mathbf{E}(t), \tag{14.1}$$

where $\mathbf{x}$ is the (time-dependent) position of the electron, b is a damping constant, K is the force constant of the spring, and $\mathbf{E}$ is the (generally time-dependent) electric field. If we assume that the electron is driven by electromagnetic radiation at frequency ω, $\mathbf{E}(t) = \mathbf{E}_0 e^{-i\omega t}$ (for mathematical simplicity, we use here the complex form of the electromagnetic field rather than the real one used in earlier chapters), then the solution to Equation (14.1) is

$$\boldsymbol{x}(t)=\frac{e\boldsymbol{E}(t)}{m\left(\omega_0^2-\omega^2-i\gamma\omega\right)}, \tag{14.2}$$

where the oscillator frequency is given by $\omega_0^2 = K/m$ and $\gamma = b/m$. Note that in the presence of frictional damping ($\gamma \neq 0$) the proportionality between $\mathbf{x}$ and $\mathbf{E}$ is a complex quantity, meaning that the electron motion is out of phase with the driving field.

The induced dipole for one oscillator is $\mu = e\mathbf{x}$, and the bulk polarization (dipole moment per unit volume) is $\mathbf{P} = N\mu = Ne\mathbf{x}$, where N is the number of oscillators per unit volume. Thus, the polarization becomes

$$\boldsymbol{P}=\frac{\varepsilon_0 \boldsymbol{E}\omega_p^2}{\omega_0^2-\omega^2-i\gamma\omega}, \tag{14.3}$$

where we define the plasma frequency as $\omega_p^2 = Ne^2/m\varepsilon_0$. Recall from Equation (2.2a) that the dielectric function ε is defined by $\mathbf{P} + \varepsilon_0\mathbf{E} = \varepsilon\mathbf{E}$ or $\mathbf{P} = (\varepsilon_r - 1)\varepsilon_0\mathbf{E}$. Therefore, the relative dielectric function is

$$\varepsilon_r(\omega)=1+\frac{\omega_p^2}{\omega_0^2-\omega^2-i\gamma\omega}, \tag{14.4}$$

with real and imaginary parts given by

$$\varepsilon_r'(\omega)=1+\frac{\omega_p^2\left(\omega_0^2-\omega^2\right)}{\left(\omega_0^2-\omega^2\right)^2+\gamma^2\omega^2} \tag{14.5a}$$

$$\varepsilon_r''(\omega)=\frac{\omega_p^2\gamma\omega}{\left(\omega_0^2-\omega^2\right)^2+\gamma^2\omega^2}. \tag{14.5b}$$

These two functions are plotted in Figure 14.1. The imaginary part, which is related to absorption, exhibits a peak at $\omega = \omega_0$, the natural oscillator resonance frequency. The real part is greater than one for frequencies below ω_0

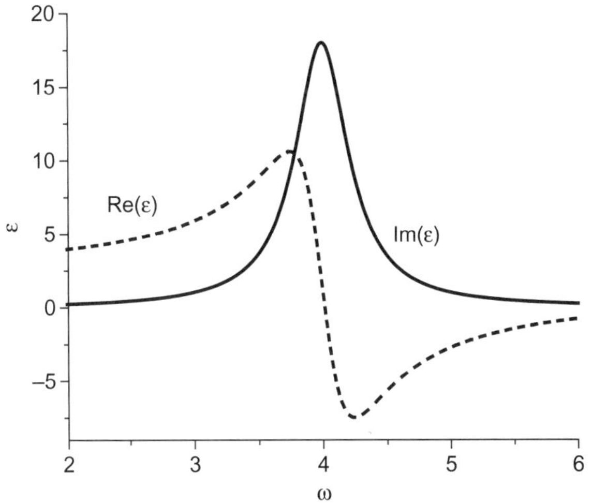

Figure 14.1. Real and imaginary parts of the complex dielectric function from Equation (14.5) for $\omega_p = 6$, $\omega_0 = 4$, and $\gamma = 0.5$ in arbitrary units.

and less than one for frequencies above ω_0. For frequencies far below the resonance frequency, the dielectric constant approaches a constant real value, $\varepsilon_r(\omega << \omega_0) \approx 1 + (\omega_p/\omega_0)^2$.

A much better model for a polarizable material is obtained by assuming that different groups of oscillators have different resonant frequencies (i.e., they are bound by springs having different force constants). A simple extension of the previous results yields for the multiple oscillator case,

$$\varepsilon_r(\omega) = \varepsilon_r^0 + \sum_j \frac{\omega_{pj}^2}{\omega_{0j}^2 - \omega^2 - i\gamma_j \omega}, \tag{14.6}$$

where the jth type of oscillator has resonant frequency ω_{0j}, plasma frequency ω_{pj}, and damping constant γ_j. The quantity ε_r^0 is introduced to include all those oscillators whose frequencies are much higher than the frequency range of interest and therefore make an approximately constant contribution to the dielectric function. If all oscillators are included in the sum over j, then $\varepsilon_r^0 = 1$.

The Lorentz model can be applied to a wide variety of materials, including insulators and semiconductors. The conduction electrons in metals are very weakly bound to the atomic cores, often to the point where the effective spring constant is essentially zero. Setting the force constant K to zero, and therefore $\omega_0 = 0$, we obtain

$$\varepsilon_r(\omega) = 1 - \frac{\omega_p^2}{\omega(\omega + i\gamma)}, \tag{14.7}$$

with real and imaginary parts given by

$$\varepsilon_r'(\omega) = 1 - \frac{\omega_p^2}{(\omega^2 + \gamma^2)} \tag{14.8a}$$

$$\varepsilon_r''(\omega) = \frac{\omega_p^2 \gamma}{\omega(\omega^2 + \gamma^2)}. \tag{14.8b}$$

This is known as the Drude model for the optical properties of a free-electron metal. Note that for frequencies below the plasma frequency ($\omega < \omega_p$) and assuming the damping is not too large, the real part of the dielectric function is negative while the imaginary part is positive. This is typical of most metals.

The two models can be combined to treat metals in which both quasi-free conduction electrons and localized electrons contribute to the optical properties:

$$\varepsilon_r(\omega) = \varepsilon_r^0 - \frac{\omega_{pe}^2}{\omega^2 + i\gamma_e\omega} + \sum_j \frac{W_j}{\omega_{0j}^2 - \omega^2 - i\gamma_j\omega}, \tag{14.9}$$

where the subscript e now designates the conduction band electrons, each group of bound electrons is designated by index j, W_j is a real, positive weighting factor that is proportional to the oscillator strength for the jth transition, and ε_r^0 contains contributions from bound electrons whose resonant frequencies are much higher than the frequency range of interest.

14.2. PLASMONS

The oscillating dipole resulting from the correlated motions of the free electrons in a metal is known as a plasma oscillation, and a single quantum of such an oscillation is called a plasmon. Plasmons manifest themselves differently depending on the geometry of the metal.

To describe plasmons in terms of the Drude model, assume that all of the electrons in a certain region of a bulk metal are displaced by a distance u in some direction (Fig. 14.2). This will generate a polarization $P = \mathrm{Ne}u$ and an electric field $E = P/\varepsilon_0 = \mathrm{Ne}u/\varepsilon_0$. The force on the electrons is $F = -\mathrm{eE} = -\mathrm{Ne}^2u/\varepsilon_0 = -m\omega_p^2 u$. The system will then obey Newton's second law as given in Equation (14.1) with $K = 0$:

$$m(d^2u/dt^2) + b(du/dt) = -\mathrm{eE} = -m\omega_p^2 u \tag{14.10a}$$

$$m(d^2u/dt^2) + \gamma m(du/dt) + m\omega_p^2 u = 0 \tag{14.10b}$$

where $\gamma = b/m$ as before. Assuming the system is underdamped ($\omega_p^2 > \gamma^2/2$), this equation has the solution

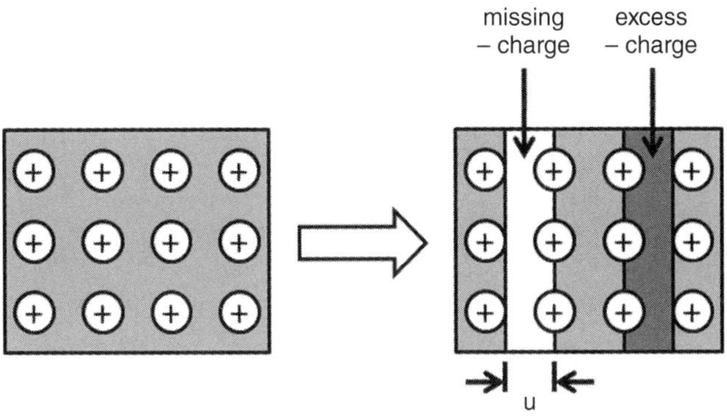

Figure 14.2. Plasmons in a bulk metal. The displacement of part of the negative charge (gray) by a constant amount u (right) sets up a damped oscillation about the equilibrium position (left) at frequency Ω_p given by Equation (14.11).

$$u(t) = e^{-\gamma t/2}[A\cos(\Omega_p t) + B\sin(\Omega_p t)] \text{ where } \Omega_p^2 = \omega_p^2 - \gamma^2/2, \qquad (14.11)$$

where A and B depend on the initial conditions of the displacement. The displaced electrons undergo damped oscillations at a frequency Ω_p that is slightly lower than the plasma frequency (equal to the plasma frequency if damping is ignored). Energy may be transferred between the plasma oscillation and electrons or electromagnetic radiation in quanta of $\hbar\Omega_p$, and these quanta are referred to as plasmons.

The above describes plasma excitations of a bulk metal. Of generally greater interest in modern spectroscopy are surface plasmons, oscillations of the conduction electrons at the interface between a large slab of metal or a small metal particle and a nonmetallic medium, often air or vacuum. It can be shown, although we would not do so here, that at an infinite planar interface between a free-electron metal and a vacuum, the plasmon oscillation occurs at a frequency (Kittel, 1996)

$$\omega_{\text{surface}}^2 = \frac{\omega_p^2}{2}, \qquad (14.12)$$

if damping is neglected. A simple handwaving way to rationalize the factor of two is that half of the positive cores that impose the restoring force on the electrons in the bulk metal are missing at the interface.

More relevant to the coming discussion of surface-enhanced spectroscopies is the surface plasmon resonance of a small metallic sphere (Fig. 14.3). This can be derived by recognizing that the electric field $\mathbf{E}$ induced by a uniform polarization $\mathbf{P}$ of a sphere is given by $\mathbf{E} = 4\pi\mathbf{P}/3(4\pi\varepsilon_0) = \mathbf{P}/3\varepsilon_0$. Then, proceeding as before to solve Newton's equation of motion, we arrive at the result, again neglecting damping (Kittel, 1996),

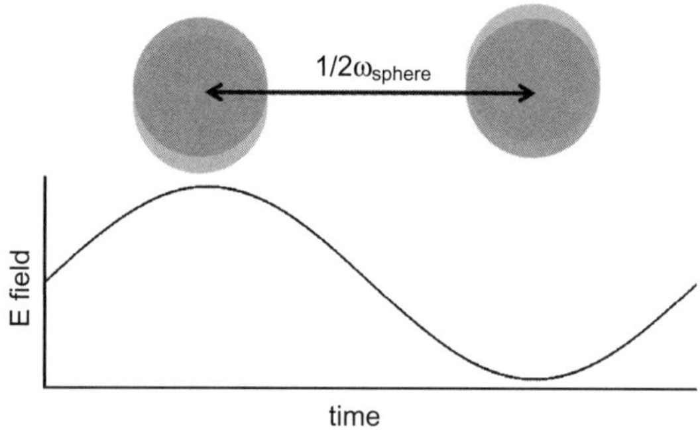

Figure 14.3. Plasmon oscillations in a small sphere. Uniform displacement of the electron cloud (e.g., by an electromagnetic field whose wavelength is large compared with the sphere) sets up oscillations of the electron cloud at a frequency ω_{sphere} given by Equation (14.13).

$$\omega_{sphere}^2 = \frac{\omega_p^2}{3} \tag{14.13}$$

Thus ω_{sphere} should be about 1.73 times lower than the plasma frequency. For main-group metals, this expectation is realized fairly closely; $\omega_p/\omega_{sphere} = 1.94$ for lithium, 1.64 for sodium, 1.58 for potassium, 1.70 for magnesium, and 1.72 for aluminum (Kittel, 1996). For transition metals, the deviations from this simple theory tend to be larger because of the influence of the localized d-electron transitions.

14.3. SPECTROSCOPY OF METAL NANOPARTICLES

Metal nanoparticles are defined here as chunks of metal ranging in size from a few atoms to somewhat less than the wavelength of visible light (hundreds of nm). Very small metal nanoparticles, up to a few dozen atoms, do not exhibit characteristically "metallic" properties. Their density of electronic states is sufficiently small that there are fairly large energy gaps between them. At reasonably low temperatures, almost all of the electrons are in the same lowest-energy state, and the electronic absorption spectrum exhibits multiple, fairly sharp peaks representing excitation to well-defined excited states. These small particles do not exhibit the plasmon resonances that arise from collective excitation of delocalized conduction band electrons.

As the particles are made larger, the density of electronic states increases and plasmon resonances become possible. Equation (10.1) gave the scattering cross section of a spherical particle whose radius a is small compared with the wavelength of the light λ (Mie theory):

$$C_{sca}(\omega_0) = \frac{8\pi k^4}{3}|\alpha_v(\omega_0)|^2, \tag{14.14}$$

where k is the wavevector of the light in the surrounding medium ($k = 2\pi n/\lambda$, where n is the real refractive index of the medium), and α_v is the polarizability volume of the sphere,

$$\alpha_v(\omega) = a^3 \frac{\varepsilon_1(\omega) - \varepsilon(\omega)}{\varepsilon_1(\omega) + 2\varepsilon(\omega)}, \tag{14.15}$$

where $\varepsilon_1(\omega)$ is the complex dielectric function of the material and $\varepsilon(\omega)$ is the dielectric constant of the medium, assumed here to be real. The corresponding expression for the absorption cross section is

$$C_{abs}(\omega_0) = 4\pi k Im[\alpha_v(\omega_0)]. \tag{14.16}$$

The scattering cross section scales as the sixth power of the particle radius while the absorption cross section scales as the cube of the radius. Thus, for small particles, the extinction is dominated by absorption, while for larger particles scattering makes an increasingly large contribution to the total extinction.

Note that we can use either actual dielectric functions or relative ones in the polarizability expression because the factors of ε_0 cancel out. Choosing relative dielectric functions, assuming $\varepsilon(\omega) = 1$ for mathematical simplicity (the environment is a vacuum), and inserting the expression for a free-electron metal from the Drude model, $\varepsilon_1(\omega) = 1 - (\omega_p^2 / \omega(\omega + i\gamma))$ [Eq. (14.7)], Equation (14.15) for the polarizability volume becomes

$$\alpha_v(\omega) = a^3 \frac{\omega_p^2(\omega_p^2 - 3\omega^2) + 3i\omega_p^2\omega\gamma}{(\omega_p^2 - 3\omega^2)^2 + 9\omega^2\gamma^2}. \tag{14.17}$$

The polarizability is maximized when $\omega^2 = \omega_p^2/3$, the same result given in Equation (14.13) for the frequency of the plasmon resonance of a sphere. It should be noted, however, that the Drude model alone does not adequately describe the optical properties even in metals such as silver and gold, which are considered to be strongly plasmonically active. Equation (14.17) predicts that small silver and gold spheres should have their absorption maxima in the ultraviolet, not in the visible as observed. A dielectric function such as Equation (14.9), which includes contributions from both bound and free electrons, is needed to adequately reproduce the optical spectra of most metals.

In gold, plasmon resonances do not become evident until the particles are about 2 nm in diameter, and the strength of the plasmon resonance increases rapidly with increasing nanoparticle size (Alvarez et al., 1997). Small spherical nanoparticles exhibit a single plasmon resonance. Larger spherical particles can have additional resonances that arise from the excitation of higher multipolar resonances in particles that are not small relative to the optical wavelength. Anisotropic particles, such as nanorods and triangular nanoprisms, exhibit multiple plasmon resonances that arise from oscillation of the conduction electrons along different directions.

14.4. SURFACE-ENHANCED RAMAN AND FLUORESCENCE

While the optical properties of bulk metals and metal nanoparticles are of interest in their own right, metal nanostructures are particularly interesting because of the ability of plasmon resonances to enhance the optical responses of other materials located in close proximity to them. The plasma oscillations that give rise to strong absorption and scattering by metallic nanoparticles also produce strong electromagnetic fields that amplify the incident field used to drive the oscillation. A molecule sitting on or near the surface of a plasmonically active nanostructured metal, upon irradiation with light, experiences an electromagnetic field that is much stronger than the field produced by the incident light alone in the absence of the metal. This has the effect of enhancing any optical response that depends on the field strength. It also enhances the spontaneous emission processes involved in Raman scattering and fluorescence (Fig. 14.4).

This effect was first seen in the context of Raman scattering. It was discovered that when silver electrodes were roughened by electrochemical cycling, thus producing nanoscale surface features, the Raman spectra of molecules adsorbed to the surfaces of those electrodes were intensified by huge factors, often orders of magnitude more than could be accounted for by the increase in surface area alone. Surface-enhanced Raman scattering or SERS was soon observed with other metals as well, most notably gold and copper, and with nanoparticles dispersed in solution as well as on electrodes. As there are enor-

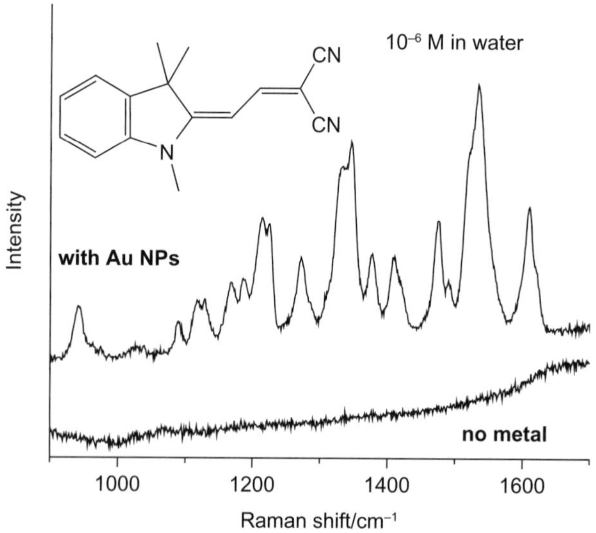

Figure 14.4. Surface plasmon enhancement of Raman scattering. The chromophore shown gives no measurable Raman spectrum in water at a concentration of 10^{-6} M. At the same concentration in the presence of gold nanoparticles, a strong Raman spectrum is observed with the same laser power and measurement time.

mous potential analytical applications for a method to amplify the highly structure-specific but normally weak Raman effect, there has been a great deal of experimental and theoretical activity in finding nanostructures that give large and reproducible Raman enhancements. Electromagnetic simulations (numerical solutions of Maxwell's equations for metal structures of specific sizes and shapes) show that the largest electric field enhancements occur at sharply pointed features and particularly at very small, molecule-sized gaps between two or more such features. Engineering such gaps in a reproducible way, and then making a molecular analyte of interest sit at an optimal position in the gap, remains challenging, but considerable progress has been made.

Part of the mechanism for SERS and other surface-enhanced spectroscopies may, at least in some cases, involve specific interactions between the analyte molecule and the metal, such as formation of charge-transfer excitations (Chapter 12). These are referred to as chemical enhancement mechanisms. However, it is generally believed that the dominant effect is the enhancement of the electromagnetic field through excitation of surface plasmons. When z-polarized light irradiates a uniform metal sphere whose radius, a, is much smaller than the wavelength of the light, the electromagnetic field outside the sphere is given by (Stiles et al., 2008)

$$\boldsymbol{E}_{\text{out}}(\omega; x, y, z) = E_0\hat{\boldsymbol{z}} - \alpha_{\text{v}}(\omega)E_0\left[\frac{\hat{\boldsymbol{z}}}{r^3} - \frac{3z}{r^5}(x\hat{\boldsymbol{x}} + y\hat{\boldsymbol{y}} + z\hat{\boldsymbol{z}})\right], \tag{14.18}$$

where $\alpha_{\text{v}}(\omega)$ is the polarizability volume given by Equation (14.15), E_0 is the electric field strength of the incident radiation, x, y, and z are the Cartesian coordinates, $r^2 = x^2 + y^2 + z^2$, and $\hat{\boldsymbol{x}}$, $\hat{\boldsymbol{y}}$, and $\hat{\boldsymbol{z}}$ are unit vectors along the Cartesian axes. If we assume that the analyte molecule is located directly on the surface of the sphere (r = a), the square of the field relative to the applied field becomes

$$\left|\frac{E_{\text{out}}}{E_0}\right|^2 = |1-g|^2 + \left[6\text{Re}(g) + 3|g|^2\right]\cos^2\theta, \tag{14.19}$$

where $g = \alpha_{\text{v}}(\omega)/a^3$. The value of $\cos^2\theta$ averaged over all (θ,φ) is 1/3, so if we assume that the sphere is coated uniformly with analyte molecules, the average value of the enhancement is

$$\left|\frac{E_{\text{out}}}{E_0}\right|^2_{\text{avg}} = 1 + 2|g|^2. \tag{14.20}$$

Inserting appropriate values for the dielectric function of silver at 514.5 nm, $\varepsilon_1(\omega) \approx -11 + 0.33i$ (Johnson and Christy, 1972), and assuming the environment is a typical solvent, such as water, with a purely real dielectric constant at optical frequencies of $\varepsilon(\omega) \approx 2$, we find $g \approx 1.86$ (the imaginary part is very small), and the average square of the field enhancement has a value of approximately 8.

The SERS enhancement factor (EF) is defined experimentally as the ratio of the Raman scattered intensity per analyte molecule in the absence and the presence of the enhancing metal. Basic electromagnetic SERS theory says that the enhancement factor should be given by the corresponding ratios of the electric field intensity enhancements at the incident (laser) and Raman-scattered frequencies (Stiles et al., 2008):

$$EF = \frac{(I_{\text{SERS}} / N_{\text{surf}})}{(I_{\text{NRS}} / N_{\text{sol}})} = |E_{\text{enh,inc}}|^2 \, |E_{\text{enh,scatt}}|^2, \tag{14.21}$$

where I_{SERS} and I_{NRS} are the measured Raman intensities with and without the metal, N_{surf} and N_{sol} are the numbers of molecules within the irradiated volume on the surface and in the unenhanced solution, and $E_{\text{enh,inc}}$ and $E_{\text{enh,scatt}}$ are the ratios between the electric field strengths with and without the metal at the incident and scattered frequencies. Neglecting the difference between the incident and scattered frequencies and inserting the previous results that $|E_{\text{enh}}|^2 = |E_{\text{out}}/E_0|^2 \approx 8$, the SERS enhancement factor is predicted to be about 64 for silver nanospheres at 514.5 nm. This is many orders of magnitude smaller than the typical reported experimental enhancement factors of 10^4–10^6 or even higher. In fact, strong SERS enhancements are typically not observed for isolated small nanospheres. In order to obtain strong enhancements from small spherical particles, it is usually necessary to aggregate the particles, producing junctions at which the electromagnetic enhancements can be much larger. Strong enhancements can also be observed with isolated particles that have sharp corners, such as triangular prisms, or for much larger spherical particles.

According to Equation (14.21), the Raman enhancement depends on the product of the squared field enhancements at the incident and scattered wavelengths. The electric field enhancement squared roughly follows the plasmon extinction spectrum which is usually quite broad, so the difference between the incident and scattered frequencies does not matter very much for low-frequency vibrations; however, for vibrations with larger Raman shifts, the enhancement profile is shifted to shorter wavelengths compared with the extinction spectrum, as shown in Fig. 14.5.

The rough calculation done previously indicates that the enhancement factor should not depend on the size of the particles as long as they are small enough to justify the approximation that the wavelength is large compared with the particle dimensions. In practice, however, larger particles (e.g., 50–100 nm) usually give greater SERS enhancements than smaller ones (10–20 nm). There are several reasons for this, two of which can be understood from the very simple theory described previously. First, while the numerical estimates made previously assumed that the analyte molecule sits directly on the particle surface, in realistic situations the analyte is slightly above the surface, and often there is a distribution of analyte molecules at different distances away from the surface. The field enhancement [Eq. (14.18)] drops off more quickly with increasing distance for small particles than for larger

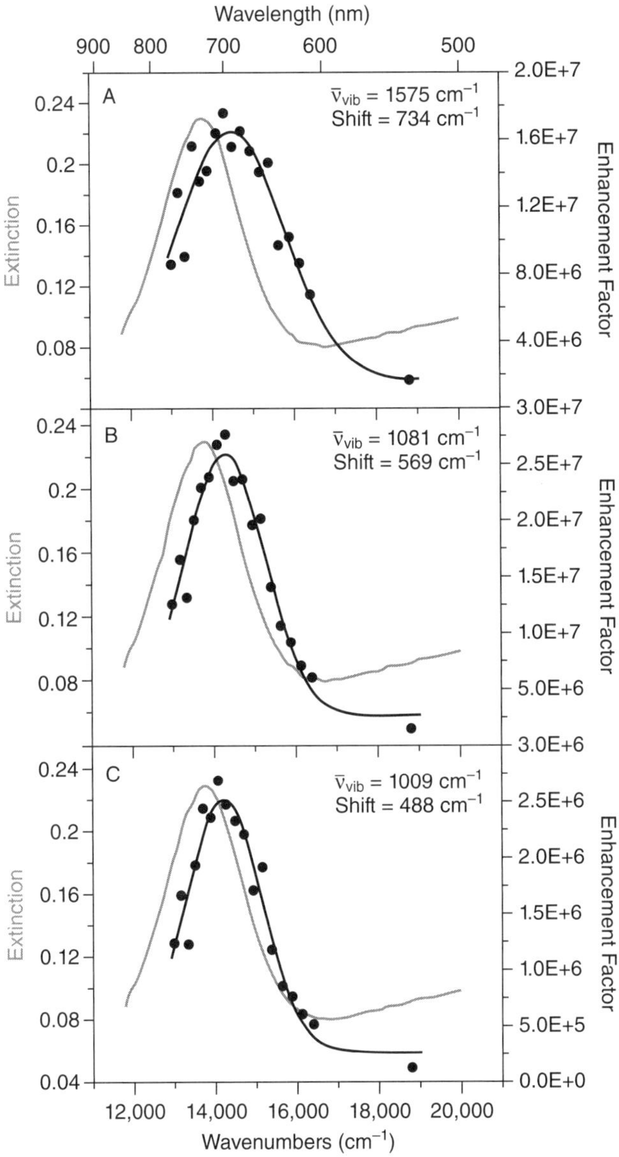

Figure 14.5. Effect of Raman shift on SERS enhancement profiles. The light solid curve is the plasmon extinction spectrum, while the dots (data) and dark curve (fits) are enhancement profiles for three Raman lines of benzenethiol. The SERS enhancement maximum occurs at about the average of the enhancement maxima for the laser light and the scattered light. Reprinted with permission from McFarland et al., J. Phys. Chem. B **109**, 11279 (2005). Copyright 2005, American Chemical Society.

ones. Second, as mentioned previously, small particles have a greater ratio of absorption to scattering than do larger ones. Thus, for a given nominal field enhancement, smaller particles will absorb more of the incident and scattered radiation than do larger ones, reducing the actual detected Raman signal.

Surface plasmon enhancement can be combined with resonance enhancement (Chapter 10) to provide extraordinarily sensitive detection of Raman signals. It has been shown that under ideal circumstances surface enhanced resonance Raman scattering (SERRS) can be sensitive enough to detect individual molecules. SERRS is often limited by the photochemical stability of the analyte at the very high field intensities provided by plasmonic enhancement. It is perhaps no surprise that SERRS is most commonly performed on molecules such as laser dyes, which have been engineered for extremely high photostability—that is, they can survive a very large number of absorption-emission or absorption-radiationless decay cycles without undergoing photochemical changes. SERRS spectra of less stable molecules are often contaminated with a variety of products of thermal and/or photochemical degradation.

Resonance Raman scattering and fluorescence each involve an excitation step and a spontaneous emission step, both of which should be subject to plasmonic field enhancement. Thus, it might appear that both processes should be enhanced to the same extent, apart from the difference in emitted wavelength between Raman and fluorescence (fluorescence typically being more strongly Stokes-shifted). In practice, however, the fluorescence enhancement is usually much smaller than the Raman enhancement. For example, in the laser dye Rhodamine 6G, it is essentially impossible to measure the spontaneous resonance Raman spectrum in solution because the underlying fluorescence is many orders of magnitude stronger, but SERRS spectra at the same excitation wavelength show strong Raman lines with only a weak fluorescence background—the Raman scattering is much more strongly enhanced than the fluorescence. This can be understood by recognizing that the metal surface also provides a pathway for increased nonradiative decay of the excited state. This nonradiative decay is fast compared with the natural radiative lifetime for fluorescence (nanoseconds), but not fast enough to compete with the dephasing processes that limit the resonance Raman intensity (subpicosecond). Thus, the ratio of Raman scattering to fluorescence is, in most cases, greatly increased by putting the molecule on a metal surface. Surface enhanced fluorescence is typically observed with the fluorophore not directly adsorbed to the surface but spaced at least a few nanometers away (Galloway et al., 2009). Furthermore, the maximum fluorescence enhancements are typically on the order of factors of 10 rather than factors of 10^6 or better for Raman.

REFERENCES AND FURTHER READING

M. M. Alvarez, J. T. Khoury, T. G. Schaaff, M. N. Shafigullin, I. Vezmar, and R. L. Whetten, J. Phys. Chem. B **101**, 3706 (1997).

C. F. Bohren and D. R. Huffman, *Absorption and Scattering of Light by Small Particles* (Wiley-VCH, Weinheim, 2004).

C. M. Galloway, P. G. Etchegoin, and E. C. Le Ru, Phys. Rev. Lett. **103**, 063003 (2009).

P. B. Johnson and R. W. Christy, Phys. Rev. B **6**, 4370 (1972).

C. Kittel, *Introduction to Solid State Physics*, 7th edition (John Wiley & Sons, New York, 1996).

P. L. Stiles, J. A. Dieringer, N. C. Shah, and R. P. Van Duyne, Annu. Rev. Anal. Chem. **1**, 601 (2008).

PROBLEMS

1. Recall that the plasma frequency ω_p of a metal is given by the Drude model as $\omega_p^2 = Ne^2/m\varepsilon_0$, where N is the density of conduction electrons and m is the electron mass. Calculate the plasma frequencies for Na, Al, and Ag and express them in eV. The densities of these metals are 0.97 g/cm^3 for Na, 2.7 g/cm^3 for Al, and 10.5 g/cm^3 for Ag; assume each Ag atom donates one conduction electron. Compare your calculation with the experimental values (5.71 eV for Na, 15.3 eV for Al, and 9.6 eV for Ag).

2. The quantity ε_r^0 in Equation (14.9) accounts for the contribution to the dielectric function from oscillators whose resonant frequencies are much higher than the frequency range of interest. Assuming those oscillators can also be modeled with the Lorentz formula [third term in Eq. (14.9)], show that ε_r^0 should be a real quantity and determine what its sign should be.

3. Equations (14.14) and (14.16) give the scattering and absorption cross sections for a spherical particle whose radius is small compared with the wavelength of the light. These quantities depend on the polarizability volume given by Equation (14.15), which involves the dielectric functions of both the sphere and the surrounding medium. Consider absorption and scattering from a gold sphere and model its dielectric function as a Drude oscillator plus one resonance due to localized electrons [Eq. (14.9)]. Use parameters from Vial et al. (Phys. Rev. B **71**, 085416 [2005]): $\omega_{pe} = 1.33 \times 10^{16}\,\text{s}^{-1}$, $\gamma_e = 1.00 \times 10^{14}\,\text{s}^{-1}$, $W_j = 1.81 \times 10^{31}\,\text{s}^{-2}$, $\gamma_j = 6.60 \times 10^{14}\,\text{s}^{-1}$, $\varepsilon_r^0 = 5.97$. Plot the scattering and absorption cross sections across the visible spectrum (400–700 nm) for a 10-nm radius Au sphere both in vacuum [$\varepsilon(\omega) = 1$] and in water [$\varepsilon(\omega) = 1.8$ at visible wavelengths].

4. Equation (14.20) gives the theoretical electric field enhancement for analyte molecules coating the surface of a metal sphere embedded in a medium. Assuming the medium is water [$\varepsilon(\omega) \approx 1.8$ at visible wavelengths] and using the Drude model, Equation (14.8), for the polarizability of the metal, determine what values of the plasma frequency ω_p and damping parameter γ should optimize the Raman enhancement for excitation near 600 nm.

5. **(a)** Derive Equation (14.19) from Equation (14.18).

(b) At what position on the surface of the sphere (in terms of r,θ,φ) will an adsorbed analyte experience the largest and smallest field enhancements? Assume that the metal dielectric function is that of silver, $\varepsilon_1(\omega) \approx -11 + 0.33i$, and the medium is water, $\varepsilon_1(\omega) \approx 1.8$.

6. In practical applications of SERS, often the analyte molecules of interest are not bound directly to the metal surface but rather are spaced a constant distance δ away from the metal surface by some type of coating. Neglecting the dielectric properties of the coating (i.e., assuming that it is a vacuum) and starting from Equation (14.18) as in Problem 5, derive the expected distance dependence of the SERS enhancement, Equation (14.21). That is, derive how the enhancement factor should depend on δ for $r = a + \delta$, where $\delta < a$. You may assume that the analyte molecules are uniformly distributed over the coated metallic sphere so that the angular averaging used to obtain Equation (14.20) may be used.

7. In SERS, it is commonly observed that when the laser excitation wavelength is chosen to give maximum enhancement of the vibrational fundamentals, the overtone transitions are enhanced by smaller factors than the fundamentals. Try to explain this in terms of the electric field enhancement factors for fundamentals and overtones.

8. Under ideal conditions, SERS can be sensitive enough to allow observation of the Raman spectrum of a single scatterer. Claims of single-molecule sensitivity were long viewed skeptically, but several lines of evidence argue in favor of single-molecule observations. One is an isotopic dilution method where measurements are made on a very dilute solution containing equal numbers of organic molecules of the normal form and of a derivative that has some of its H atoms replaced by deuterium (D). H/D substitution has essentially no effect on the electronic or chemical properties of the molecule, but does shift some of its vibrations (recall Chapter 8) so the two species have distinct Raman spectra. In an experiment using rhodamine 6G and its tetradeuterated analog, 50 different spectra were observed under presumed "single-molecule" conditions. Twenty-two showed vibrations of only the all-H containing molecule, 24 showed only the all-D, and four showed both. If the distribution of observed molecules is random, then it should follow a Poisson distribution:

$$P(N) = e^{-\rho}\frac{\rho^N}{N!},$$

where $P(N)$ is the probability of observing N molecules, and ρ is a positive-valued parameter. Use the previous information to estimate what fraction of observations in the previous experiment involved a single scatterer.

CHAPTER 15

CRYSTALS

In a perfect crystal, every atom has a precisely defined position relative to every other atom. Although real crystals are not perfect, theoretical results derived for perfect crystals often describe the properties of real crystals quite well. This chapter describes some of the basic concepts, terminology, and formalism used to describe the structure and spectroscopy of crystals, and briefly addresses the vibrations (phonons) of crystals. The electronic states and spectroscopy of semiconductors are discussed in the following chapter.

15.1. CRYSTAL LATTICES

The fundamental property that distinguishes a crystal from a large molecule, liquid, or amorphous solid is periodicity. The entire crystal can be built up by taking one or more atoms (the unit cell), repeatedly translating it by specified distances along specified directions, and replicating it. Different kinds of crystalline solids can be differentiated by the nature of their building blocks. A molecular solid, such as naphthalene, is composed of discrete, covalently bound molecules which interact with one another through weaker dipolar or van der Waals forces. An ionic solid, such as sodium chloride, is composed of individual cations and anions that interact mainly through Coulombic forces. An atomic solid, such as neon, consists of discrete, neutral atoms that interact

Condensed-Phase Molecular Spectroscopy and Photophysics, First Edition. Anne Myers Kelley.

through van der Waals forces. A network solid, such as a diamond, consists of atoms that interact through covalent bonds. A metallic solid, such as iron, contains discrete, positively charged atomic cores that interact through the attraction of the cores to the delocalized conduction electrons. There are many intermediate cases: for example, the cation and/or anion in an ionic solid may be a charged polyatomic molecule (e.g., SO_4^{2-}), and ionic solids may be only partially ionic in that the effective charges on the ions in the crystal may be noninteger. The unit cell for a molecular solid consists of an integer number of molecules, usually one or two. The unit cell for an ionic solid contains an integer number of formula units, often one or two but sometimes more. The unit cell for an atomic, network, or metallic solid consists of one or more atoms.

A crystal lattice (sometimes known as the direct lattice to distinguish it from the reciprocal lattice defined below) is specified by the number, type, and arrangement of the atoms making up the unit cell, and a set of three unit vectors, $\mathbf{a}_1$, $\mathbf{a}_2$, and $\mathbf{a}_3$. The unit vectors may be, but are not necessarily, mutually perpendicular. Once the unit cell has been defined, the entire crystal can be built up by replicating the unit cell at positions given by integer multiples of the unit vectors: $\mathbf{R} = i\mathbf{a}_1 + j\mathbf{a}_2 + k\mathbf{a}_3$, where i, j, and k are any integers. The lengths of the unit vectors are designated a, b, and c, respectively, and these are known as the lattice constants. Each crystal lattice type is defined by the angles between the unit vectors and by which unit vectors, if any, are constrained to have the same length. It is useful to also define a reciprocal lattice by the vectors

$$\mathbf{b}_1 = 2\pi(\mathbf{a}_2 \times \mathbf{a}_3)/V \tag{15.1a}$$

$$\mathbf{b}_2 = 2\pi(\mathbf{a}_3 \times \mathbf{a}_1)/V \tag{15.1b}$$

$$\mathbf{b}_3 = 2\pi(\mathbf{a}_1 \times \mathbf{a}_2)/V. \tag{15.1c}$$

where $V = \mathbf{a}_1 \cdot \mathbf{a}_2 \times \mathbf{a}_3$ is the volume of the unit cell. Here $\times$ and $\cdot$ refer to the cross product and dot product, respectively, of two vectors.

In an infinite crystal, every unit cell is identical and sits in an identical environment. Thus, the Hamiltonian is the same at corresponding positions in every unit cell in the structure, that is, at every point $\mathbf{R} = i\mathbf{a}_1 + j\mathbf{a}_2 + k\mathbf{a}_3$, where i, j, and k are any integers. In other words, if we define a translation operator $T_\mathbf{R}$ such that $T_\mathbf{R}[f(\mathbf{r})] = f(\mathbf{r} + \mathbf{R})$, then the Hamiltonian commutes with $\mathrm{T}_\mathbf{R}$, and all eigenfunctions of the Hamiltonian can be expressed as eigenfunctions of $T_\mathbf{R}$. These functions are known as Bloch wavefunctions, and they have the form $\Phi_\mathbf{k}(r) = \exp(i\mathbf{k}\cdot\mathbf{r})\, u_\mathbf{k}(\mathbf{r})$ where $u_\mathbf{k}(\mathbf{r})$ is a periodic function that has the same periodicity as the lattice, $u_\mathbf{k}(\mathbf{r}) = u_\mathbf{k}(\mathbf{r} + \mathbf{R})$ (the Bloch lattice function). The wavefunction can differ by only a phase factor at equivalent lattice points in the crystal, so the absolute square of the wavefunction is the same at corresponding lattice points. The constant $\mathbf{k}$ is the wavevector associated with a particular Bloch function; it has the same meaning as the wavevector for an electromagnetic wave introduced in Chapter 2.

In general, the wavevector **k** can take any value from zero to infinity. However, if only the lattice points are considered, all physically distinguishable wavefunctions can be obtained by limiting **k** to a certain finite range. Consider (Fig. 15.1) a one-dimensional lattice along the x-direction in which the lattice points are spaced by a distance R, and consider the two wavevectors k and $k' = k + (2\pi n/R)$, where n is an integer. The function $\exp(ik'x) = \exp(ikx)$ $\exp(2\pi nix/R)$ has the same value as $\exp(ikx)$ when x is any integer multiple of R, that is, at the lattice points. Therefore, we need to consider only those k-values ranging from zero to $2\pi/R$, or equivalently from $-\pi/R$ to $+\pi/R$. This region of k-space in one dimension is known as the first Brillouin zone.

In a three-dimensional crystal, the definition of the first Brillouin zone is a little more involved. It can be defined as the smallest polyhedron contained by the six planes perpendicularly bisecting the reciprocal lattice vectors. This is easiest to demonstrate with the simplest three-dimensional example, a simple cubic lattice whose lattice vectors are mutually perpendicular and all the same length: $\mathbf{a}_1 = a\hat{\mathbf{x}}$, $\mathbf{a}_2 = a\hat{\mathbf{y}}$, and $\mathbf{a}_3 = a\hat{\mathbf{z}}$, where $\hat{\mathbf{x}}$, $\hat{\mathbf{y}}$, and $\hat{\mathbf{z}}$ are the unit vectors along the three Cartesian directions (Fig. 15.2). The volume is $V = a^3$,

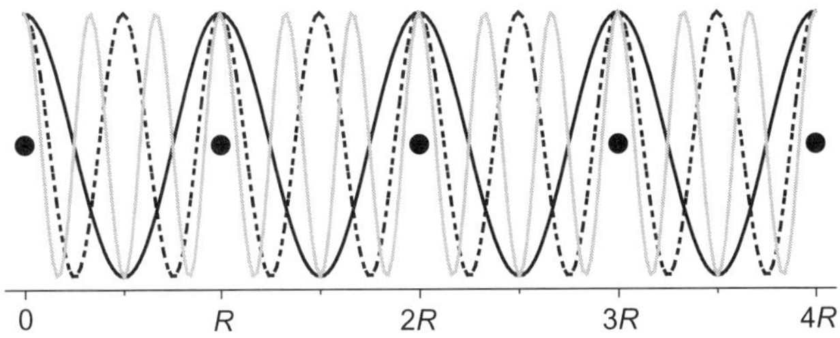

Figure 15.1. Real part of the function $\exp(ikx)$ for $k = (2\pi n/R)$, where $n = 1, 2$, and 3. All three functions have the same value at each lattice point.

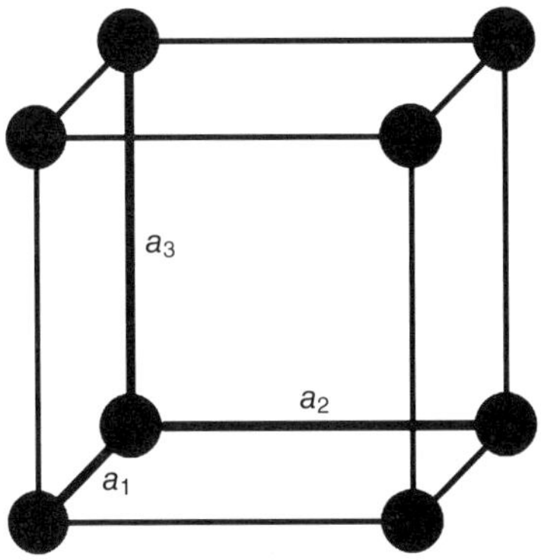

Figure 15.2. The unit cell for a simple cubic lattice. The three lattice vectors are mutually perpendicular and all have the same length.

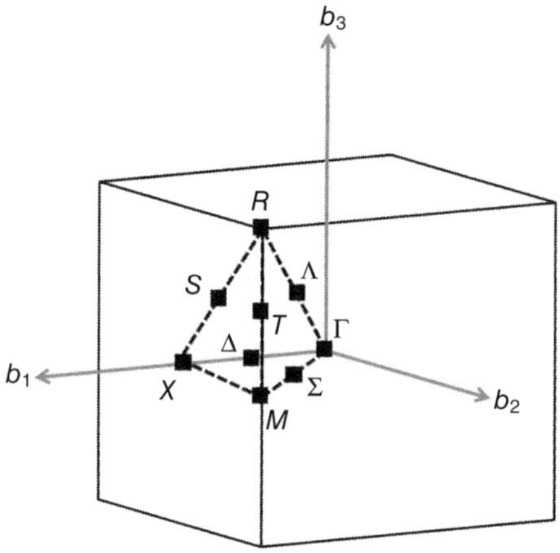

Figure 15.3. Definition of points in the first Brillouin zone of the simple cubic reciprocal lattice.

and the three reciprocal lattice vectors defined as above become $\mathbf{b_1} = (2\pi/a)\mathbf{x}$, $\mathbf{b_2} = -(2\pi/a)\mathbf{y}$, and $\mathbf{b_3} = (2\pi/a)\mathbf{z}$. The reciprocal lattice in this case is also a simple cubic, and the first Brillouin zone is a box whose sides extend from $-\pi/a$ to $+\pi/a$ along x, y, and z. Thus, we only need to consider those wavevectors $\mathbf{k}$ whose lengths vary between $-\pi/a$ and $+\pi/a$ along each of the three spatial directions. This is a simple case for demonstration; for most lattices, the reciprocal lattice does not have the same symmetry as the direct lattice, and the first Brillouin zone may be a more complicated looking polyhedron.

Special symbols are used to represent particular points in the first Brillouin zone. The center of the Brillouin zone, that is, the point at the origin, is always denoted Γ. The other special points are different for different lattices, but in general points on the surfaces of the Brillouin zone are denoted by Roman letters, and points inside the Brillouin zone are named with Greek letters. For the specific example just discussed of a simple cubic lattice (Fig. 15.3), the point at the surface in the center of one of the faces is denoted X and the point halfway between Γ and X is called Δ. The point bisecting the intersection of two faces is denoted M and the point halfway between Γ and M is called Σ. The point at the intersection of all three faces is called R and the point halfway between Γ and R is called Λ. The directions Γ to X, Γ to M, and Γ to R are designated [1 0 0], [1 1 0], and [1 1 1], respectively, according to their components along $\mathbf{b_1}$, $\mathbf{b_2}$, and $\mathbf{b_3}$. Bear in mind that since this is the reciprocal lattice, the Γ point corresponds to the longest wavelength (infinitely long wavelength), and the points on the surface of the Brillouin zone correspond to the shortest physically meaningful wavelengths.

15.2. PHONONS IN CRYSTALS

The vibrations and vibrational spectra of molecules were discussed in some detail in Chapter 7. A crystal is really just a very large molecule, and the physics governing its vibrational spectroscopy is fundamentally the same. Just as for a molecule, one begins by writing down the classical equation of motion and finding the classical vibrational frequencies and normal modes of vibrational motion. These are then equated to the frequencies and normal coordinates of the corresponding quantum mechanical harmonic oscillators to obtain the quantized energy levels and the functional form of the wavefunctions. The collective vibrations of the atoms in a crystal are referred to as phonons. The obvious difficulty in applying the techniques described in Chapter 7 to an infinite crystal is that the number of atoms is infinite! However, this turns out not to be a problem in a crystal because of the translational symmetry it possesses.

In order to get a feel for how this works and to simplify the calculations, we will begin with a one-dimensional example (Fig. 15.4). Consider first a lattice of identical atoms of mass M, in an array along the x-axis. Let the nth atom be located at position $x_n = na$ along the x-axis and let its displacement from equilibrium be denoted u_n. Assume that each atom is connected to its nearest neighbors by a harmonic spring with force constant K, as discussed in Chapter 7. The force felt by the nth atom is then given by

$$F_n = K(u_{n+1} - u_n) + K(u_{n-1} - u_n) \tag{15.2}$$

According to Newton's law,

$$M(d^2u_n/\mathrm{d}t^2) = F_n = K(u_{n+1} - u_n) + K(u_{n-1} - u_n) = -K(2u_n - u_{n+1} - u_{n-1}). \tag{15.3}$$

Note that we could also include smaller harmonic terms between non-nearest-neighbor atoms if desired. For a small finite number of atoms N, one can write down the N equations and solve them simultaneously. For an effectively infinite number of atoms, we instead seek solutions of the form $u_n = A\exp(iqx_n)\exp(-i\omega t)$, where ω is the frequency of the oscillation. That is, each atom oscillates with the same frequency and amplitude but a different phase. This has the form of a Bloch function multiplied by an oscillatory factor in time.

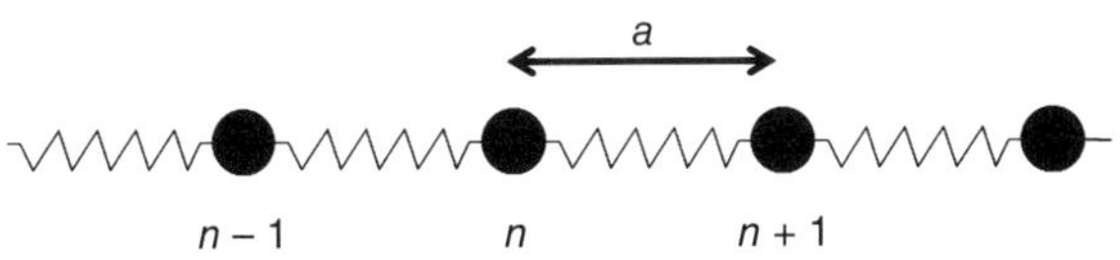

Figure 15.4. Model one-dimensional lattice of identical atoms with lattice constant a.

Substituting this functional form into the above equation and taking the second time derivative leaves

$$-\omega^2 MAe^{iqna}e^{-i\omega t} = -KAe^{-i\omega t}\left[2e^{iqna} - e^{iq(n+1)a} - e^{iq(n-1)a}\right], \tag{15.4}$$

which simplifies to

$$M\omega^2 = K\left[2 - e^{iqa} - e^{-iqa}\right] = 2K\left[1 - \cos(qa)\right] = 4K\sin^2(qa/2). \tag{15.5}$$

Thus we get the relationship between the frequency of the vibration and its wavevector,

$$\omega = \sqrt{\frac{4K}{M}}\left|\sin(qa/2)\right|. \tag{15.6}$$

This is an example of a dispersion relation, which relates the frequency or energy of a state to its wavevector. When the wavevector is zero, the frequency is zero; all of the atoms move together, corresponding to an overall translation of the lattice through space with no changes in the interatomic distances. When the wavevector has the value $q = \pm\pi/a$ (the ends of the first Brillouin zone), the frequency has its maximum value of $(4K/M)^{1/2}$. Here, each atom is completely out of phase with its nearest neighbor, giving maximum changes in the nearest-neighbor distances. For intermediate values of the wavevector, the frequency has an intermediate value, and the variation in nearest-neighbor distances is intermediate.

This is a simple and rather boring example. An example that is still quite simple but more interesting is a one-dimensional diatomic lattice consisting of alternating atoms with masses m and M, respectively (Fig. 15.5). Now there are two different equations of motion,

$$m(d^2u_n/dt^2) = -K(2u_n - u_{n+1} - u_{n-1}) \tag{15.7a}$$

$$M(d^2u_{n+1}/dt^2) = -K(2u_{n+1} - u_{n+2} - u_n), \tag{15.7b}$$

and we seek solutions having the form $u_n = A_1\exp(iqna)\exp(-i\omega t)$ and $u_{n+1} = A_2\exp(iq[n+1]a)\exp(-i\omega t)$. Substituting these into the differential equation yields

$$(2K - m\omega^2)A_1 - 2K\cos(qa)A_2 = 0 \tag{15.8a}$$

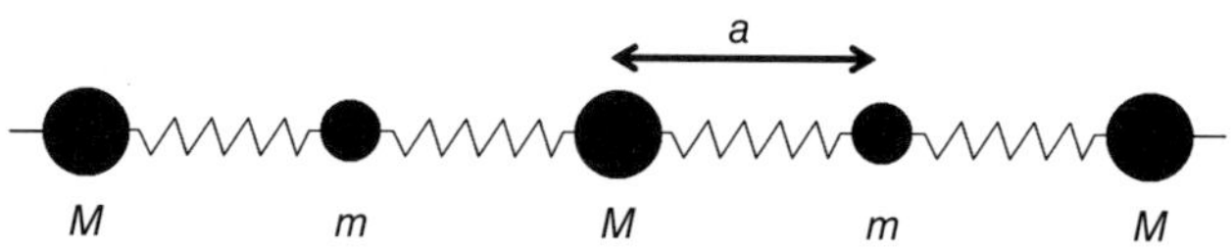

Figure 15.5. A one-dimensional diatomic lattice.

$$-2K\cos(qa)A_1 + (2K - M\omega^2)A_2 = 0. \tag{15.8b}$$

This is a system of linear homogeneous equations for which a solution exists only if the determinant of the corresponding matrix is zero:

$$\begin{vmatrix} 2K - m\omega^2 & -2K\cos(qa) \\ -2K\cos(qa) & 2K - M\omega^2 \end{vmatrix} = 0. \tag{15.9}$$

Solution yields

$$\omega^2 = K\left\{\left(\frac{1}{m}+\frac{1}{M}\right) \pm \sqrt{\left(\frac{1}{m}+\frac{1}{M}\right)^2 - \frac{4\sin^2(qa)}{mM}}\right\}.$$

There are two different solutions corresponding to the two sign choices. The corresponding dispersion curves in the first Brillouin zone are plotted in Figure 15.6 (note that the lattice constant is now $2a$ rather than a). The lower-frequency branch (negative sign) is called the acoustic branch. It corresponds to solutions where the atoms in the same unit cell move together. The zero-wavevector limit corresponds to the overall translation of the whole lattice as in the identical-atoms case, and the frequency goes to zero. The higher-frequency branch (positive sign) is called the optical branch. It corresponds to solutions where the two atoms in the same unit cell move in opposite directions, so the zero-wavevector limit corresponds to the greatest relative motion of neighboring atoms and the highest vibrational frequency.

Real three-dimensional crystals have more complicated phonon dispersion relations, but crystals that have more than one atom per unit cell still have distinguishable acoustic and optical phonon branches. In general, the optical

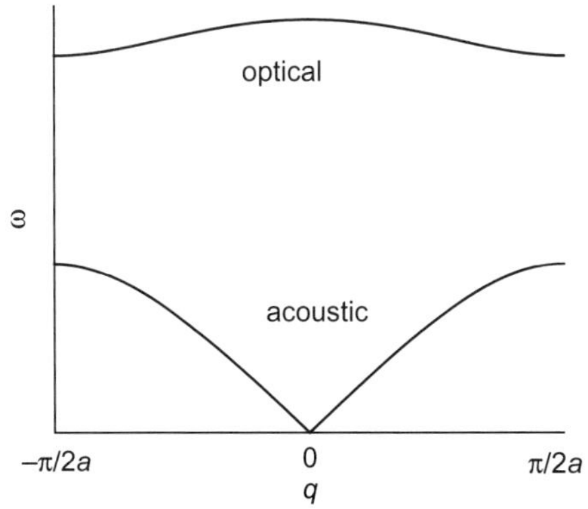

Figure 15.6. Optical and acoustic branches of the phonon dispersion curve for a one-dimensional diatomic lattice, calculated for a mass ratio of $M/m = 5$.

phonons have higher frequencies than acoustic phonons, and optical phonons usually have much flatter dispersion curves, that is, less dependence of the frequency on the wavevector. This is because the frequency of an optical phonon is determined more strongly by the relative motions of neighboring atoms in the unit cell than by the relative motions in different unit cells. Each phonon mode is characterized by its wavevector **q** and amplitude **A**, both vector quantities. **q** contains both the wavelength and the direction of propagation of the phonon. If **A** and **q** are parallel, the phonon is designated as longitudinal; if they are perpendicular, the mode is transverse. If there are s atoms per unit cell, there are a total of 3s dispersion curves of which three branches are acoustic (the frequency goes to zero at $q = 0$) and the remainder are optical.

In the one-dimensional examples discussed above, q is a scalar quantity and one need consider only the dependence of the phonon frequencies on the magnitude of q. In a three-dimensional crystal, **q** is a vector quantity and one must consider both its magnitude and its direction in calculating and plotting the phonon dispersion curves. It is conventional to plot these curves along a few "high-symmetry" directions, for example, [1 0 0], [1 1 0], and [1 1 1], as described above. As an example, we plot phonon dispersion curves along three directions for the wurtzite crystal form of CdSe. This is an anisotropic crystal structure with unit vectors defined by

$$\boldsymbol{a}_1 = \frac{\sqrt{3}a}{2}\hat{\boldsymbol{x}} + \frac{a}{2}\hat{\boldsymbol{y}},\ \boldsymbol{a}_2 = -\frac{\sqrt{3}a}{2}\hat{\boldsymbol{x}} + \frac{a}{2}\hat{\boldsymbol{y}},\ \text{and}\ \boldsymbol{a}_3 = c\hat{\boldsymbol{z}},$$

where a and c are two different unit vectors needed to define the structure. Notice that $\boldsymbol{a}_1$ and $\boldsymbol{a}_2$ are not perpendicular to one another. The reciprocal lattice vectors calculated from Equation (15.1) are

$$\boldsymbol{b}_1 = \frac{2\pi}{\sqrt{3}a}\hat{\boldsymbol{x}} + \frac{2\pi}{a}\hat{\boldsymbol{y}},\ \boldsymbol{b}_2 = -\frac{2\pi}{\sqrt{3}a}\hat{\boldsymbol{x}} + \frac{2\pi}{a}\hat{\boldsymbol{y}},\ \text{and}\ \boldsymbol{b}_3 = \frac{2\pi}{c}\hat{\boldsymbol{z}}.$$

The corresponding first Brillouin zone is displayed in Figure 15.7, with just four of the high-symmetry points (the zone center, Γ, and three other points) labeled. Figure 15.8 displays the phonon dispersion curves along these three directions.

The energy levels of phonons are subject to the same quantization conditions as the vibrations of polyatomic molecules. The energy in a given phonon mode is constrained to have the values $E = \hbar\omega(n + ½)$, where $n = 0, 1, 2, \ldots$. The total phonon energy is given by $E = \Sigma\, \hbar\omega_p(n_p + ½)$, where n_p is the phonon quantum number in mode p having frequency ω_p.

15.3. INFRARED AND RAMAN SPECTRA

The basics of infrared and Raman spectroscopy of molecules were discussed in Chapters 7 and 10, respectively. There, it was shown that within the harmonic

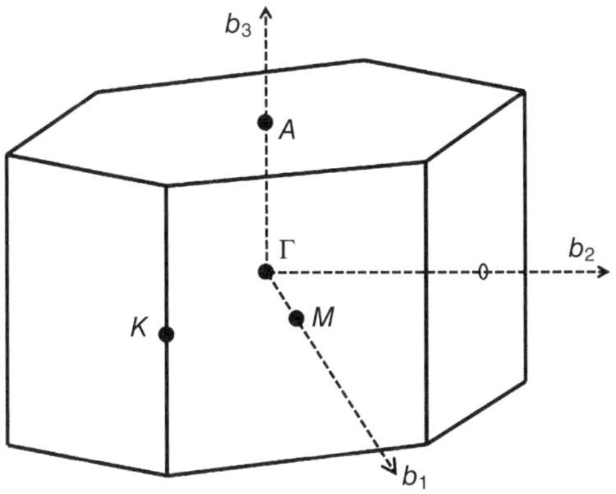

Figure 15.7. First Brillouin zone of the wurtzite crystal lattice with four of the high-symmetry points labeled.

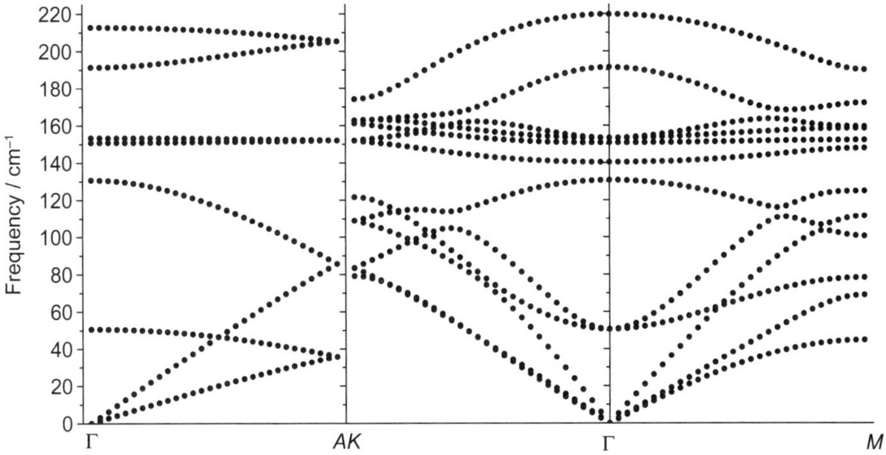

Figure 15.8. Phonon dispersion curves calculated for CdSe in its wurtzite crystal structure, along the Γ to A direction (left) and the Γ to K and Γ to M directions (middle and right). There are four atoms in the unit cell, leading to three acoustic and nine optical phonon modes at each lattice point, some of which are degenerate. The calculations use an empirical force field (Rabani, 2002) and the Generalized Lattice Utility Program (GULP) (Gale and Rohl, 2003).

oscillator approximation, direct vibrational transitions arising from absorption or emission of infrared radiation, or vibrational transitions resulting from nonresonant Raman scattering, normally occur with a change of only one vibrational quantum (the fundamental transition). Transitions involving more than one vibrational quantum (combination bands and overtones) do occur in real molecules but are normally much weaker than the fundamentals. We also showed that group theory can be used to determine which vibrational modes may be infrared or Raman allowed based on their symmetries, while recognizing that transitions that are symmetry allowed may be too weak to be detected.

Corresponding rules apply to infrared and nonresonant Raman transitions of phonons in crystals. Usually, only fundamentals have high intensities; overtones and combination bands are observed weakly because of anharmonicities. As with molecules, group theory may be used to determine which symmetries of phonons may have nonzero intensities; however, for crystals, space groups rather than point groups must be used. Space groups are discussed elsewhere (Yu and Cardona, 2010; Tinkham, 1964). Some of the other qualitative results found for molecules also apply to crystals. For example, if the crystal structure is centrosymmetric, all infrared-active phonons have odd symmetry with respect to inversion, while all Raman-active phonons have even symmetry, so there is complete exclusion between IR and Raman modes just as in centrosymmetric molecules. In most noncentrosymmetric crystals, there are some phonons that are both IR and Raman active, some that are active in only one, and some that are seen in neither.

There are also some qualitative differences between IR and Raman spectroscopy in crystals and in molecules. In molecular spectroscopy, one is most often working with unoriented samples; the molecule-fixed axes are randomly distributed relative to a space-fixed axis system. Thus, if one uses light that is polarized along, say, the z-axis in the laboratory, it can cause an infrared transition in any IR-allowed vibration in the molecule. In crystal spectroscopy, on the other hand, if large single crystals are available, one can accurately define the orientation of the crystal axes relative to the polarization direction(s) of the incident and (for Raman) scattered fields, providing additional control over which vibrations are observed. A second important difference relevant to semiconductor crystals is that because many semiconductors absorb light in the visible region of the spectrum, Raman scattering is more often performed under conditions of electronic resonance in semiconductors than in small molecules. In fact, in semiconductors with small band gaps, it is nearly impossible to carry out a nonresonant Raman experiment. Because the $\Delta n = 1$ selection rule in ordinary Raman scattering is relaxed in resonance Raman as discussed in Chapter 10, Raman spectra of semiconductors often show long overtone progressions.

15.4. PHONONS IN NANOCRYSTALS

The discussion of crystals given above assumes that the crystal is infinite in extent. While infinite crystals do not actually exist, single crystals of macroscopic size may reasonably be treated as infinite. As the crystal becomes smaller, the finite extent of the crystal and the presence of surfaces become important for describing the spectroscopy. The size at which finite size effects become important depends on the material and the property of interest but typically ranges from microns to tens of nanometers. For the purposes of this chapter and the next, "large enough" crystals are referred to as "bulk" and smaller ones as "nanocrystals." Finite size effects on electronic properties of semiconductor crystals are discussed in the next chapter. Here, we touch briefly on finite size effects on phonons and associated spectroscopies.

Phonons of nanocrystals are generally divided into two categories: those that mainly involve motions of atoms in the interior of the crystal and fairly closely resemble the phonons of infinite crystals, and those that mainly involve motions of atoms on the surface of the crystal. Nanocrystals that are at least several unit cells long in each direction, that is, diameters of a few nanometers or larger, usually have some phonons that correspond quite closely to those of the infinite crystal. They can usefully be classified as acoustic or optical, longitudinal or transverse, and so on, although the distinctions are not as clear as in a bulk crystal. The phonon frequencies of interior modes of nanocrystals also tend to be quite similar to those of bulk crystals, often within ±5% or so. The deviations may be to either higher or lower frequencies depending on the material and the nature of the mode. Because the full translational symmetry of the bulk is not present in a nanocrystal, the bulk selection rules are not followed exactly. In addition to these bulklike modes, nanocrystals also have phonon modes that are localized on surface atoms and do not have any counterpart in the bulk crystal.

REFERENCES AND FURTHER READING

J. D. Gale and A. L. Rohl, Mol. Simul. **29**, 291 (2003).

C. Kittel, *Introduction to Solid State Physics*, 7th edition (John Wiley & Sons, New York, 1996).

J. D. Patterson and B. C. Bailey, *Solid-State Physics* (Springer, Berlin Heidelberg, 2007).

E. Rabani, J. Chem. Phys. **116**, 258 (2002).

M. Tinkham, *Group Theory and Quantum Mechanics* (McGraw-Hill, New York, 1964).

P. Y. Yu and M. Cardona, *Fundamentals of Semiconductors*, 4th edition (Springer-Verlag, Berlin Heidelberg, 2010).

PROBLEMS

1. Consider a two-dimensional lattice of identical atoms arranged in a hexagonal structure with an atom at the vertex of each hexagon. This could represent, for example, a single sheet of graphene. The distance between each atom and its nearest neighbors is 1.4 Å.

(a) Make a rough drawing of such a lattice and define an x,y-coordinate system. Indicate on your drawing the smallest unit cell that can reproduce the lattice. How many atoms does it contain? Give the coordinates of the two basis vectors of this lattice in units of Å.

(b) Find the reciprocal lattice vectors, in units of Å^{-1}, and sketch the reciprocal lattice in the x,y-coordinate system. Sketch the first Brillouin zone and calculate its area.

2. The three-dimensional face-centered cubic lattice can be described by the basis vectors

$$\boldsymbol{a}_1 = \frac{a}{2}(\hat{x}+\hat{y}) \quad \boldsymbol{a}_2 = \frac{a}{2}(\hat{y}+\hat{z}) \quad \boldsymbol{a}_3 = \frac{a}{2}(\hat{z}+\hat{x}).$$

Find the reciprocal lattice.

3. In a three-dimensional crystal having two atoms per unit cell, how many phonon dispersion curves are there? How many correspond to longitudinal acoustic, transverse acoustic, longitudinal optical, and transverse optical modes, respectively?

4. In the text, we derived the phonon dispersion relations for infinite one-dimensional lattices containing either one or two atoms per unit cell. This problem addresses the related situation of an impurity in an otherwise homogeneous lattice. We assume an infinite one-dimensional lattice in which one atom has a different mass and a different nearest-neighbor force constant from the others. We will assume that all interatomic spacings are the same. In principle, we would like to consider the case where the fraction of impurity atoms is very small; however, for mathematical tractability, we will consider here that every fifth atom is an impurity, so our effective unit cell has a length of $5a$, where a is the internuclear separation. The approach to this problem follows that done in Equations (15.7)–(15.9) of the text, but requires a couple of additional assumptions.

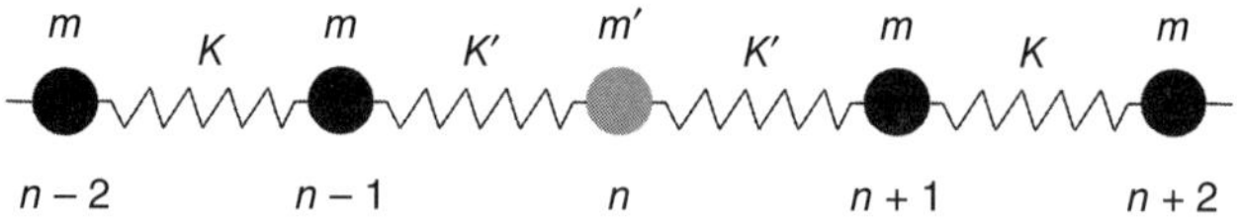

(a) Write down the classical equations of motion for the five atoms in the unit cell.

(b) As before, assume $u_n = A_n\exp(iqna)\exp(-i\omega t)$, insert into the equations of motion and cancel out the time-dependent terms.

(c) Rewrite these equations in terms of the parameters $K/m = \alpha$, $K'/m = \beta$, and $K'/m' = \delta$. At this point, you should have five equations in seven unknowns, the amplitudes A_{n-3} through A_{n+3}. Now recognize that since A_{n-3} and A_{n+2} are in adjacent unit cells, we can equate their amplitudes, so replace A_{n-3} by A_{n+2}, and, similarly, A_{n+3} by A_{n-2} (this is an example of "cyclic boundary conditions"). You now have five equations in five unknowns. Write this as a matrix whose determinant has to equal zero for a nontrivial solution to exist.

(d) Finding a general analytic solution to this determinant is a lot of work. Rather, choose the longest-wavelength ($q = 0$) solution and explore the behavior of the five eigenvalues (ω^2) as a function of the relative values of the three parameters α, β, and δ, referring back to what they mean physically.

5. For phonons of a one-dimensional diatomic lattice, for what relative masses is the dispersion in the optical phonon branch minimized? What about the acoustic branch?

6. CdSe in its wurtzite crystal structure has four atoms per unit cell and therefore 12 phonon modes, but at many points in the Brillouin zone, some of these modes are degenerate. By examining Figure 15.8, determine the frequencies and degeneracies of each of the phonons at the K point.

7. In deriving the dispersion relation for the phonons of a one-dimensional lattice (Fig. 15.4), the chapter assumed that only nearest-neighbor harmonic forces are involved. Derive the dispersion relation analogous to Equation (15.6) with the addition of second nearest-neighbor harmonic forces (e.g., between atom n and atom $n + 2$) with a force constant K'.

8. The chapter derived the phonon dispersion relation for a one-dimensional diatomic lattice:

$$\omega^2 = K\left\{\left(\frac{1}{m}+\frac{1}{M}\right) \pm \sqrt{\left(\frac{1}{m}+\frac{1}{M}\right)^2 - \frac{4\sin^2(qa)}{mM}}\right\}.$$

Find the corresponding numerical values of the frequencies by inserting appropriate numbers for a hypothetical one-dimensional NaCl lattice. Estimate the force constant K by assuming that the Na^+ and Cl^- ions interact through a Coulomb plus Lennard–Jones type potential (Alejandre et al., J. Chem. Phys. **130**, 174505 [2009]):

$$V(r) = 4\varepsilon\left[\left(\frac{\sigma}{r}\right)^{12} - \left(\frac{\sigma}{r}\right)^{6}\right] - \frac{e^2}{4\pi\varepsilon_0 r},$$

where $\varepsilon = 0.351\,\text{kJ mol}^{-1}$ and $\sigma = 3.275$ Å. Find the equilibrium Na-Cl distance by setting the derivative of the potential to zero, and then calculate the second derivative of the potential at the minimum to get the force constant. For the actual three-dimensional crystal, the highest fundamental phonon frequency of NaCl is $\omega \sim 4.8 \times 10^{13}\ \text{s}^{-1}$. Compare this experimental 3D result with your simple 1D estimate.

CHAPTER 16

ELECTRONIC SPECTROSCOPY OF SEMICONDUCTORS

The previous chapter discussed the structure and the vibrations of crystals in a fairly general manner that can be applied to many different types of crystalline solids. Electronic states of crystals are not so easy to describe in a general way; there are fundamental, qualitative differences between the electronic states and electronic spectroscopy of semiconductors, metals, and insulating molecular solids. Therefore, this chapter will focus on just one important class of crystalline solids, namely semiconductors.

16.1. BAND STRUCTURE

A traditional description of the electronic structure of molecules starts with a set of atomic orbitals or atomic orbital-like basis functions localized on the individual atoms. The variational principle is then used to find the set of molecular orbitals, formed from linear combinations of these basis orbitals, which, when populated with the available valence electrons according to the *aufbau* principle, minimize the ground-state energy. The result is a set of orbitals ordered according to their energy. Because the number of atomic orbital basis functions is finite, so is the number of molecular orbitals. The highest-energy orbital that is filled with electrons in the ground state is often referred to as the HOMO (highest occupied molecular orbital) and the lowest-energy

Condensed-Phase Molecular Spectroscopy and Photophysics, First Edition. Anne Myers Kelley.

orbital that is not occupied in the ground state is referred to as the LUMO (lowest unoccupied molecular orbital). The lowest-energy electronic transition can be approximated as the transition that moves one electron from the HOMO to the LUMO (refer also to Section 13.4). In small molecules, the energy spacings between orbitals are usually quite large. In larger molecules, both the occupied orbitals and the unoccupied orbitals become more closely spaced in energy, although there usually remains a gap between the HOMO and the LUMO. In the limit of an infinitely large molecule (a crystal), the energy levels become continuous rather than discrete, although in a semiconductor there remains a separation (the band gap) between those levels that are occupied by electrons in the ground state and those that are unoccupied. The occupied orbitals are referred to as the valence band and the unoccupied ones as the conduction band (Fig. 16.1).

The ground state has all of the valence band levels filled and the conduction band levels empty. Thermal or optical excitation can move an electron out of the valence band and into the conduction band. In the standard terminology of solid state physics, an extra electron in the conduction band is referred to as an electron, and a vacancy (missing electron) in the valence band is referred to as a hole. Electrons and holes are treated as different particles with opposite charges, usually different effective masses (see later section), and different wavefunctions. Those accustomed to working with atoms or molecules may find it difficult to grasp the concept of a missing electron as a particle, but when the total number of electrons in the system approaches Avogadro's number,

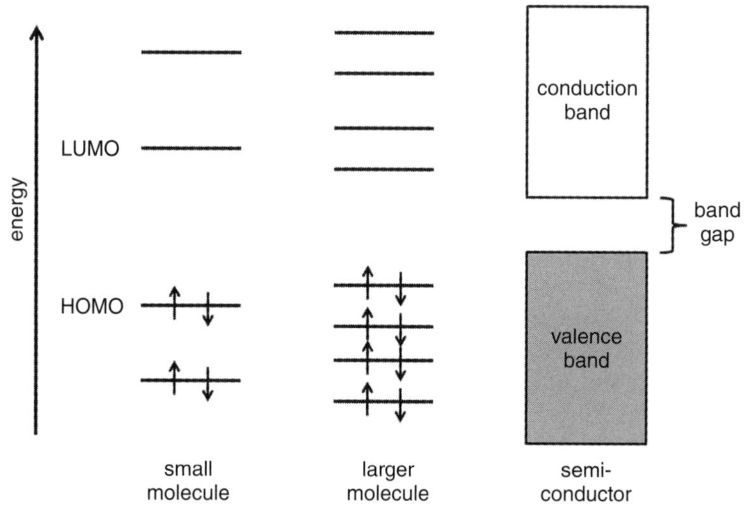

Figure 16.1. Evolution of the electronic energy levels from discrete filled and empty orbitals to continuous bands with increasing size.

it makes a lot of sense to focus on the one that is missing from a band rather than all the others that are present.

Usually, we are most interested in the behavior of the electronic energy levels near the top of the valence band and near the bottom of the conduction band, as these levels make the dominant contribution to the electronic and optical properties. These levels can be considered to originate from a small number of basis functions located on the atoms of a single unit cell and duplicated throughout the crystal. The resulting "molecular orbitals" are delocalized throughout the crystal. The situation is somewhat analogous to that of the pi-electron wavefunctions in conjugated organic molecules. Each atom in the conjugated system contributes a p_z orbital and one pi electron. The linear combinations of these N atomic orbitals produce N delocalized molecular orbitals. In the ground state, the electrons are spin-paired in the lowest-energy $N/2$ pi orbitals (the "valence band"), while the other $N/2$ pi orbitals remain empty ("conduction band"). Recall that another way of treating the electronic wavefunctions of pi-conjugated molecules is to consider that the electrons move in a periodic potential produced by the atoms in the conjugated chain. The simplest such potential is the one-dimensional particle in a box, where the attraction of the pi electrons to the nuclei is not explicitly considered, and the electrons are simply assumed to move freely over the length of the conjugated chain. The electronic wavefunctions and energy levels of semiconductors may be treated in similar ways, by considering either explicit linear combinations of atomic orbitals that are repeated in each unit cell, or by considering motion of a quasi-free particle in a periodic potential having the periodicity of the lattice.

The simplest possible model is the free-electron ("empty lattice") model, where the periodicity of the lattice potential is ignored completely. This is analogous to the simple particle-in-a-box model for pi-electron wavefunctions of conjugated organic molecules. Since the crystal is assumed to be infinite in extent, the wavefunctions are those of a free particle:

$$\Phi(x, y, z) = \exp(i[k_x x + k_y y + k_z z]),\ E(k_x, k_y, k_z) = \hbar^2(k_x^2 + k_y^2 + k_z^2)/2m. \quad (16.1)$$

Recall when we discussed phonons in crystals in Chapter 15, we argued that only those wavevectors that lie within the first Brillouin zone are physically meaningful. For electronic wavefunctions, this restriction does not apply because the electrons are delocalized over the entire crystal and not restricted to the lattice points. That is, electronic wavefunctions can have shorter wavelengths (larger wave vectors) that contained within the first Brillouin zone. Recall from Chapter 15 that for a simple cubic lattice, the first Brillouin zone extends from $-\pi/a$ to $+\pi/a$ along each of x, y, and z. Changing k_x to $(k_x + 2\pi/a)$, for example, does not change the *phonon* frequency or the displacement of the atoms, but it does change the *electronic* energy. It is still traditional to plot the electronic energies as a function of wave vector only within the first Brillouin zone, but the energy functions are multiple valued because k_x, $(k_x + 2\pi/a)$,

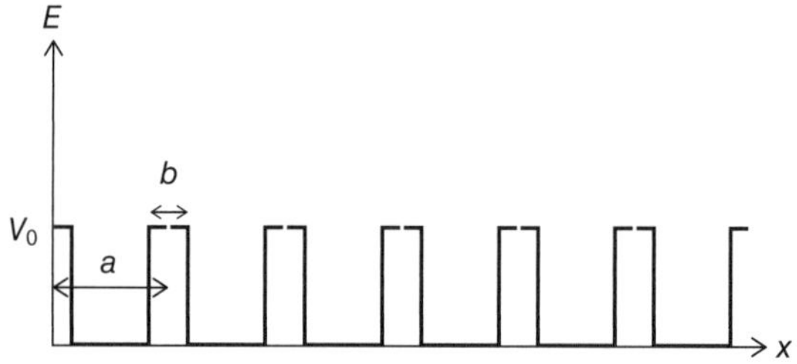

Figure 16.2. Potential energy function for the Kronig–Penney model.

$(k_x + 4\pi/a)$, and so on have different energies yet correspond to the same point in the first Brillouin zone.

The empty lattice is a very crude approximation to the dispersion curves of semiconductors. The most obvious deficiency is that it does not predict the existence of a band gap; there is a continuous distribution of allowed energy levels, as in a metal. In order to obtain semiconducting behavior, it is necessary to assume, at a minimum, that the electrons move in a periodic potential imposed by the lattice. The simplest such model that has an analytic solution is the Kronig–Penney model. Although it is only one-dimensional, it does capture some of the basic physics responsible for the semiconducting band gap. Like the free electron model, the Kronig–Penney model treats the electrons as moving independently—that is, electron-electron repulsion and quantum-mechanical exchange effects are neglected. The electron is considered to move in a potential defined by an infinite array of square barriers (alternatively, an array of square wells for V_0 negative) (Fig. 16.2). The potential energy is defined by

$$V(x) = V_0 \text{ for } 0 < x < b/2 \text{ or } a - b/2 < x < \text{a} \tag{16.2a}$$

$$V(x) = 0 \text{ for } b/2 \le x \le a - b/2 \tag{16.2b}$$

$$V(x + a) = V(x). \tag{16.2c}$$

The third condition simply defines the periodicity of the lattice. As discussed in Chapter 15, the solutions to the time-independent Schrödinger equation for such a periodic potential must have the form $\psi_q(x) = e^{iqx}u_q(x)$, where $u_q(x + a) = u_q(x)$, that is, a Bloch function. Inserting this $\psi_q(x)$ into the Schrödinger equation yields

$$\frac{d^2u_q}{dx^2} + 2iq\frac{du_q}{dx} - \left\{\frac{2m}{\hbar^2}[V(x) - E] + q^2\right\}u_q = 0. \tag{16.3}$$

This equation is simplest to solve in the limit $E < V_0$. Defining the two auxiliary variables $\alpha^2 = 2mE/\hbar^2$ and $\beta^2 = 2m(V_0 - E)/\hbar^2$, the Schrödinger equation becomes

$$\frac{d^2u_q}{dx^2} + 2iq\frac{du_q}{dx} + \{\alpha^2 - q^2\}u_q = 0 \text{ for } V = 0 \tag{16.4a}$$

$$\frac{d^2u_q}{dx^2} + 2iq\frac{du_q}{dx} - \{\beta^2 + q^2\}u_q = 0 \text{ for } V = V_0. \tag{16.4b}$$

These two equations have solutions of the form $u_q(x) = Ae^{i(\alpha-q)x} + Be^{-i(\alpha-q)x}$ for $V = 0$ and $u_q(x) = Ce^{(\beta-iq)x} + De^{-(\beta+iq)x}$ for $V = V_0$. Using the periodicity of the potential and requiring that both the wavefunction and its derivative be continuous leads to the result

$$(ab\beta^2/2\alpha a)\sin(\alpha a) + \cos(\alpha a) = \cos(qa). \tag{16.5}$$

Now, $\cos(qa)$ can only take values between −1 and 1; however, depending on the value of the parameter $ab\beta^2/2$, there may be certain values of αa for which the left-hand side of the equation is greater than 1 or less than −1. That means that those values of αa are not allowed—that is, there are regions of the energy spectrum where no solutions to the Schrödinger equation exist (Fig. 16.3). These are the band gaps in the Kronig–Penney model. Although this model is very simple and actual solids are also three-dimensional, it illustrates the basic physics behind the existence of band gaps in semiconductors.

For a free electron, the dependence of energy on wavevector is given by $d^2E/dk^2 = \hbar^2/m$. More sophisticated theories of electronic band structure give

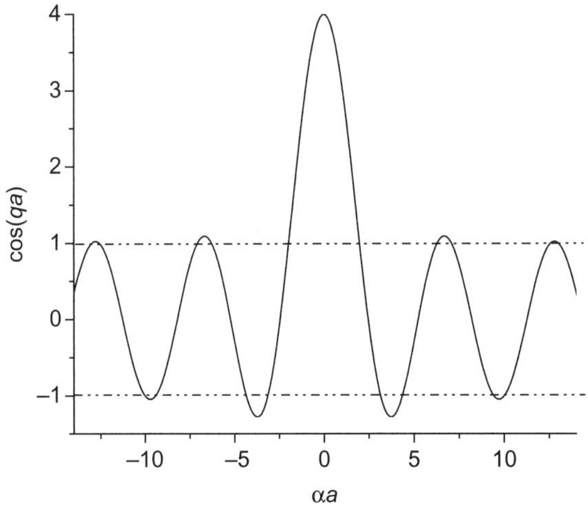

Figure 16.3. Solution to the Kronig–Penney Hamiltonian, Equation (16.5), for the case $ab\beta^2/2 = 3$. The dash-dotted lines indicate the range of allowed values of $\cos(qa)$. For values of αa outside this range, there is no allowed solution—that is, there is no choice of wavevector q corresponding to that energy α ($\alpha^2 = 2mE/\hbar^2$, where E is the energy).

different values for the energies, but often the dependence on k is still approximately quadratic near the top of the valence band and the bottom of the conduction band. When the actual dispersion relations are known or can be calculated, it is conventional to define a quantity known as the effective mass m^*, defined by $m^* = \hbar^2(d^2E/dk^2)^{-1}$. Usually the effective masses are different for the energy levels at the top of the valence band (hole effective mass) and at the bottom of the conduction band (electron effective mass). Typical effective masses for electrons in semiconductors range from $0.01m_e$ to $0.3m_e$ where m_e is the free electron mass, while effective masses for holes range from $0.3m_e$ to m_e. In most materials, the hole effective mass is at least twice the electron effective mass. This means that the hole energy varies more slowly with wavevector than does the electron energy. It also means that the wavefunctions for holes tend to be more localized than those for electrons.

16.2. DIRECT AND INDIRECT TRANSITIONS

In many semiconductors, the highest-energy point in the valence band and the lowest point in the conduction band occur at the same value of the wavevector, specifically the Γ point. Such semiconductors are known as direct gap semiconductors, meaning that the lowest-energy optical transition that is possible occurs between an electron and a hole with the same (zero) momentum. In a direct gap semiconductor, the optical transition at the bandgap energy may be, and often is, electric dipole allowed and may be quite strong. There also exist indirect gap semiconductors in which the top of the valence band and the bottom of the conduction band occur at different values of the wavevector. In an indirect gap material, the purely electronic optical transition at the bandgap energy cannot occur because of the momentum mismatch between initial and final states; the bandgap transition has to be accompanied by absorption or emission of a phonon to satisfy momentum conservation. This is roughly analogous to electronic transitions in molecules that are electronically forbidden but vibronically induced. Indirect gap materials typically have much weaker absorption at the bandgap energy, and, in particular, much longer radiative lifetimes than direct gap materials.

16.3. EXCITONS

Most approaches to band structure theory begin by assuming that the electrons behave independently apart from the Pauli exclusion principle that limits the number of electrons with the same wavevector to two. As long as electron–electron interactions are neglected, the wavefunctions of a perfect crystal are completely delocalized over the crystal. Thus, when optical or thermal excitation promotes an electron from the valence band to the conduction band, both the electron and the hole are completely delocalized. This picture changes when electron–electron interactions are included, usually as a perturbation on the independent-electron wavefunctions. Most notably, the electron in the conduction band and the hole in the valence band are oppositely charged, and thus attract each other. This provides a potential function V(**r**) that tends to keep the electron and the hole localized to the same region of the crystal. The resulting coupled electron-hole wavefunction is referred to as an exciton (Fig. 13.5). It has properties in common with the wavefunction for the hydrogen atom, which also involves a spherically symmetric interaction potential between oppositely charged particles. Confining the electron and hole lowers the potential energy of the system but raises the kinetic energy of each particle, and the average separation at which the total energy is minimized is known as the Bohr radius of the exciton, in homage to the hydrogen atom. The Bohr radius is a measure of the average extent to which the electron and hole are confined to the same region of space. The energy needed to separate the electron and hole from the Bohr radius to an infinite distance is referred to as the exciton binding energy. Recall that for the hydrogen atom the Bohr radius, a_0, is the most probable value for the separation between the electron and the nucleus in the lowest-energy (1s) wavefunction. It is given by

$$a_0 = \frac{4\pi\varepsilon_0\hbar^2}{m_e e^2},$$

where m_e is the electron mass. In a semiconductor, the exciton Bohr radius has the same meaning of the most probable separation between electron and hole. It is given by the same expression except that ε_0 must be replaced by the dielectric constant of the semiconductor, $\varepsilon = \varepsilon_0\varepsilon_r$, and m_e must be replaced by the reduced mass μ_{eh}, where

$$\frac{1}{\mu_{eh}} = \frac{1}{m_e^*} + \frac{1}{m_h^*},$$

with m_e^* and m_h^* being the effective masses of the electron and hole, respectively. Recall that for the hydrogen atom, $a_0 = 0.529$ Å. Most semiconductors have relative dielectric constants in the range 5–15, so the shielding of the electron and hole charges by the lattice increases the Bohr radius by about an order of magnitude. Furthermore, most semiconductors have effective masses for both the hole and, particularly, the electron that are considerably smaller, sometimes by factors of 10 or more, than the bare electron mass. Taken together, these two factors combine to give typical exciton Bohr radii in the range from 1 to 100 nm for different semiconductors.

There are two principal types of excitons distinguished by their sizes. In materials with small dielectric constants, where opposite charges attract each other quite strongly, excitons tend to be on the order of or smaller than a unit cell. These are referred to as Frenkel excitons, and their exciton binding energies tend to be on the order of 0.1–1 eV. Most semiconductors have sufficiently high dielectric constants that there is a strong screening of the electron–hole interaction and the exciton spans multiple unit cells. These Wannier excitons tend to have small binding energies, on the order of 0.01 eV. There is not always a clear distinction between the two types, and some materials can support both types of excitons.

16.4. DEFECTS

An important factor that tends to localize charges in semiconductors is the presence of defects. Real crystals are never perfect—they do not have every atom located at exactly the position expected from the nominal lattice structure. The structure may include misplaced atoms, extra atoms, or missing atoms, or atoms of the wrong type (chemical impurities). There may be discontinuities in the crystal structure such that some atoms are translated relative to others. All crystals have a finite size, and their surfaces constitute a type of defect. Because the atoms of the crystal are not fully coordinated at the surface and the surface atoms are exposed to the environment, surfaces are a prime site for the formation of chemical impurities. For some applications, semiconductor crystals are deliberately produced with some level of defects. For example, they may be doped with atoms having either more or fewer valence electrons to produce an excess of either electrons or holes, increasing electrical conductivity.

16.5. SEMICONDUCTOR NANOCRYSTALS

Much of the spectroscopic work carried out on semiconductors since the 1990s has involved not bulk materials but nanocrystals. The phonon modes of nanocrystals were discussed briefly in the previous chapter; here, we consider their electronic states. There are important ways in which the electronic band structure of a nanocrystal differs from that of a bulk crystal. Because the number of atoms in a nanocrystal is large but decidedly finite (typically hundreds to tens of thousands), the available energy levels are discrete rather than continuously distributed, and there can be fairly large energy gaps between available states near the top of the valence band and the bottom of the conduction band. In this regard, a nanocrystal behaves somewhat more like a large molecule than a bulk crystal (see Fig. 16.1). For sufficiently small crystals, the band gap also becomes larger than that of the bulk, and continues to increase with decreasing crystal size. The size at which this becomes important is given roughly by the Bohr radius of the exciton. If the radius of the nanocrystal significantly exceeds the Bohr radius, then the energy of the exciton is relatively unaffected by the finite size of the crystal, but as the crystal is made smaller than the Bohr radius, the exciton is confined to a smaller region of space and its energy goes up. This effect is known as quantum confinement, and it can be used to tune the bandgap energy over a broad region by chang-

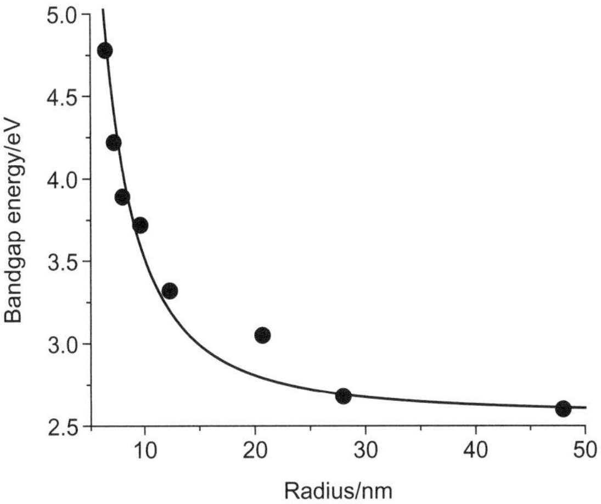

Figure 16.4. Experimental bandgap energy of CdS nanoclusters as a function of particle radius. The solid curve is a fit to the $1/r^2$ dependence predicted by a simple particle in a sphere model (see below). Data from Vossmeyer et al., J. Phys. Chem. **98**, 7665 (1994).

ing the size of the nanocrystal (Fig. 16.4). Another important factor affecting the optical properties of nanocrystals is the fact that a large fraction of the atoms are on the surface, making effects of surface defects and traps potentially much more important. Finally, nanocrystals can occur in a wide variety of shapes, from approximately spherical to rods and disks to more complicated branched structures.

Recall that for an infinite crystal, the electronic wavefunctions can be expressed as linear combinations of Bloch wavefunctions, $\Phi_{\mathbf{k}}(\mathbf{r}) = \exp(i\mathbf{k}\cdot\mathbf{r})u_{\mathbf{k}}(\mathbf{r})$, where $u_{\mathbf{k}}(\mathbf{r})$ is a function that has the same periodicity as the lattice. The function $\exp(i\mathbf{k}\cdot\mathbf{r})$ is an eigenstate of a free particle and can be considered as an "envelope function" that modulates the Bloch lattice function $u_{\mathbf{k}}(\mathbf{r})$. The free particle eigenstates are not appropriate envelope functions for a confined nanocrystal; one must instead use functions that have the correct boundary conditions and symmetry. The simplest approximation to the wavefunctions for a spherical nanocrystal of radius a are the solutions to the "particle in a sphere" problem: $V(\mathbf{r}) = 0$ for $r \leq a$ and $V(r) = \infty$ for $r > a$. This is the spherical symmetry analog to the particle in a box problem summarized in Chapter 1. The solutions to the particle in a sphere problem have the form

$$\begin{aligned}\psi_{n\ell m}(r,\theta,\varphi) &= R_{n\ell}(r)Y_{\ell m}(\theta,\varphi)\\ n = 1, 2, 3\ldots; \ \ell &= 0, 1, 2, \ldots; \ m = -\ell, -\ell+1, \ldots \ell.\end{aligned} \tag{16.6}$$

The wavefunctions separate into a radial part and an angular part in the same way as the solutions to the hydrogen atom, and $Y_{\ell m}(\theta,\varphi)$ are the same spherical harmonics (the eigenfunctions of angular momentum) that appear in the hydrogen atom wavefunctions. Note that unlike the hydrogen atom, here the angular momentum quantum number ℓ may have any non-negative integer value regardless of the value of the radial quantum number n. Because the potential energy is constant (zero) rather than Coulombic, the radial functions for the particle in a sphere have a different functional form than the hydrogen atom solutions:

$$R_{n\ell}(r) = N_{n\ell} j_\ell(k_{n\ell} r), \tag{16.7}$$

where $N_{n\ell}$ is a normalization factor, $j_\ell(z)$ is a spherical Bessel function, the first few of which are

$$j_0(z) = \frac{\sin z}{z} \tag{16.8a}$$

$$j_1(z) = \frac{\sin z}{z^2} - \frac{\cos z}{z} \tag{16.8b}$$

$$j_2(z) = \left(\frac{3}{z^3} - \frac{1}{z}\right)\sin z - \frac{3}{z^2}\cos z, \tag{16.8c}$$

and $k_{n\ell}$ must satisfy $k_{n\ell} = z_{n\ell}/a$, where $z_{n\ell}$ is the nth zero of $j_\ell(z)$. The first few zeros are given below:

	$n = 1$	$n = 2$	$n = 3$
$\ell = 0$	3.142	6.283	9.425
$\ell = 1$	4.493	7.725	10.904
$\ell = 2$	5.763	9.095	12.323
$\ell = 3$	6.988	10.417	13.698

Finally, the allowed energy levels are $E_{n\ell} = z_{n\ell}^2 \hbar^2 / 2ma^2$. This approach can describe either the electron or the hole wavefunction, with the appropriate effective mass (see above) used for each.

These wavefunctions have much in common with the hydrogen atom wavefunctions. The functions with $\ell = 0, 1, 2, 3$ are referred to as S, P, D, and F functions and have the same shapes as the corresponding hydrogen atom functions. The radial quantum number n determines the number of radial nodes in the wavefunction.

The simplest approximation to the electron or hole wavefunction of a semiconductor nanocrystal, then, is a particle in a sphere envelope function multiplying a Bloch lattice function whose details depend on the material. This is a fairly crude approximation. For example, the discontinuity in the potential at r = a causes the wavefunctions to be strictly zero beyond that point. A more realistic treatment gives the spherical well a finite depth, allowing the bound wavefunctions to leak out slightly beyond the boundaries of the crystal. However, even the simplest treatment produces useful results. For example, the lowest energy wavefunction for both electron and hole is the 1S type function, which has no nodes in the interior of the crystal. The lowest energy optical transition (the "band gap" of the nanoparticle) is the $1S_h$–$1S_e$ transition (the subscripts h and e refer to hole and electron, respectively), and the energy is expected to scale as the inverse square of the radius.

The next lowest energy transition should be between the 1P hole and the 1S electron; since the effective mass of the hole is larger than that of the electron, the energy separations between hole levels are smaller than between electron levels. However, this transition is not optically allowed. To a first approximation, the transition dipole operator can be considered to operate only on the unit cell portion of the wavefunction and not the envelope function. Because of the orthogonality of the spherical harmonics and the spherical Bessel functions, only transitions between states with the same n and ℓ are allowed: $1S_h$–$1S_e$, $1P_h$–$1P_e$, and so on. Therefore, the second lowest-energy transition should be $1P_h$–$1P_e$. In real nanocrystals, the selection rules $\Delta n = 0$ and $\Delta\ell = 0$ are only approximate. Transitions between states of different n are normally observed, but transitions that change ℓ are usually not. Therefore, the next two transitions in the optical spectrum should be $2S_h$–$1S_e$ and $1P_h$–$1P_e$, and which one is lower in energy depends on the electron and hole effective masses and other factors not considered in this simple envelope function approximation (Fig. 16.5).

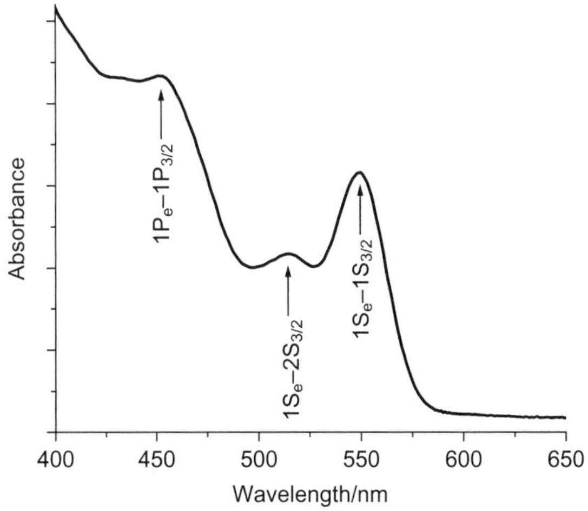

Figure 16.5. Optical absorption spectrum of 3.0-nm diameter CdSe nanocrystals (zinc blende crystal structure) with the first three transitions assigned as particle in a sphere functions. The subscript e stands for electron and the 3/2 and 1/2 subscripts refer to spin-orbit components of the hole wavefunction not considered within the simple particle in a sphere treatment. Data courtesy Professor David F. Kelley, University of California, Merced.

FURTHER READING

M. G. Bawendi, M. L. Steigerwald, and L. E. Brus, Annu. Rev. Phys. Chem. **41**, 477 (1990).

C. Kittel, *Introduction to Solid State Physics*, 7th edition (John Wiley & Sons, New York, 1996).

P. Y. Yu and M. Cardona, *Fundamentals of Semiconductors*, 4th edition (Springer-Verlag, Berlin Heidelberg, 2010).

PROBLEMS

1. In the II–VI semiconductors containing Zn or Cd with S, Se, or Te, the valence band is composed primarily of the filled 3p, 4p, or 5p atomic orbitals of the anion, while the conduction band is composed mainly of the empty 4s or 5s atomic orbitals of the cation.

(a) Which should have the largest and smallest band gap: CdS, CdSe, or CdTe?

(b) Which should have the largest band gap: CdSe or ZnSe?

2. In the "free-electron" model, the electrons are actually not completely free; they are confined to a "box" the size of the crystal. For electrons confined to a three-dimensional box, calculate the density of states (states per unit energy interval), $\rho(E)$. Note that for electrons, there are two available spin states for each particle-in-a-box spatial function.

3. Consider the simplest possible two-dimensional lattice in which all of the atoms are of the same type and are located on a square grid. Let a be the interatomic spacing along x and y. The unit vectors are thus $\mathbf{a}_1 = a\hat{x}$ and $\mathbf{a}_2 = a\hat{y}$.

(a) Find the reciprocal lattice.

(b) In the free-electron approximation, plot the electronic energy as a function of q within the first Brillouin zone, in the (1 0) and (1 1) directions.

4. For the Kronig–Penney model and assuming $\beta^2 = 2m(V_0 - E)/\hbar^2 << 1$ (i.e., the periodic potential is weak), find the energy of the lowest energy band at $q = 0$.

5. **(a)** Calculate the Bohr radius for InSb, for which $m_e^* = 0.013\, m_e$, $m_h^* = 0.6\, m_e$, and $\varepsilon_r = 18$.

(b) Calculate the Bohr radius for Si, for which $m_e^* = 0.33\, m_e$, $m_h^* = 0.5\, m_e$, and $\varepsilon_r = 12$.

(c) For InSb and Si, comment on whether you would expect the optical properties of a 10-nm radius nanocrystal to differ significantly from those of the bulk—that is, do these materials exhibit quantum confinement at this size?

6. ZnO has a density of 5.606 g cm^{-3} at 300 K. Neglecting any details of crystal structure, approximately what fraction of the atoms are on the surface in a spherical ZnO nanocrystal with a diameter of 2 nm? What fraction are on the surface if the diameter is 20 nm?

7. The radial probability density for the particle in a sphere wavefunctions defined by Equations (16.6)–(16.8) is given by $dP = |R_{n\ell}|^2\ 4\pi r^2 dr$. Plot this quantity for the 1S, 2S, and 1P particle in a sphere functions from $r = 0$ to $r = a$, omitting normalization (i.e., you may leave out any constants), and comment on how they differ.

8. A central potential is one for which, when written in spherical polar coordinates, the potential energy depends only on the radius and not on the angles. The hydrogen atom, the three-dimensional isotropic harmonic oscillator, and the particle in a sphere are three examples of central potentials.

(a) Give the expression for the radial probability distribution, defined in Problem 7 above, for the lowest-energy eigenfunction of each of these three potentials. (You will have to generalize the result for the one-dimensional harmonic oscillator to three dimensions and write the solutions in spherical polar coordinates.) Discuss how the radial probability distributions depend on the mass of the particle (assume that for the hydrogen atom, the charge is the normal electron charge, but the mass could be different from m_e).

(b) The particle in a sphere envelope functions given in Equations (16.6)–(16.8) are oversimplified models for the wavefunctions of a semiconductor nanocrystal because they assume that the potential energy is infinitely high outside the sphere. A better model, but more complicated to solve, assumes a large but finite potential energy outside the sphere. Qualitatively, how do you expect the electron and hole wavefunctions to differ when the finite depth of the potential well is considered? Recall that the effective mass is usually much smaller for the electron than for the hole.

APPENDIX A

PHYSICAL CONSTANTS, UNIT SYSTEMS, AND CONVERSION FACTORS

UNIT SYSTEMS USED IN CHEMISTRY AND PHYSICS

Most physical constants or quantities may be expressed in more than one set of units. For example, length may be expressed in meters, or millimeters, or inches. Although English or Imperial units (pounds, feet, etc.) are still used in commerce and everyday life in the United States and a few other countries, scientific quantities are almost universally expressed in the metric system (grams, meters, etc.). However, even within the metric system, there are different conventions for the standard units of various quantities.

For purely mechanical systems, in which all quantities can be expressed in terms of mass, length, and time, two competing metric systems used to be in common use: the centimeter-gram-second system (cgs) system and the meter-kilogram-second system (MKS). The modern preferred unit system for scientific quantities, and the system used in commerce and industry in most of the world, is the International System of Units, abbreviated SI. This is equivalent to MKS for mechanical systems, but includes four additional base units. The base units for the SI and cgs systems are summarized in Table A.1.

A number of other useful units are derived, that is, defined in terms of the base units. Some of the common ones are given in Table A.2.

The most common prefixes used to indicate decimal powers of quantities are given in Table A.3.

Condensed-Phase Molecular Spectroscopy and Photophysics, First Edition. Anne Myers Kelley.

TABLE A.1. Base Units in SI and cgs Systems

Quantity	SI Unit	cgs Unit
Length	Meter (m)	Centimeter (cm)
Mass	Kilogram (kg)	Gram (g)
Time	Second (s)	Second (s)
Electric current	Ampere (A)	
Temperature	Kelvin (K)	
Amount of substance	Mole (mol)	
Luminous intensity	Candela (cd)	

TABLE A.2. Derived Units in SI and cgs Systems

Quantity	SI Unit	Definition in Terms of Other SI Units	cgs Unit	Definition in Terms of Other cgs Units
Frequency	Hertz (Hz)	s^{-1}	Hertz (Hz)	s^{-1}
Force	Newton (N)	$kg \cdot m \cdot s^{-2}$	dyne (dyn)	$g \cdot cm \cdot s^{-2}$
Pressure	Pascal (Pa)	$N \cdot m^{-2}$		
Energy	Joule (J)	N·m	erg (erg)	dyn cm
Power	Watt (W)	$J \cdot s^{-1}$		
Electric charge	Coulomb (C)	A s		
Electric potential difference	Volt (V)	$W \cdot A^{-1}$		

TABLE A.3. Prefixes

Factor	Name
Tera (T)	10^{12}
Giga (G)	10^{9}
Mega (M)	10^{6}
Kilo (k)	10^{3}
Deci (d)	10^{-1}
Centi (c)	10^{-2}
Milli (m)	10^{-3}
Micro (μ)	10^{-6}
Nano (n)	10^{-9}
Pico (p)	10^{-12}
Femto (f)	10^{-15}

TABLE A.4. Maxwell's Equations and Coulomb's Law in SI and cgs Units

SI	cgs-Gaussian
$\nabla \cdot \mathbf{D} = \rho$	$\nabla \cdot \mathbf{D} = 4\pi\rho$
$\nabla \times \mathbf{E} = -\partial \mathbf{B}/\partial t$	$\nabla \times \mathbf{E} = -\frac{1}{c}\frac{\partial \boldsymbol{B}}{\partial t}$
$\nabla \cdot \mathbf{B} = 0$	$\nabla \cdot \mathbf{B} = 0$
$\nabla \times \mathbf{H} - (\partial \mathbf{D}/\partial t) = \mathbf{J}$	$\nabla \times \mathbf{H} = \frac{1}{c}\frac{\partial \boldsymbol{D}}{\partial t} + \frac{4\pi}{c}\mathbf{J}$
$F = \frac{q_1 q_2}{4\pi\varepsilon_0 r^2}$	$F = \frac{q_1 q_2}{r^2}$

For purely mechanical systems, the values of quantities or physical constants in cgs and SI (or MKS) units are related by powers of 10 through straightforward unit conversions. When electromagnetic phenomena are involved, unit conversions are less straightforward because there exist several different conventions for electromagnetism within cgs, and they result in different forms for the relevant equations. Table A.4 gives Maxwell's equations and Coulomb's law in SI units as well as Gaussian units, probably the most common convention in the cgs system.

This book uses SI units throughout, but cgs-Gaussian units are often encountered in older papers and textbooks.

VALUES OF PHYSICAL CONSTANTS

Planck's constant $\mathrm{h} = 6.62606957 \times 10^{-34}\ \mathrm{J \cdot s}$

$\hbar = \mathrm{h}/2\pi = 1.054571726 \times 10^{-34}\ \mathrm{J \cdot s}$

Boltzmann's constant $\mathrm{k_B} = 1.3806488 \times 10^{-23}\ \mathrm{J \cdot K^{-1}}$

Speed of light in vacuum $\mathrm{c} = 2.99792458 \times 10^{8}\ \mathrm{m \cdot s^{-1}}$

Electron charge $\mathrm{e} = 1.602176565 \times 10^{-19}\ \mathrm{C}$

Electron mass $\mathrm{m_e} = 9.10938291 \times 10^{-31}\ \mathrm{kg}$

Permittivity of free space $\varepsilon_0 = 8.854187817 \times 10^{-12}\ \mathrm{C^2 \cdot J^{-1} \cdot m^{-1}}$

Avogadro's number $\mathrm{N_A} = 6.02214129 \times 10^{23}\ \mathrm{mol^{-1}}$.

USEFUL CONVERSION FACTORS

$1\ \mathrm{J} = 1\ \mathrm{kg \cdot m^2 \cdot s^{-2}}$

$1\ \mathrm{N} = 1\ \mathrm{J \cdot m^{-1}}$

$$1\ \mathrm{J} = 6.242 \times 10^{18}\ \mathrm{eV} = 5.034 \times 10^{22}\ \mathrm{cm}^{-1}$$
$$1\ \mathrm{eV} = 8065.5\ \mathrm{cm}^{-1} = 1.602 \times 10^{-19}\ \mathrm{J}$$
$$1\ \mathrm{cm}^{-1} = 1.240 \times 10^{-4}\ \mathrm{eV} = 1.986 \times 10^{-23}\ \mathrm{J}$$
$$1\ \mathrm{D} = 3.336 \times 10^{-30}\ \mathrm{C \cdot m}$$
$$1\ \text{Å} = 10^{-10}\ \mathrm{m}.$$

APPENDIX B

MISCELLANEOUS MATHEMATICS REVIEW

VECTORS

A scalar is a quantity that has a single value. A scalar function $F(x,y,z)$ may have a different value at each point in space, but at each point, the value is a single (real or complex) number. A vector has both a magnitude and a direction. A vector function $\mathbf{F}(x,y,z)$ has a value at each point in space that can be written as $\hat{x}F_x(x, y, z) + \hat{y}F_y(x, y, z) + \hat{z}F_z(x, y, z)$, where $\hat{x}, \hat{y}$, and $\hat{z}$ are unit vectors in the x-, y-, and z-directions.

The dot product of two vectors $\mathbf{F}$ and $\mathbf{G}$ is a scalar quantity given by $\mathbf{F} \cdot \mathbf{G} = F_x G_x + F_y G_y + F_z G_z = |\mathbf{F}||\mathbf{G}|\cos\theta$, where $|\mathbf{F}|$ is the length of vector $\mathbf{F}$ $\left(|\mathbf{F}| = \sqrt{F_x^2 + F_y^2 + F_z^2}\right)$, and θ is the angle between the two vectors. The cross product of two vectors is another vector that is perpendicular to each of them. The cross product is given by $\mathbf{F} \times \mathbf{G} = \hat{x}(F_y G_z - F_z G_y) + \hat{y}(F_z G_x - F_x G_z) + \hat{z}(F_x G_y - F_y G_x)$, or more compactly in the form of a determinant (Appendix C),

$$\mathbf{F} \times \mathbf{G} = \begin{vmatrix} \hat{x} & \hat{y} & \hat{z} \\ F_x & F_y & F_z \\ G_x & G_y & G_z \end{vmatrix}.$$

Condensed-Phase Molecular Spectroscopy and Photophysics, First Edition. Anne Myers Kelley.
© 2013 John Wiley & Sons, Inc. Published 2013 by John Wiley & Sons, Inc.

VECTOR OPERATORS

The del or grad operator is defined as

$$\boldsymbol{\nabla} = \hat{\boldsymbol{x}}\frac{\partial}{\partial x} + \hat{\boldsymbol{y}}\frac{\partial}{\partial y} + \hat{\boldsymbol{z}}\frac{\partial}{\partial z}.$$

It operates on a scalar function to return a vector quantity:

$$\boldsymbol{\nabla}\mathrm{F}(\mathrm{x}, \mathrm{y}, \mathrm{z}) = \hat{\boldsymbol{x}}\frac{\partial F}{\partial x} + \hat{\boldsymbol{y}}\frac{\partial F}{\partial y} + \hat{\boldsymbol{z}}\frac{\partial F}{\partial z}.$$

The divergence (div) operator acts on a vector function to return a scalar:

$$\boldsymbol{\nabla}\cdot\mathbf{F}(x, y, z) = \frac{\partial F_x}{\partial x} + \frac{\partial F_y}{\partial y} + \frac{\partial F_z}{\partial z}.$$

The curl of a vector function is another vector that can be written as

$$\boldsymbol{\nabla}\times\mathbf{F} = \hat{\boldsymbol{x}}\left(\frac{\partial F_z}{\partial y} - \frac{\partial F_y}{\partial z}\right) + \hat{\boldsymbol{y}}\left(\frac{\partial F_x}{\partial z} - \frac{\partial F_z}{\partial x}\right) + \hat{\boldsymbol{z}}\left(\frac{\partial F_y}{\partial x} - \frac{\partial F_x}{\partial y}\right),$$

or more compactly as a determinant,

$$\boldsymbol{\nabla}\times\mathbf{F} = \begin{vmatrix} \hat{\boldsymbol{x}} & \hat{\boldsymbol{y}} & \hat{\boldsymbol{z}} \\ \frac{\partial}{\partial x} & \frac{\partial}{\partial y} & \frac{\partial}{\partial z} \\ F_x & F_y & F_z \end{vmatrix}.$$

Finally, the Laplacian, $\boldsymbol{\nabla}^2 = \boldsymbol{\nabla}\cdot\boldsymbol{\nabla}$, acts on a scalar function to give a scalar result:

$$\boldsymbol{\nabla}^2\mathrm{F} = \frac{\partial^2 F}{\partial x^2} + \frac{\partial^2 F}{\partial y^2} + \frac{\partial^2 F}{\partial z^2}.$$

Three other useful properties are that for any vector $\mathbf{A}$ and scalar $\boldsymbol{\Phi}$,

$\boldsymbol{\nabla}\cdot(\boldsymbol{\nabla}\times\mathbf{A})=0$ (divergence of the curl)
$\boldsymbol{\nabla}\times(\boldsymbol{\nabla}\boldsymbol{\Phi})=0$ (curl of the gradient)
$\boldsymbol{\nabla}\times(\boldsymbol{\nabla}\times\mathbf{A})=\boldsymbol{\nabla}(\boldsymbol{\nabla}\cdot\mathbf{A}) - \boldsymbol{\nabla}^2\mathbf{A}$ (curl of the curl).

SERIES EXPANSIONS

It is often convenient to express a function in the form of an infinite series. One particularly useful type of series expansion is the Taylor series, which approximates a function near a particular point in terms of the derivatives of the function evaluated at that point:

$$f(x) = f(a) + \frac{f'(a)}{1!}(x-a) + \frac{f''(a)}{2!}(x-a)^2 + \frac{f'''(a)}{3!}(x-a)^3 + \cdots \text{ (Taylor series)},$$

where the primes signify derivatives of the function with respect to x, evaluated at $x = a$. A Maclaurin series is a special case of a Taylor series when the expansion is about the origin, although the more general term "Taylor series" is often used when expansion about the origin is implied.

$$f(x) = f(0) + \frac{f'(0)}{1!}x + \frac{f''(0)}{2!}x^2 + \frac{f'''(0)}{3!}x^3 + \cdots \text{ (Maclaurin series)}.$$

When the argument of the function is known to be small, the first few terms of the Maclaurin series expansion often give a very good approximation to the value of the function. A few of the series expansions most useful in quantum mechanics and spectroscopy are summarized below, along with the range of values for which each expansion is valid.

$$(1 \pm x)^n = 1 \pm nx + \frac{n(n-1)x^2}{2!} \pm \frac{n(n-1)(n-2)x^3}{3!} + \cdots \quad (x^2 < 1)$$

$$(1 \pm x)^{-n} = 1 \mp nx + \frac{n(n+1)x^2}{2!} \mp \frac{n(n+1)(n+2)x^3}{3!} + \cdots \quad (x^2 < 1)$$

$$e^x = 1 + x + \frac{x^2}{2!} + \frac{x^3}{3!} + \frac{x^4}{4!} + \cdots \text{ (all real } x\text{)}$$

$$\sin x = x - \frac{x^3}{3!} + \frac{x^5}{5!} - \frac{x^7}{7!} + \cdots \text{ (all real } x\text{)}$$

$$\cos x = 1 - \frac{x^2}{2!} + \frac{x^4}{4!} - \frac{x^6}{6!} + \cdots \text{ (all real } x\text{)}$$

APPENDIX C

MATRICES AND DETERMINANTS

MATRICES

A matrix is an array of numbers. If the matrix **A** has n rows and m columns, it is called an $m \times n$ matrix. The element in the ith row and jth column is referred to as element a_{ij}.

$$\mathbf{A} = \begin{pmatrix} a_{11} & a_{12} & \cdots & a_{1m} \\ a_{21} & \cdots & \cdots & a_{2m} \\ \vdots & \vdots & \ddots & \vdots \\ a_{n1} & a_{n2} & \cdots & a_{nm} \end{pmatrix}$$

Two matrices can be added or subtracted only if they have the same dimensions, in which case one simply adds or subtracts each of the corresponding elements.

A square matrix has $m = n$. A diagonal matrix is a square matrix in which only the diagonal elements are nonzero, for example, $a_{ij} = 0$ unless $i = j$. The unit matrix, **I**, is a diagonal matrix having 1's on the diagonal, $a_{ij} = 1$ for $i = j$ and 0 everywhere else. A matrix for which one of the dimensions is 1 is also known as a vector. A column vector is a matrix having dimension $n \times 1$, while a row vector has dimension $1 \times n$.

The product of two matrices, $\mathbf{AB} = \mathbf{C}$, is defined only if the number of columns in **A** is the same as the number of rows in **B**. If **A** has dimension

Condensed-Phase Molecular Spectroscopy and Photophysics, First Edition. Anne Myers Kelley.

$m \times k$ and **B** has dimension $k \times n$, then **C** has dimension $m \times n$. The elements of **C** are given by $c_{ij} = \sum_{s=1}^{k} a_{is} b_{sj}$. In general, matrix multiplication is not commutative; that is, $\mathbf{AB} \neq \mathbf{BA}$ except in special cases. If **A** is a $1 \times n$ row vector, and **B** is a $n \times 1$ column vector, then their product **AB** is a 1×1 matrix, that is, a scalar, while their product **BA** is a square matrix of dimension n.

DETERMINANTS

The determinant of a square matrix, abbreviated det, is a number (not a matrix) and is symbolized by putting the matrix between vertical lines: det $\mathbf{A} = |\mathbf{A}|$. The determinant of a 2×2 matrix has a simple formula:

$$\begin{vmatrix} a & b \\ c & d \end{vmatrix} = ad - bc.$$

The determinant of a 3×3 matrix also has a reasonably simple formula:

$$\begin{vmatrix} a & b & c \\ d & e & f \\ g & h & i \end{vmatrix} = aei + bfg + cdh - gec - hfa - idb.$$

Determinants of larger $n \times n$ matrices are most simply evaluated by expanding them as a sum of lower-order determinants: $|\mathbf{A}| = \sum_{j=1}^{n} (-1)^{i+j} a_{ij} A^{ij}$ for any row i, where A^{ij}, the minor of element a_{ij}, is the determinant of the matrix that is left when row i and column j are deleted (the equivalent formula using columns rather than rows is also valid). If any two rows or any two columns of a matrix are interchanged, the determinant changes sign; if any two rows or two columns are identical, the determinant is zero.

Determinants appear naturally in the solution of simultaneous linear equations. Assume we have a set of equations of the form

$$\begin{aligned} a_{11}x_1 + a_{12}x_2 + \cdots + a_{1n}x_n &= b_1 \\ a_{21}x_1 + a_{22}x_2 + \cdots + a_{2n}x_n &= b_2 \\ &\vdots \\ a_{n1}x_1 + a_{n2}x_2 + \cdots + a_{nn}x_n &= b_n \end{aligned}$$

These n equations can be written compactly as the matrix equation

$$\begin{pmatrix} a_{11} & a_{12} & \cdots & a_{1n} \\ a_{21} & \cdots & \cdots & a_{2n} \\ \vdots & \vdots & \ddots & \vdots \\ a_{n1} & a_{n2} & \cdots & a_{nn} \end{pmatrix} \begin{pmatrix} x_1 \\ x_2 \\ \vdots \\ x_n \end{pmatrix} = \begin{pmatrix} b_1 \\ b_2 \\ \vdots \\ b_n \end{pmatrix}$$

There always exists a trivial solution (all of the $x_i = 0$), but that is usually not interesting. It turns out that there exists a nontrivial solution (at least some of the $x_i \neq 0$) only if the determinant of the coefficients is zero: $|\mathbf{A}| = 0$.

EIGENVALUE PROBLEMS

In quantum chemistry and spectroscopy, determinants are most often used in the solution of eigenvalue problems. In this case, the matrix equation has the form $\mathbf{AX} = \lambda\mathbf{X}$, where $\mathbf{A}$ is the matrix of a quantum-mechanical operator in some basis, λ is an eigenvalue (a scalar constant) and $\mathbf{X}$ is an eigenvector containing the coefficients of each of the basis states. If there are n basis states, $\mathbf{A}$ is an $n \times n$ matrix, and there are n eigenvalues and eigenvectors. The eigenvalue equation can be written in matrix form as

$$\begin{pmatrix} a_{11}-\lambda & a_{12} & \cdots & a_{1n} \\ a_{21} & a_{22}-\lambda & \cdots & a_{2n} \\ \vdots & \vdots & \ddots & \vdots \\ a_{n1} & a_{n2} & \cdots & a_{nn}-\lambda \end{pmatrix} \begin{pmatrix} x_1 \\ x_2 \\ \vdots \\ x_n \end{pmatrix} = \begin{pmatrix} 0 \\ 0 \\ \vdots \\ 0 \end{pmatrix}.$$

and a nontrivial solution exists only if

$$\begin{vmatrix} a_{11}-\lambda & a_{12} & \cdots & a_{1n} \\ a_{21} & a_{22}-\lambda & \cdots & a_{2n} \\ \vdots & \vdots & \ddots & \vdots \\ a_{n1} & a_{n2} & \cdots & a_{nn}-\lambda \end{vmatrix} = 0.$$

Expanding this determinant gives an nth order polynomial in λ, which can be solved to give the n eigenvalues. Each eigenvalue is then separately inserted back into the original matrix equation to solve for the eigenvector $\mathbf{X}$ which corresponds to that eigenvalue. This procedure gives the x_i only to within an overall multiplicative constant which is determined by requiring that the eigenvector be normalized, $\Sigma_i |x_i|^2 = 1$.

APPENDIX D

CHARACTER TABLES FOR SOME COMMON POINT GROUPS

$C_{\infty v}$	E	$2C(\varphi)$	σ_v		
$A_1 = \Sigma^+$	1	1	1	z	$x^2 + y^2, z^2$
$A_2 = \Sigma^-$	1	1	−1	R_z	
$E_1 = \Pi$	2	$2\cos(\varphi)$	0	(x, y) (R_x, R_y)	(xz, yz)
$E_2 = \Delta$	2	$2\cos(2\varphi)$	0		$(x^2 - y^2, xy)$
$E_3 = \Phi$	2	$2\cos(3\varphi)$	0		

$D_{\infty h}$	E	$2C(\varphi)$	σ_v	i	$2S(\varphi)$	C_2		
Σ_g^+	1	1	1	1	1	1		$x^2 + y^2, z^2$
Σ_g^-	1	1	−1	1	1	−1	R_z	
Π_g	2	$2\cos(\varphi)$	0	2	$-2\cos(\varphi)$	0	(R_x, R_y)	(xz, yz)
Δ_g	2	$2\cos(2\varphi)$	0	2	$2\cos(2\varphi)$	0		$(x^2 - y^2, xy)$
Σ_u^+	1	1	1	−1	−1	−1	z	
Σ_u^-	1	1	−1	−1	−1	1		
Π_u	2	$2\cos(\varphi)$	0	−2	$2\cos(\varphi)$	0	(x, y)	
Δ_u	2	$2\cos(2\varphi)$	0	−2	$-2\cos(2\varphi)$	0		

Condensed-Phase Molecular Spectroscopy and Photophysics, First Edition. Anne Myers Kelley.

C_s	E	σ_h		
A'	1	1	x, y, R_z	x^2, y^2, z^2, xy
A"	1	−1	z, R_x, R_y	xz, yz

C_i	E	i		
A_g	1	1	R_x, R_y, R_z	$x^2, y^2, z^2, xy, xz, yz$
A_u	1	−1	x, y, z	

C_2	E	C_2		
A	1	1	z, R_z	x^2, y^2, z^2, xy
B	1	−1	x, y, R_x, R_y	xz, yz

C_{2v}	E	C_2	$\sigma_v(xz)$	$\sigma'_v(yz)$		
A_1	1	1	1	1	z	x^2, y^2, z^2
A_2	1	1	−1	−1	R_z	xy
B_1	1	−1	1	−1	x, R_y	xz
B_2	1	−1	−1	1	y, R_x	yz

C_{3v}	E	$2C_3$	$3\sigma_v$		
A_1	1	1	1	z	$x^2 + y^2, z^2$
A_2	1	1	−1	R_z	
E	2	−1	0	(x, y) (R_x, R_y)	$(x^2 - y^2, xy)$ (xz, yz)

C_{2h}	E	C_2	i	σ_h		
A_g	1	1	1	1	R_z	x^2, y^2, z^2, xy
B_g	1	−1	1	−1	R_x, R_y	xz, yz
A_u	1	1	−1	−1	z	
B_u	1	−1	−1	1	x, y	

D_{2h}	E	$C_2(z)$	$C_2(y)$	$C_2(x)$	i	$\sigma(xy)$	$\sigma(xz)$	$\sigma(yz)$		
A_g	1	1	1	1	1	1	1	1		x^2, y^2, z^2
B_{1g}	1	1	–1	–1	1	1	–1	–1	R_z	xy
B_{2g}	1	–1	1	–1	1	–1	1	–1	R_y	xz
B_{3g}	1	–1	–1	1	1	–1	–1	1	R_x	yz
A_u	1	1	1	1	–1	–1	–1	–1		
B_{1u}	1	1	–1	–1	–1	–1	1	1	z	
B_{2u}	1	–1	1	–1	–1	1	–1	1	y	
B_{3u}	1	–1	–1	1	–1	1	1	–1	x	

D_{3h}	E	$2C_3$	$3C_2$	σ_h	$2S_3$	$3\sigma_v$		
A_1'	1	1	1	1	1	1		$x^2 + y^2, z^2$
A_2'	1	1	–1	1	1	–1	R_z	
E'	2	–1	0	2	–1	0	(x, y)	$(x^2 - y^2, xy)$
A_1''	1	1	1	–1	–1	–1		
A_2''	1	1	–1	–1	–1	1	z	
E''	2	–1	0	–2	1	0	(R_x, R_y)	(xz, yz)

D_{4h}	E	$2C_4$	C_2	$2C_2'$	$2C_2''$	i	$2S_4$	σ_h	$2\sigma_v$	$2\sigma_d$		
A_{1g}	1	1	1	1	1	1	1	1	1	1		$x^2 + y^2, z^2$
A_{2g}	1	1	1	–1	–1	1	1	1	–1	–1	R_z	
B_{1g}	1	–1	1	1	–1	1	–1	1	1	–1		$x^2 - y^2$
B_{2g}	1	–1	1	–1	1	1	–1	1	–1	1		xy
E_g	2	0	–2	0	0	2	0	–2	0	0	(R_x, R_y)	(xz, yz)
A_{1u}	1	1	1	1	1	–1	–1	–1	–1	–1		
A_{2u}	1	1	1	–1	–1	–1	–1	–1	1	1	z	
B_{1u}	1	–1	1	1	–1	–1	1	–1	–1	1		
B_{2u}	1	–1	1	–1	1	–1	1	–1	1	–1		
E_u	2	0	–2	0	0	–2	0	2	0	0	(x, y)	

D_{6h}	E	$2C_6$	$2C_3$	C_2	$3C_2'$	$3C_2''$	i	$2S_3$	$2S_6$	σ_h	$3\sigma_d$	$3\sigma_v$		
A_{1g}	1	1	1	1	1	1	1	1	1	1	1	1		x^2+y^2, z^2
A_{2g}	1	1	1	1	−1	−1	1	1	1	1	−1	−1	R_z	
B_{1g}	1	−1	1	−1	1	−1	1	−1	1	−1	1	−1		
B_{2g}	1	−1	1	−1	−1	1	1	−1	1	−1	−1	1		
E_{1g}	2	1	−1	−2	0	0	2	1	−1	−2	0	0	(R_x, R_y)	(xz, yz)
E_{2g}	2	−1	−1	2	0	0	2	−1	−1	2	0	0		(x^2-y^2, xy)
A_{1u}	1	1	1	1	1	1	−1	−1	−1	−1	−1	−1		
A_{2u}	1	1	1	1	−1	−1	−1	−1	−1	−1	1	1	z	
B_{1u}	1	−1	1	−1	1	−1	−1	1	−1	1	−1	1		
B_{2u}	1	−1	1	−1	−1	1	−1	1	−1	1	1	−1		
E_{1u}	2	1	−1	−2	0	0	−2	−1	1	2	0	0	(x, y)	
E_{2u}	2	−1	−1	2	0	0	−2	1	1	−2	0	0		

D_{3d}	E	$2C_3$	$3C_2$	i	$2S_6$	$3\sigma_d$		
A_{1g}	1	1	1	1	1	1		x^2+y^2, z^2
A_{2g}	1	1	−1	1	1	−1	R_z	
E_g	2	−1	0	2	−1	0	(R_x, R_y)	$(x^2-y^2, xy), (xz, yz)$
A_{1u}	1	1	1	−1	−1	−1		
A_{2u}	1	1	−1	−1	−1	1	z	
E_u	2	−1	0	−2	1	0	(x, y)	

APPENDIX E

FOURIER TRANSFORMS

A Fourier transform is a mathematical operation that converts a function of one variable to a function of another, conjugate variable. Most often these variables are frequency and time, although in quantum mechanics, the position and the momentum are also conjugate variables related through Fourier transformation. If $f(t)$ is some integrable function of time t, then its Fourier transform is a function of angular frequency ω and is given by

$$F(\omega) = \frac{1}{\sqrt{2\pi}} \int_{-\infty}^{\infty} f(t) e^{-i\omega t} dt.$$

With this definition of the Fourier transform, we can also define the inverse Fourier transform as

$$f(t) = \frac{1}{\sqrt{2\pi}} \int_{-\infty}^{\infty} F(\omega) e^{i\omega t} d\omega.$$

That is, given either the function of time or the function of frequency, we can recover the other. (In an alternative convention, the Fourier transform itself omits the $1/\sqrt{2\pi}$ prefactor, and the inverse Fourier transform contains a prefactor of $1/2\pi$.) As a rule, the more $f(t)$ is concentrated within some small region of time, the more $F(\omega)$ is spread out over a broad range of frequencies, and vice versa.

Condensed-Phase Molecular Spectroscopy and Photophysics, First Edition. Anne Myers Kelley.

Fourier transformation is linear: for any complex numbers α and β, if $h(t) = \alpha f(t) + \beta g(t)$, then its Fourier transform is $H(\omega) = \alpha F(\omega) + \beta G(\omega)$. In addition, for any nonzero real number a, if $h(t) = f(at)$, then its Fourier transform is

$$H(\omega) = \frac{1}{|a|} F\left(\frac{\omega}{a}\right).$$

Most functions do not have a closed-form Fourier transform, and the integral must be evaluated numerically. Below are a few functions that do have simple analytical transforms.

$f(t)$	$F(\omega)$
1	$\sqrt{2\pi}\delta(\omega)$
$e^{-\alpha t^2}$ $[\mathrm{Re}(\alpha) > 0]$	$\frac{1}{\sqrt{2\alpha}} e^{-\omega^2/4\alpha}$
$e^{-a\lvert t\rvert}$	$\sqrt{\frac{2}{\pi}} \frac{a}{a^2 + \omega^2}$
$\delta(t)$	$\frac{1}{\sqrt{2\pi}}$
e^{iat}	$\sqrt{2\pi}\delta(\omega - a)$
sech (at)	$\frac{1}{a}\sqrt{\frac{\pi}{2}}\,\mathrm{sech}\left(\frac{\pi}{2a}\omega\right)$
1 for $\lvert t\rvert < a$, 0 for $\lvert t\rvert > a$ $(a > 0)$	$\sqrt{\frac{2}{\pi}} \frac{\sin(a\omega)}{\omega}$

INDEX

absorbance 70, 220
absorption cross-section 69, 94, 97, 192, 218, 223, 269
accepting mode 172
acoustic phonon 283
adiabatic approximation 26
adiabatic electron transfer 246
ammonia (NH_3) 106, 125
angle tuning 229
angular momentum 14, 18, 58, 109, 141, 172, 182, 300
angular momentum quantum number 14
anharmonicity 9, 118, 127, 146, 154, 174, 205, 286
anisotropy (of light emission) 184
annihilation operator 37, 60, 196
anomalous polarization 207
antibonding orbital 24, 104
anti-Stokes Raman scattering 194
antisymmetric function 20, 24, 103
associated Laguerre polynomials 14
atomic solid 277
atomic units 20
aufbau process 17, 291
azulene 164

band gap 286, 292, 295, 301
band structure 291
basis 2, 17, 25, 86, 106, 110, 121, 150, 170, 174, 246, 252, 255, 291, 315
bath 86, 91, 215
Beer-Lambert Law 70
benzene 106
blackbody radiation 41, 58, 68
Bloch function 278, 281, 294, 299
blue shift 155, 253
$B(OH)_3$ 106
Bohr radius 13, 297
Boltzmann distribution/factor 68, 79, 82, 90, 146, 154, 180, 219, 242
bonding orbital 24
Born-Oppenheimer approximation 22, 118, 155, 168, 173
boson 19, 37
bright state 174

Condensed-Phase Molecular Spectroscopy and Photophysics, First Edition. Anne Myers Kelley.

Brillouin zone 279, 293
butadiene 106, 257, 258
2-butene 154

$C_2H_2Cl_2Br_2$ 106
C_2H_2ClBr 106
cadmium selenide (CdSe) 284, 301
cadmium sulfide (CdS) 299
center of mass 9, 25, 55, 115, 121
CFClBrI 106
character/character table 107, 129, 142, 317
charge-transfer transition 131, 239, 271
chemical enhancement 271
chloroform 131
chromophore 74, 155, 179, 251, 255
closure 3, 83, 91, 144, 200, 204, 208
CO_2 124
coherence 90, 97, 222, 233
coherent antistokes Raman scattering (CARS) 232
combination band 127, 205, 209, 285
commutator 2, 5, 91
compatible observables 7
Condon approximation 141, 148, 152, 155, 209, 244
conduction band 268, 292
conformers 154
conjugated polymer 7, 256
conjugation length 259
conservative system 49, 87
constants of the motion 48
continuous basis 3
continuous-wave laser 44, 218
copper (Cu) 270
correlation function 84, 132
Coulomb gauge 35
Coulomb integral 23
Coulomb potential/interaction 12, 256, 277, 300, 307
creation operator 37, 60, 196
crystal lattice 278
CS_2 124, 210
cyanine dye 254

d orbital 14, 151
dark state 174
defect 298
degenerate/degeneracy 4, 8, 14, 21, 124, 141, 151, 175, 207, 246, 252, 255, 285
del operator (∇) 6
density matrix 86, 133, 215, 221, 233
density operator 86, 93
density of states 41, 54, 61, 71, 201, 210, 244
dephasing 91, 133, 191, 218, 222, 233, 274
depolarization ratio 184, 206
determinant 4, 17, 20, 23, 123, 283, 309, 313
Dexter energy transfer 179
diagonal anharmonicity 127
diamond 278
diatomic molecule 10, 24, 117, 127, 130, 141, 147
dichroic beamsplitter 220
dielectric constant 32, 180, 191, 194, 243, 263, 269, 297
dielectric continuum 180, 243
dielectric function 32, 193, 263, 269
differential cross section 201
dihedral angle 123
dimensionless coordinate 10, 147
dimer 252
dipole correlation function 84
dipole moment 55, 59, 84, 89, 93, 117, 129, 132, 141, 155, 180, 193, 205, 215, 252, 254
Dirac delta function 3
Dirac notation 2, 104
direct gap 296
direct lattice 278
discrete basis 3
dispersion relation 282, 294
dispersive interaction 155
dodecahexaene 258
doped semiconductor 296
Doppler broadening 115
Drude model 266, 269
Duschinsky rotation/effect 145, 149, 154, 171, 209
dynamic Stokes shift 157

effective mass 296, 301
Ehrenfest's theorem 48
eigenstate 4, 12, 19, 47, 67, 79, 82, 119, 144, 167, 173, 180, 205, 252, 255, 299

eigenvalue 4, 9, 12, 18, 49, 79, 119, 170, 174, 315
eigenvector 4, 121, 174, 315
Einstein coefficients 42, 67, 222
elastic scattering 193
electrical anharmonicity 118
electric dipole approximation 55, 116, 196
electric dipole transition 57, 59, 82, 109, 126, 141, 149, 176, 296
electric displacement 31
electric field 31, 43, 54, 74, 91, 193, 215, 222, 228, 264, 271
electric quadrupole 59, 109, 142
electromagnetic field enhancement 271
electromagnetic radiation 31, 52, 73, 191, 194, 264
electron correlation 24
electronic origin 145, 258
electronic Raman scattering 203
electron spin 18
electron transfer 166, 239
empty lattice model 293
energy density 34, 58, 61, 67, 74, 216, 222
energy gap law 172
energy transfer 133, 166, 179
envelope function 299
ethane 106
ethylene 105, 110, 257
exchange integral 23
exciton 256, 297
exciton binding energy 297
expectation value 6, 16, 18, 48, 83, 87, 93, 140, 227
extinction 192, 269, 272

Fermi's Golden Rule 54, 60, 69, 167, 170, 180, 243
fermion 19
fluorescence 55, 155, 165, 176, 179, 209, 218, 225, 231, 253, 270, 274
fluorescence line-narrowing 158
fluorescence upconversion 157
formaldehyde 173
force constant 9, 22, 119, 121, 130, 242, 254, 265, 281
Förster energy transfer 179
Fourier transform 40, 84, 99, 133, 226, 321
four-wave mixing 232
Franck-Condon factor 144, 152, 180, 209, 243
free-electron model 293
free energy 245
Frenkel exciton 297
frequency-time uncertainty principle 39
full-width at half maximum 81
fundamental (vibration) 126, 179, 205, 209, 285

gamma point 280
gauge transformation 34
Gaussian function/distribution 10, 39, 81, 134, 242
Gaussian type orbital 25
Gaussian units 307
gold (Au) 269
grad operator (∇) 6, 310
ground-state recovery 224
group 104
group theory 104, 127, 141, 286

H-aggregate 253
H-dimer 253
H_2 molecule 24, 104, 106
H_2^+ molecule 22
H_2O_2 106
H_2O (water) 106, 124, 128, 271
half-width at half maximum 81, 85
Hamiltonian 5, 8, 11, 16, 22, 36, 47, 56, 60, 79, 83, 88, 96, 150, 167, 174, 180, 195, 252, 257, 278
harmonic oscillator 9, 36, 117, 125, 144, 154, 178, 194, 205, 209, 281
HCN 106
heavy-atom effect 157, 172
Heisenberg picture 83
Heisenberg uncertainty relation 7
Hermite polynomials 9
Hermitian operator 4, 17, 87
Herzberg-Teller coupling 150
heterodimer 252
hexatriene 257
highest occupied molecular orbital (HOMO) 240, 247, 257, 291
hole 256, 292, 296

homodimer 252
homogeneous broadening 132, 158, 233
Hooke's Law 9
horizontal reflection plane 105
Huang-Rhys factor 147
Hückel model 257
Hund's cases 143
hydrogen atom 12, 297, 299
hydrogen bonding 124, 131, 134

identity operation 104, 108
improper rotation 105
impulsive stimulated Raman scattering 225
incoherent light 42
indirect gap 296
infrared spectroscopy 42, 118
inhomogeneous broadening 132, 158, 180, 233
inner product 2, 86, 168
inner-sphere reorganization 156
intensity of light 34, 38, 43, 70, 82, 94
intensity borrowing 151
interferences in scattering 195, 203, 210
internal conversion 164, 170, 184, 243, 253
internal coordinate 123
intersystem crossing 165, 172
intramolecular vibrational redistribution/relaxation 133, 164, 174, 251
inversion operation 105, 141, 206, 219, 229, 286
inversion mode 124
ionic solid 277
ion pair 240
iron (Fe) 288
irreducible representation 105, 108, 111
isolated binary collision 178
isotopic shift/substitution 130, 255

J-aggregate 253
J-dimer 253
Jablonski diagram 163
Jahn-Teller effect 151

Kasha's Rule 165
Kronecker delta 3
Kronig-Penney model 294

laboratory-fixed coordinates 129, 206, 217
Laplacian (∇^2) 6, 13, 310
laser 37, 40, 42, 57, 158, 182, 204, 218, 224, 230
lattice constant 278, 283
libration 183, 255
lifetime 72, 80, 85, 146, 165, 181, 202, 217, 225
lifetime broadening 132, 157
light scattering 191
linear combination of atomic orbitals 24, 107
linear polyenes 7, 257
lineshape 80, 84, 94, 132, 135, 234, 242
linewidth 43, 81, 85, 132, 153, 158, 203, 234
local field 73, 157
locally excited transition 240, 247
longitudinal phonon 284
Lorentzian distribution/function 81, 85, 94, 133, 135, 147, 178
Lorentz local field 74
Lorentz model 263
lowering operator (harmonic oscillator) 12, 37
lowest unoccupied molecular orbital (LUMO) 240, 247, 257, 292

magic angle 186, 220
magnetic dipole matrix element 59, 109, 142
magnetic field 14, 21, 31
magnetic induction 31
magnetic polarization 32
magnetic quantum number 14
Marcus inverted region 246
Marcus theory 243
maria 218
mass-weighted Cartesian coordinate 122
matrix element 4, 17, 52, 89, 97, 104, 109, 116, 120, 128, 140, 150, 167, 171, 177, 196, 203, 240, 244, 255
matrix representation 3, 86
Maxwell's equations 31, 56, 193, 271, 307
mean value of observable 6
mechanical anharmonicity 118, 127

metallic solid 278
Mie scattering 192, 268
mixed case density matrix 87
mode-locked laser 44
molecular orbital 7, 24, 103, 107, 239, 247, 257, 291
molecular solid 277, 291
molecule-fixed coordinates 129, 142, 184, 208, 286
molar absorptivity 69
molar extinction coefficient 69
momentum basis 2
momentum operator 5, 12, 56
Morse potential 119, 127
multiphoton absorption 216
multipole expansion 55

Na_2 139
nanocrystal 7, 255, 287, 298
nanoparticle 56, 268, 301
nanoprism 269
nanorod 269
naphthalene 277
natural radiative lifetime 72, 166, 274
neon (Ne) 277
network solid 278
NH_3 (ammonia) 106, 125
nonadiabatic electron transfer 246
non-Condon terms 149, 157
nondegenerate eigenvalue 4, 120, 141, 209, 246, 255
nonlinear optical responses 216
nonphotochemical hole burning 158
nonresonant Raman scattering 204, 225, 285
normal coordinate 121, 127, 150, 281
normal mode 110, 121, 144, 149, 169, 173, 178, 206, 281
normalization 10, 17, 23, 255, 300
number operator 12, 37

observable 4, 48, 87, 93
octahedral complex 151
octatetraene 257, 258
off-diagonal anharmonicity 127
operator 2, 11, 17, 19, 22, 32, 36, 48, 55, 83, 86, 104, 107, 128, 141, 168, 196, 219, 310, 315
optical phonon 283
optical gating 230
orthogonal 2, 13, 15, 19, 49, 117, 120, 141, 144, 173, 182, 196, 205, 301
orthonormal 3, 15, 20, 60
oscillator strength 73, 240, 252, 266
outer-sphere reorganization 156
overlap integral 17, 3
overtone 120, 127, 206, 209, 285

p orbital 14, 25, 107,
parallel transition 142
partial trace of matrix 88
particle in a box 7, 35, 127, 257, 293, 299
particle in a sphere 299
Pauli exclusion principle 18, 296
permanent dipole 117, 129, 140, 155
permeability 32
permittivity 32, 307
perpendicular transition 142
perturbation theory (time-independent) 17, 150
phase matching 230
phase velocity 33
phonon 225, 277, 281, 293, 296
phosphorescence 55, 165, 184
photochemical hole-burning 158
photochemistry 165, 274
photophysical hole-burning 158
photon 31, 37, 60, 163, 166, 179, 191, 195, 201, 210, 215, 232,
photon echo 232
photoselection 182
plasma frequency 264
plasma oscillation 266
plasmon 266
point group 104, 124, 128, 131, 139, 148, 206, 286, 317
polarizability 155, 194, 205, 251, 269
polarizability tensor 205, 209
polarizability volume 193, 269
polarization
  of light 35, 42, 61, 82, 109, 129, 142, 182, 184, 196, 201, 206, 220, 286
  of matter 32, 74, 94, 97, 184, 216, 228, 232, 264, 266
polaron 256
polyethylene 257
population (density matrix) 89

population decay 97, 132, 186
position basis 2
position operator 5
potential energy surface 26, 115, 140, 144, 148, 167, 225, 246
Poynting vector 34
principal axis 104
principal quantum number 14
probability 2
probability amplitude 2
promoting mode 150, 172
propagation direction 33, 115, 182, 284
pump-probe experiment 219
pure case density matrix 86
pure dephasing 133, 158, 233

quantized radiation field 35, 69
quantum beats 176
quantum confinement 298
quantum yield 166

Rabi frequency 97
radial wavefunction 14, 299
radiationless transition 158, 167, 172, 219
radiative lifetime 72, 296
raising operator (harmonic oscillator) 12, 37
Raman excitation profile 210
Raman polarizability 194, 209
Raman scattering 38, 54, 130, 194, 203, 216, 270, 285
Raman shift 130, 194, 272
Raman spectrum 131, 194, 210, 270
Rayleigh scattering 193, 203
reciprocal lattice 278, 284
red shift 155, 259
reduced coordinates (harmonic oscillator) 11
reduced mass 9, 124, 130, 169, 297
reducible representation 108
redundant coordinate system 123
reflection operation 105
reflection plane 104
refractive index 33, 41, 73, 180, 193, 201, 216, 222, 229, 243, 252, 263, 269
relaxation (density matrix) 91
reorganization energy 156, 242, 245
representation 3
resonance enhancement 210, 274
resonance Raman 203, 207, 217, 225, 274, 286
root-mean-square deviation 6
rotating wave approximation 96
rotation operation 105
rotation axis 104
rotational diffusion 183
rotational motion 115, 143, 182, 220
Rydberg transition 155

s orbital 14, 23, 107
scalar potential 34
scalar product 2
scattering cross-section 193, 203, 268
Schr ödinger equation
  time-dependent 47, 79, 87
  time-independent 5, 9, 13, 16, 22, 168, 294
Schr ödinger picture 83
second harmonic generation 227
second Legendre polynomial (P_2) 184
selection rule 117, 120, 126, 141, 149, 165, 205, 219, 286, 301
self-phase modulation 221
semiconductor 286, 292
silver (Ag) 269, 270
single-mode laser 44
singlet state 21, 141, 164
Slater determinant 21, 24
Slater type orbital 25
sodium chloride (NaCl) 277
solvent effects 154
solvent reorganization 156, 219, 241, 246
space-fixed coordinates 129
space group 104, 286
spectral congestion 153
spectral hole burning 158
spherical Bessel function 192, 300
spherical Hankel function 192
spherical harmonics 13, 301
spherical polar coordinates 2, 13
spin angular momentum 19
spin echo 232
spin selection rules 141
spinorbital 19
spin-orbit coupling 143, 157, 170, 177
spin states 3, 18

spontaneous emission 35, 38, 55, 62, 67, 72, 132, 178, 201, 222, 231, 270, 274
spontaneous light scattering 195, 232, 274
square law detector 38
state function 1, 80
state vector 1, 48, 86, 175
stationary state 47, 49, 79, 167, 175
statistical case density matrix 87
stimulated emission 42, 67, 95, 98, 215, 219, 232
stimulated photon echo 234
stochastic theories 135
Stokes Raman scattering 194
Stokes shift 157, 274
sulfate ion (SO_4^{2-}) 278
sum frequency generation 227, 230
supersonic free-jet expansion 251
surface enhanced fluorescence 274
surface enhanced Raman scattering (SERS) 270
surface enhanced resonance Raman scattering (SERRS) 274
surface phonon 287
surface plasmon 267, 271, 274
susceptibilities 216
symmetrically equivalent atoms 103
symmetry coordinates 128
symmetry element 104, 206
symmetry operation 104, 128, 229
symmetry species 105, 124, 128, 139, 149, 172, 206

T_1 133
T_2 133
T_2^* 133
Taylor series 9, 58, 99, 117, 126, 140, 148, 194, 205, 310
temperature tuning 229
thermal equilibrium 41, 67, 90, 97, 146, 154, 163, 221, 242
third harmonic generation 232
three-pulse photon echo 234
three-wave mixing 232
tight-binding model 257
time-correlated single photon counting 157
time-dependent perturbation theory 50, 69, 80, 196, 202, 215
torsion 123, 125, 154
total dephasing time 133
totally symmetric 107, 124, 128, 141, 150, 206, 209
trace (of matrix) 86
trans-butadiene 106
trans-dichloroethylene 125
transient absorption 225
transient bleaching 224
transient hole burning 159
transition dipole 61, 63, 72, 140, 150, 179, 202, 217, 233, 252, 301
transition rate 58, 61, 82, 116, 199, 216, 243
translation operator 278
transmittance 220
transverse phonon 284, 287
transverse wave 33
triplet state 21, 141, 158, 165, 177, 219
tunneling 10, 244
turning point (harmonic oscillator) 10
two-level (two-state) system 67, 91, 95, 174
two-photon absorption 216, 227
two-photon cross section 218

umbrella mode 124
unimolecular processes 166
unit cell 277, 283, 293, 301
unit vector (crystal lattice) 278, 284

vacuum state 37
valence band 292, 296
van der Waals molecule 251
variational method/principle 16
vector potential 34, 37, 56, 59
vertical reflection plane 104
vibrational overlap 144, 148, 172, 209, 244
vibrational Raman scattering 203, 205
vibrational relaxation 179, 219
vibrational reorganization 156
vibrational series 146
vibrational spectroscopy 116, 132, 231, 281
vibronic coupling 148
vibronic transition 140, 153, 243

vibronic structure 143, 153, 243, 259
viscosity 185
Voigt profile 135

wag 123, 125
Wannier exciton 298
water 106, 124, 128, 271
wavefunction 2, 5, 7, 10, 13
wavenumber 43, 73, 120, 127
wavevector 33, 35, 42, 62, 193, 227, 232, 269, 278, 282, 293, 295
wavevector mismatch 229
white light 221
Wilson FG method 124
wurtzite crystal structure 284

zeaxanthin 254
zero-point energy 10, 172
zero-zero transition 145, 158